Statistical Modeling by Wavelets

BRANI VIDAKOVIC
Duke University

A Wiley-Interscience Publication
JOHN WILEY & SONS, INC.
New York / Chichester / Weinheim / Brisbane / Singapore / Toronto

This book is printed on acid-free paper. ∞

Copyright © 1999 by John Wiley & Sons, Inc. All rights reserved.

Published simultaneously in Canada.

No part of this publication may be reproduced, stored in a retrieval system or transmitted in any form or by any means, electronic, mechanical, photocopying, recording, scanning or otherwise, except as permitted under Sections 107 or 108 of the 1976 United States Copyright Act, without either the prior written permission of the Publisher, or authorization through payment of the appropriate per-copy fee to the Copyright Clearance Center, 222 Rosewood Drive, Danvers, MA 01923, (978) 750-8400, fax (978) 750-4744. Requests to the Publisher for permission should be addressed to the Permissions Department, John Wiley & Sons, Inc., 605 Third Avenue, New York, NY 10158-0012, (212) 850-6011, fax (212) 850-6008, E-Mail: PERMREQ@WILEY.COM.

For ordering and customer service, call 1-800-CALL-WILEY.

Library of Congress Cataloging-in-Publication Data:

Vidakovic, Brani, 1955–
    Statistical modeling by wavelets / Brani Vidakovic.
       p. cm.
    Includes bibliographical references and index.
    ISBN 0-471-29365-2 (alk. paper)
    1. Wavelets (Mathematics).   2. Mathematical statistics.   I. Title.
QA403.3.V53   1999
515'.2433—dc21   98-48449

Printed in the United States of America

10 9 8 7 6 5 4 3 2 1

Contents

Preface		*xi*
Acknowledgments		*xiii*
1	**Introduction**	1
	1.1 Wavelet Evolution	1
	1.2 Wavelet Revolution	6
	1.3 Wavelets and Statistics	16
	1.4 An Appetizer: California Earthquakes	19
2	**Prerequisites**	23
	2.1 General	23
	2.2 Hilbert Spaces	24
	2.2.1 Projection Theorem	25
	2.2.2 Orthonormal Sets	26
	2.2.3 Reproducing Kernel Hilbert Spaces	29
	2.3 Fourier Transformation	29
	2.3.1 Basic Properties	30
	2.3.2 Poisson Summation Formula and Sampling Theorem	32

		2.3.3	Fourier Series	*33*
		2.3.4	Discrete Fourier Transform	*34*
	2.4	Heisenberg's Uncertainty Principle		*35*
	2.5	Some Important Function Spaces		*36*
	2.6	Fundamentals of Signal Processing		*39*
	2.7	Exercises		*40*
3	Wavelets			*43*
	3.1	Continuous Wavelet Transformation		*43*
		3.1.1	Basic Properties	*44*
		3.1.2	Wavelets for Continuous Transformations	*48*
	3.2	Discretization of the Continuous Wavelet Transform		*50*
	3.3	Multiresolution Analysis		*51*
		3.3.1	Derivation of a Wavelet Function	*57*
	3.4	Some Important Wavelet Bases		*60*
		3.4.1	Haar's Wavelets	*60*
		3.4.2	Shannon's Wavelets	*63*
		3.4.3	Meyer's Wavelets	*65*
		3.4.4	Franklin's Wavelets	*70*
		3.4.5	Daubechies' Compactly Supported Wavelets	*73*
	3.5	Some Extensions		*80*
		3.5.1	Regularity of Wavelets	*80*
		3.5.2	The Least Asymmetric Daubechies' Wavelets: Symmlets	*86*
		3.5.3	Approximations and Characterizations of Functional Spaces	*87*
		3.5.4	Daubechies-Lagarias Algorithm	*89*
		3.5.5	Moment Conditions	*91*
		3.5.6	Interpolating (Cardinal) Wavelets	*93*
		3.5.7	Pollen-Type Parameterization of Wavelets	*94*
	3.6	Exercises		*96*
4	Discrete Wavelet Transformations			*101*
	4.1	Introduction		*101*
	4.2	The Cascade Algorithm		*104*
	4.3	The Operator Notation of DWT		*109*
		4.3.1	Discrete Wavelet Transformations as Linear Transformations	*115*
	4.4	Exercises		*117*

Statistical Modeling by Wavelets

WILEY SERIES IN PROBABILITY AND STATISTICS
APPLIED PROBABILITY AND STATISTICS SECTION

Established by WALTER A. SHEWHART and SAMUEL S. WILKS

Editors: *Vic Barnett, Noel A. C. Cressie, Nicholas I. Fisher, Iain M. Johnstone, J. B. Kadane, David G. Kendall, David W. Scott, Bernard W. Silverman, Adrian F. M. Smith, Jozef L. Teugels; Ralph A. Bradley, Emeritus, J. Stuart Hunter, Emeritus*

A complete list of the titles in this series appears at the end of this volume.

5	Some Generalizations		119
	5.1	Coiflets	120
		5.1.1 Construction of Coiflets	121
	5.2	Biorthogonal Wavelets	122
		5.2.1 Construction of Biorthogonal Wavelets	125
		5.2.2 B-Spline Wavelets	128
	5.3	Wavelet Packets	133
		5.3.1 Basic Properties of Wavelet Packets	135
		5.3.2 Wavelet Packet Tables	138
	5.4	Best Basis Selection	140
		5.4.1 Some Cost Measures and the Best Basis Algorithm	141
	5.5	ϵ-Decimated and Stationary Wavelet Transformations	145
		5.5.1 ϵ-Decimated Wavelet Transformation	145
		5.5.2 Stationary (Non-Decimated) Wavelet Transformation	147
	5.6	Periodic Wavelet Transformations	150
	5.7	Multivariate Wavelet Transformations	153
	5.8	Discussion	160
	5.9	Exercises	163
6	Wavelet Shrinkage		167
	6.1	Shrinkage Method	168
	6.2	Linear Wavelet Regression Estimators	170
		6.2.1 Wavelet Kernels	171
		6.2.2 Local Constant Fit Estimators	172
	6.3	The Simplest Non-Linear Wavelet Shrinkage: Thresholding	175
		6.3.1 Variable Selection and Thresholding	177
		6.3.2 Oracular Risk for Thresholding Rules	177
		6.3.3 Why the Wavelet Shrinkage Works	179
		6.3.4 Almost Sure Convergence of Wavelet Shrinkage Estimators	184
	6.4	General Minimax Paradigm	184
		6.4.1 Translation of Minimaxity Results to the Wavelet Domain	186
	6.5	Thresholding Policies and Thresholding Rules	188
		6.5.1 Exact Risk Analysis of Thresholding Rules	188
		6.5.2 Large Sample Properties of $\hat{f}$	189

	6.5.3 Some Other Shrinkage Rules	191
6.6	How to Select a Threshold	194
	6.6.1 Mallat's Model and Induced Percentile Thresholding	194
	6.6.2 Universal Threshold	195
	6.6.3 A Threshold Based on Stein's Unbiased Estimator of Risk	199
	6.6.4 Cross-Validation	200
	6.6.5 Thresholding as a Testing Problem	202
	6.6.6 Lorentz Curve Thresholding	203
	6.6.7 Block Thresholding Estimators	206
6.7	Other Methods and References	207
6.8	Exercises	213

7 Density Estimation — 217

7.1	Orthogonal Series Density Estimators	217
7.2	Wavelet Density Estimation	219
	7.2.1 δ-Sequence Density Estimators	219
	7.2.2 Bias and Variance of Linear Wavelet Density Estimators	223
	7.2.3 Linear Wavelet Density Estimators in a More General Setting	224
7.3	Non-Linear Wavelet Density Estimators	225
	7.3.1 Global Thresholding Estimator	227
7.4	Non-Negative Density Estimators	228
	7.4.1 Estimating the Square Root of a Density	229
	7.4.2 Density Estimation by Non-Negative Wavelets	234
7.5	Other Methods	236
	7.5.1 Multivariate Wavelet Density Estimators	236
	7.5.2 Density Estimation as a Regression Problem	239
	7.5.3 Cross-Validation Estimator	240
	7.5.4 Multiscale Estimator	240
	7.5.5 Estimation of a Derivative of a Density	241
7.6	Exercises	243

8 Bayesian Methods in Wavelets — 247

8.1	Motivational Examples	247
8.2	Smooth Shrinkage	250
8.3	Bayesian Thresholding	255

8.4	MAP-Principle		257
8.5	Density Estimation Problem		259
8.6	Full Bayesian Model		262
8.7	Discussion and References		265
8.8	Exercises		267

9 Wavelets and Random Processes — 271

9.1	Stationary Time Series		272
9.2	Wavelets and Stationary Processes		273
	9.2.1	Wavelet Transformations of Stationary Processes	273
	9.2.2	Whitening of Stationary Processes	274
	9.2.3	Karhunen-Loève-Like Expansions	276
9.3	Estimation of Spectral Densities		279
	9.3.1	Gao's Algorithm	280
	9.3.2	Non-Gaussian Stationary Processes	284
9.4	Wavelet Spectrum		286
	9.4.1	Wavelet Spectrum of a Stationary Time Series	286
	9.4.2	Scalogram and Periodicities	289
9.5	Long-Memory Processes		290
	9.5.1	Wavelets and Fractional Brownian Motion	290
	9.5.2	Estimating Spectral Exponents in Self-Similar Processes	292
	9.5.3	Quantifying the Whitening Property of Wavelet Transformations for fBm Processes	295
9.6	Discussion and References		296
9.7	Exercises		297

10 Wavelet-Based Random Variables and Densities — 299

10.1	Scaling Function as a Density		299
10.2	Wavelet-Based Random Variables		300
10.3	Random Densities via Wavelets		308
	10.3.1	Tree Algorithm	308
10.4	Properties of Wavelet-Based Random Densities		310
10.5	Random Densities With Constraints		312
	10.5.1	Smoothness Constraints	312
	10.5.2	Constraints on Symmetry	312
	10.5.3	Constraints on Modality	313
	10.5.4	Skewed Random Densities	313

    10.6 Exercises   *316*

11 Miscellaneous Statistical Applications   *319*
    11.1 Deconvolution Problems   *319*
    11.2 Wavelet-Vaguelette Decompositions   *323*
    11.3 Pursuit Methods   *325*
    11.4 Moments of Order Statistics   *329*
    11.5 Wavelets and Statistical Turbulence   *333*
          11.5.1 K41 Theory   *335*
          11.5.2 Townsend's Decompositions   *336*
    11.6 Software and WWW Resources for Wavelet Analysis   *338*
          11.6.1 Commercial Wavelet Software   *338*
          11.6.2 Free Wavelet Software   *340*
          11.6.3 Some WWW Resources   *341*
    11.7 Exercises   *342*

References   *345*

Notation Index   *371*

Author Index   *373*

Subject Index   *379*

Preface

> Just two months ago astronomers did not know about it. But now they are giving good odds that Hyakutake will be the most impressive comet since the invention of telescope 400 years ago. *(Herald Sun, Durham, NC, March 24, 1996.)*

One can trace the origins of wavelets back to the beginning of this century; however, wavelets, understood as a systematic way of producing local orthogonal bases, are a recent unification of existing theories in various fields and some important "discoveries." They are mathematical objects that have interpretation and application in many scientific fields, most notably in the fields of signal processing, nonparametric function estimation, and data compression. In the early 1990s, a series of papers by Donoho and Johnstone and their coauthors demonstrated that wavelets are appropriate tools in problems of denoising, regression, and density estimation. The subsequent burgeoning wavelet research broadened to a wide range of statistical problems.

This book is aimed at graduate students in statistics and mathematics, practicing statisticians, and statistically curious engineers. It can serve as a text for an introductory wavelet course concerned with an interface of wavelet methods and statistical inference. The necessary mathematical background is proficiency in advanced calculus and algebra; consequently, this book should be useful to advanced undergraduate students as well as to graduate students in statistics, mathematics, and engineering.

This book originated from the class notes supporting the Special Topics Course on Multiscale Methods at Duke University. The content can be divided into two parts:

an introduction to wavelets (Chapters 1–5) and statistical modeling (Chapters 6–11). An introduction and some mathematical prerequisites are presented in Chapters 1 and 2. Continuous and discrete wavelet transformations are covered in Chapters 3 and 4. Some important generalizations (coiflets, biorthogonal wavelets, wavelet packets, stationary, periodized and multivariate wavelets) are covered in Chapter 5.

Chapters 6–11 are data-oriented. Chapter 6 is the crux of the book, covering the theory and practice of wavelet shrinkage. Important theoretical aspects of wavelet density estimation are covered in Chapter 7. Chapter 8 discusses Bayesian modeling in the wavelet domain. Time series are covered in Chapter 9, while Chapter 10 contains several probabilistic and simulational properties of wavelet-based random functions and densities. Chapter 11 gives some novel and important wavelet applications in statistics.

Instead of providing appendices with data sets and programs used in the book, I opted for a more modern style. The web page:

http://www.isds.duke.edu/~brani/wiley.html

is associated with the book. This page contains all data sets, functions, and programs referred to.

I hope the reader will find this book useful. All comments, suggestions, updates, and critiques will be appreciated.

BRANI VIDAKOVIC

Institute of Statistics and Decision Sciences
Duke University
Durham, February 1999

Acknowledgments

This book was made possible by the support of many individuals. First, I am grateful to colleagues from the Institute of Statistics and Decision Sciences (ISDS) – Duke University for supporting a wavelet-based statistics course. The attending students were patient and understood the difficulties of transferring research papers to a working copy of course notes. Their industry and enthusiasm is treasured. Duke University helped with this project through the Grant of Arts and Sciences Council 1997, and partial support was provided by the National Science Foundation Award DMS-9626159 at Duke University. Figure 1.10 is reproduced with the permission of the Salvador Dalí Museum, Inc. in St. Petersburg, Florida.

Many colleagues contributed to this project in different ways: Anestis Antoniadis, Tony Cai, Merlise Clyde, Lubo Dechevsky, Iain Johnstone, Gabriel Katul, Eric Kolaczyk, Pedro Morettin, Peter Müller, Giovanni Parmigiani, Marianna Pensky, David Rios, Fabrizio Ruggeri, Rainer von-Sacks, Naoki Saito, Yazhen Wang, and Gilbert Walter, to list a few. Collaboration with software gurus Hong-Ye Gao [TeraLogic Inc.] and Andrew Bruce [MathSoft Inc.] was fruitful. The S+Wavelets module (for S-Plus) was used for almost all of the computer examples, figures, and calculations. I am grateful to Alison Bory, Angioline Loredo, and Steve Quigley from Wiley, for their enthusiastic assistance, and to Courtney Johnson, Michael Kozdron, and Kathy Zhou, doctoral students at Duke University, for their help in proofreading the manuscript.

And most of all, I am grateful to my family for their love and strong and continuous support.

1
Introduction

In this chapter, we give a brief overview of the history of wavelets, make a case for their use in statistics, and provide a real-life example that emphasizes specificities of wavelets in data processing problems. The wavelet method in this example is compared with its counterpart traditional approaches. The reader may encounter unfamiliar jargon or undefined objects. Some of these notions will be defined later and some are used to illustrate the general picture.

1.1 WAVELET EVOLUTION

Wavelets are developed not only from a couple of bright discoveries, but from concepts and theories that already existed in various fields. In this section, we will give a brief historic tour of some important milestones in the development of wavelets.

Functional series have a long history that can be traced back to the early nineteenth century. French mathematician (and politician!) Jean-Baptiste-Joseph Fourier [Fig. 1.1(a)] in 1807 [1] decomposed a continuous, periodic on $[-\pi, \pi]$ function $f(x)$ into

---

[1] Jean-Baptiste-Joseph Fourier's *Theorie analitique de la chaleur* (The Mathematical Theory of Heat) inaugurated simple methods for the solution of boundary value problems occurring in the conduction of heat.

the series

$$\frac{a_0}{2} + \sum_{n=1}^{\infty} a_n \cos nx + b_n \sin nx,$$

where the coefficients a_n and b_n are defined as

$$a_n = \frac{1}{\pi} \int_{-\pi}^{\pi} f(x) \cos nx \, dx, \quad n = 0, 1, 2, \ldots$$
$$b_n = \frac{1}{\pi} \int_{-\pi}^{\pi} f(x) \sin nx \, dx, \quad n = 1, 2, \ldots .$$

It is interesting that, at the time of Fourier's discovery, the notion of a function was not yet precisely defined.

Fig. 1.1 (a) Jean-Baptiste-Joseph Fourier 1768-1830 and (b) Alfred Haar 1885-1933.

The first "wavelet basis" was discovered in 1910 when Alfred Haar [Fig. 1.1(b)] showed that any continuous function $f(x)$ on $[0, 1]$ can be approximated by

$$f_n(x) = \langle \xi_0, f \rangle \xi_0(x) + \langle \xi_1, f \rangle \xi_1(x) + \cdots + \langle \xi_n, f \rangle \xi_n(x), \tag{1.1}$$

and that, when $n \to \infty$, f_n converges to f uniformly ([181]). The coefficients $\langle \xi_i, f \rangle$ are given by $\int \xi_i(x) f(x) dx$. The Haar basis is very simple:

$$\begin{aligned}
\xi_0(x) &= \mathbf{1}(0 \le x \le 1), \\
\xi_1(x) &= \mathbf{1}(0 \le x \le 1/2) - \mathbf{1}(1/2 \le x \le 1), \\
\xi_2(x) &= \sqrt{2}[\mathbf{1}(0 \le x \le 1/4) - \mathbf{1}(1/4 \le x \le 1/2)], \\
&\ldots \\
\xi_n(x) &= 2^{j/2}[\mathbf{1}(k \cdot 2^{-j} \le x \le (k+1/2) \cdot 2^{-j}) \\
&\quad - \mathbf{1}((k+1/2) \cdot 2^{-j} \le x \le (k+1) \cdot 2^{-j})], \\
&\ldots
\end{aligned}$$

where n is uniquely decomposed as $n = 2^j + k$, $j \ge 0$, $0 \le k \le 2^j - 1$, and $\mathbf{1}(A)$ is the indicator of a set A, i.e., $\mathbf{1}(A) = 1$, if $x \in A$, and $\mathbf{1}(A) = 0$, if $x \in A^c$.

The approximation in (1.1) is equivalent to an approximation by step functions whose values are the averages (mean values) of the function over appropriate dyadic intervals.

Fig. 1.2 gives an exemplary function, $f(x) = \sin \pi x + \cos 2\pi x + 0.6 \cdot \mathbf{1}(x > 1/2)$, and three different levels of approximation: f_3, f_{15}, and f_{63}. Basis functions ξ_1, ξ_2, ξ_{14}, and ξ_{25} are shown in Fig. 1.3. Since $\int \xi_n^2(x)\,dx = 1$ for an arbitrary n, there is a trade-off between the magnitude and the support of the basis functions in the Haar system.

Notice that for any $n \ge 1$ the basis function ξ_n can be expressed as a scale-shift transformation of a single function ξ_1,

$$\xi_n(x) = 2^{j/2}\xi_1(2^j \cdot x - k), \quad n = 2^j + k,$$

a property shared by critically sampled wavelets, as we will see later. The function $\xi_0(x)$ is different in nature than the functions ξ_n, $n \ge 1$; while the functions ξ_n, $n \ge 1$ describe the details in the decomposition, the function $\xi_0(x)$ is responsible for the "average" of the decomposed function.

The Schauder basis on $[0, 1]$ (Schauder [369]) consists of the primitives of the Haar basis functions, the triangle functions. Let $\Delta(x) = 2x\,\mathbf{1}(0 \le x \le 1/2) + 2(1-x)\,\mathbf{1}(1/2 \le x \le 1)$, and let $\Delta_n(x) = \Delta(2^j x - k)$, $n = 2^j + k$, $j \ge 0$, $0 \le k \le 2^j - 1$. Then $\{1, \Delta(x), \Delta_1(x), \ldots\}$ constitutes a Schauder basis on $[0, 1]$ and, as in the case of Haar's basis, any continuous function $f(x)$ on $[0, 1]$ can be approximated by

$$f_N(x) = a + bx + \sum_{n=1}^{N} s_n \Delta_n(x). \tag{1.2}$$

Coefficients a and b are solutions of the system $f(0) = a$ and $f(1) = a + b$, while the coefficients s_n can be obtained by the simple relation

4 INTRODUCTION

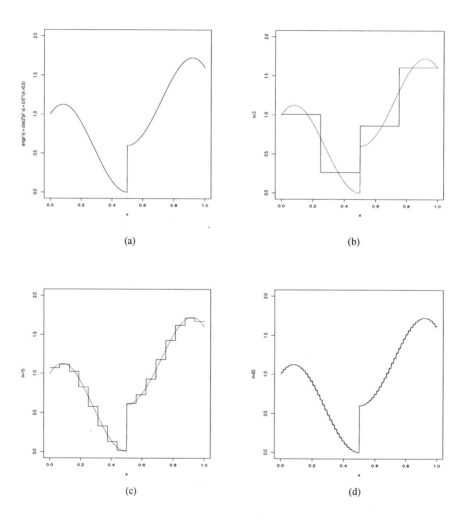

Fig. 1.2 Panels (a)-(d) show the original function $f(x) = \sin \pi x + \cos 2\pi x + 0.6 \cdot \mathbf{1}(x > \frac{1}{2})$, $0 \leq x \leq 1$, and three different levels of approximation in the Haar basis. Using the notation of (1.1), approximations f_3, f_{15}, and f_{63} are plotted.

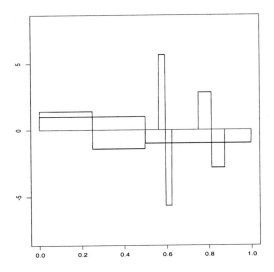

Fig. 1.3 Functions ξ_1, ξ_2, ξ_{14}, and ξ_{25} from the Haar basis of $\mathbb{L}_2([0,1])$.

$$s_n = f\left(\frac{k+1/2}{2^j}\right) - \frac{1}{2}\left[f\left(\frac{k}{2^j}\right) + f\left(\frac{k+1}{2^j}\right)\right],$$
$$n = 2^j + k,\ j \geq 0,\ 0 \leq k \leq 2^j - 1.$$

The convergence $f_N(x) \to f(x)$ is uniform and the coefficients are unique; however, the Schauder system is not orthogonal. We will see later that its orthogonalization leads to a family of wavelets, known as Franklin wavelets.

In the mid-1930s, Littlewood-Paley techniques (based on Fourier methods) [264] were broadly used in research on Fourier summability and in investigation of the behavior of analytic functions.

Prototypes of wavelets first appeared in Lusin's work in the 1930s. A standard characterization of Hardy's spaces can be given in terms of Lusin's "area" functions.

In the 1950s and 1960s, techniques by Littlewood-Paley and Lusin were developed into powerful tools for studying physical phenomena describable by solutions of differential and integral equations. Researchers realized that these techniques could be unified by the Calderón-Zygmund theory [292], now a branch of harmonic analysis.

Strömberg [390] was the first to construct an orthonormal basis of $\mathbb{L}_2(\mathbb{R})$ of the form $\{\psi_{jk}(x) = 2^{j/2}\psi(2^j x - k),\ j,k \in \mathbb{Z}\}$, a wavelet-like basis more general than Haar's basis. Stromberg's construction uses Franklin systems which are Gram-Schmidt orthogonalized Schauder basis functions $\Delta_n(x)$.

For more information about the historical roots of wavelets, we direct the reader to monographs by Meyer [294, 295] and Daubechies [104].

1.2 WAVELET REVOLUTION

The first definitions of wavelets can be attributed to Morlet et al. [300] and Morlet and Grossmann [179] and it is given in the Fourier domain: A wavelet is an $\mathbb{L}_2(\mathbb{R})$ function for which the Fourier transformation $\Psi(\omega)$ satisfies

$$\int_0^\infty |\Psi(t\omega)|^2 \frac{dt}{t} = 1, \text{ for almost all } \omega.$$

The definition of Morlet and Grossmann is quite broad and over time the meaning of the term *wavelet* became narrower. Currently, the term wavelet is usually associated with a function $\psi \in \mathbb{L}_2(\mathbb{R})$ such that the translations and dyadic dilations of ψ,

$$\psi_{jk}(x) = 2^{j/2} \psi(2^j x - k), \ j, k \in \mathbb{Z} \tag{1.3}$$

constitute an orthonormal basis of $\mathbb{L}_2(\mathbb{R})$.

Calculating wavelet expansions directly is a computationally expensive task, moreover, most interesting wavelets are without a closed form. In the mid-1980s, Mallat [274, 275, 276] connected quadrature-mirror filtering and pyramidal algorithms from the signal processing theory with wavelets. He demonstrated that discrete wavelet transformation can be calculated very rapidly via cascade-like algorithm. This link was of paramount importance for the practice of wavelets. Daubechies' discovery of compactly supported wavelet bases represents another important milestone in the development of wavelet theory. Daubechies' bases are versatile in smoothness and locality and represent a starting point for much of the subsequent generalizations and theoretical advances.

Wavelet theory has developed now into a methodology used in many disciplines: mathematics, geophysics, astronomy, signal processing, numerical analysis, and statistics, to list a few. Wavelets are providing a rich source of useful and sometimes intriguing tools for applications in "time-scale" types of problems. In analyses of signals, the wavelet representations allow us to view a time-domain evolution in terms of scale components. In this respect, wavelet transformations behave similarly to Fourier transformations. The Fourier transform extracts details from the signal frequency, but all information about the location of a particular frequency within the signal is lost. Time localization can be achieved by first windowing the signal, and then by taking its Fourier transform. The problem with windowing is that slices of the processed signal are of a fixed length, which is determined by the window. Slices of the same length are used to resolve both high and low frequency components. For nonstationary signals, this lack of adaptivity may lead to a local under- or over-fitting.

Table 1.1 Coefficients of the doppler function in the Haar basis plotted in levels determined by the length of support of corresponding basis functions, ξ_n, $n \geq 0$.

level in Fig. 1.4	coefficients of	support	j and k in the notation: $n = 2^j + k$ j	k
...	...	...	...	...
d1	ξ_{512} - ξ_{1023}	$1/2^9$	$j = 9$	$0 \leq k \leq 2^9 - 1$
d2	ξ_{256} - ξ_{511}	$1/2^8$	$j = 8$	$0 \leq k \leq 2^8 - 1$
...	...	...	...	...
d8	ξ_4 - ξ_7	$1/4$	$j = 2$	$0 \leq k \leq 3$
d9	ξ_2 - ξ_3	$1/2$	$j = 1$	$0 \leq k \leq 1$
d10	ξ_1	1	$j = 0$	$k = 0$
s10	ξ_0	1		
...	...	...	...	...

In contrast to windowed Fourier transforms, wavelets select widths of time slices according to the local frequency in the signal. This adaptivity property of wavelets is very important, and we will make it more precise later in the discussion of Heisenberg's uncertainty principle. Two panels in Fig. 1.11, on page 18, depict slicing the time–scale plane for a windowed Fourier (left) and a wavelet transformation (right).

Now we give several examples: The first example views the Haar decomposition as a wavelet decomposition and discusses connections between "levels" and resolutions of the decomposition. The subsequent four examples demonstrate important properties of wavelets: the ability to filter, "disbalance", and "whiten" signals as well as to detect self-similarity within a signal.

Example 1.2.1 The Haar basis as a wavelet basis. To illustrate the time and scale adaptivity of wavelets, and to introduce some necessary wavelet notations and jargon, let us consider a decomposition of the function

$$y(x) = \sqrt{x(1-x)} \sin \frac{2.1\pi}{x + 0.05}, \quad 0 \leq x \leq 1, \qquad (1.4)$$

in Haar's basis. This function is known as the doppler test-function. Notice that frequency in the function increases as x decreases.

In Table 1.1, n is represented as $2^j + k$ where j is a *level* and k is a *shift* within the level. Notice that all functions ξ within a level have supports of the same length. The support of a function is defined as closure of the set at which the function differs from zero.

When $j \to \infty$, the number of coefficients in the level increases and the length of support of the corresponding basis functions decreases. For example, the level indexed by $j = 5$ has $2^5 = 32$ coefficients and the supports are of length 2^{-5}. The shifts within a level are indexed by k, where k ranges from 0 to $2^j - 1$. For

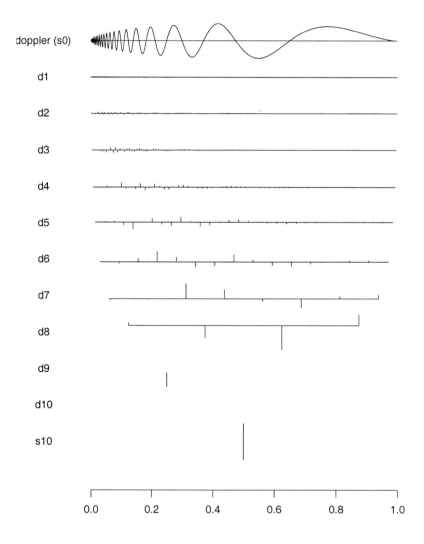

Fig. 1.4 The doppler function and its Haar basis decomposition.

an exact description of an arbitrary function, the number of levels is infinite. The coefficient corresponding to ξ_0 is called the "smooth" coefficient ($s0$ in Fig. 1.4) and the coefficients corresponding to ξ_n, $n > 1$ are called "detail" coefficients. Level $j = 9$ ($d1$ in Fig. 1.4) contains coefficients corresponding to "fine" details.

When dealing with functions that are given by their sampled values, it is customary to set the sampled values to be "smooth" coefficients at the level $j = J$. The subsequent "detail" levels denoted by $d1, d2, \ldots$, correspond to $j = J-1, J-2, \ldots$.

We provide four more examples that emphasize the most interesting features of wavelet transformations. Occasionally, we will use terms like "fine and coarse levels", "wavelet domain", and "energy", which have not been previously defined and will be defined in the subsequent chapters. However, the intended messages of the examples should be clear even without precise definitions of such terms.

Example 1.2.2 Wavelets generate local bases. Classical orthonormal bases (Fourier, Hermite, Legendre, etc.) have been used with great success in applied mathematics for decades. However, there is a serious limitation shared by many classical bases, which is *non-locality*. A basis is non-local when many basis functions are substantially contributing at any value of a decomposition. The convergence of non-local classical decompositions often relies on a multitude of cancellations.

Local bases are desirable since they are more adaptive and parsimonious. In 1946, Gabor [161] suggested localizing Fourier bases by modulating and translating an appropriate "window" function g. More precisely, Gabor suggested bases in the form

$$\{g_{m,n}(x) = e^{2\pi m i x} g(x - n)\},$$

where m and n are integers and g is a square-integrable function. An example of a function g that produces an orthonormal basis of $\mathbb{L}_2(\mathbb{R})$ is $\sin(\pi x)/(\pi x)$.

The Balian-Low theorem stipulates limitations of Gabor bases. If the Gabor basis is orthogonal and $\hat{g}(\omega)$ is the Fourier transformation of the window $g(x)$ then, by the Balian-Low theorem, either $\int x^2 |g(x)|^2 \, dx = \infty$ or $\int \omega^2 |\hat{g}(\omega)|^2 \, d\omega = \infty$. In other words, orthogonal Gabor bases are non-local either in time or in scale (frequency). Modulations and translations of the Gaussian window $g(x) = \frac{1}{\sqrt{2\pi}} e^{-x^2/2}$ (which is well localized in both time and frequency, and for which the above integrals are finite) will not produce an orthonormal basis.

Locality of wavelet bases comes from their construction. Most of the wavelets that are used in statistics now are either compactly supported or decay exponentially. An exception are Meyer-type wavelets (with a polynomial decay) used in deconvolution problems.

Example 1.2.3 Wavelets filter data. To illustrate the action of wavelets as a filtering device, we generate two periodic functions with different frequencies, $y_1 = \sin x + \cos 2x$, and $y_2 = \frac{1}{5} \arcsin(\sin 20x)$, where $x \in [-2\pi, 2\pi]$. These are shown in panel (a) in Fig. 1.5. Our goal is to filter out the component y_2 from the given sum $y_1 + y_2$

[Fig. 1.5(b)]. Since the periods of y_1 and y_2 are different, the functions are described by wavelets with different supports (and whose coefficients belong to different levels). Fig. 1.5(c), depicts the level-wise energies (sums of squares of wavelet coefficients). The support of wavelets associated with level 1 is 32 times larger than the support of wavelets associated with level 5. This means that almost all the energy in levels 0,1, and 2 comes from signal y_1, and the energy in level 5 comes from y_2, thus allowing an easy separation. The filtered components are depicted in Fig. 1.5(d).

Example 1.2.4 Wavelets "disbalance" energy in data. The term "disbalance" is coined and it relates to an uneven distribution of energy in a signal. Disbalancing is desirable since a signal can be well described by only a few energetic components.

To illustrate the disbalancing action that is typical of wavelets, we first introduce some necessary notation. Given a vector $\underline{a} = (a_1, a_2, \ldots, a_n)$ let $||\underline{a}||^2 = \sum_i a_i^2$ be the total energy of $\underline{a}$ and let a_i^2 be the ith energy component. Let $a_{(1)}^2, a_{(2)}^2, \ldots, a_{(n)}^2$ be increasingly ordered energy components. The standard measure of disbalance used in economics is the Lorentz curve. The Lorentz curve was introduced at the beginning of the century. It was used by economics researchers to assess inequality of distribution of wealth in a country, region, or among people within a particular population group.

One definition of the Lorentz curve, in terms of energy components, is

$$L(p) = \frac{1}{||\underline{a}||^2} \cdot \sum_{i=1}^{\lfloor np \rfloor} a_{(i)}^2, \ p \in [0, 1],$$

where $\lfloor x \rfloor$ is the largest integer smaller than x. In Fig. 1.6(a), an observed time series (turbulence data set) is given. Below is its wavelet transformation represented in a vector form beginning with coarse coefficients. Orthogonality of the transformation preserves the total energy, $||\underline{a}||^2$. However, the energy in the wavelet domain is more disbalanced, as indicated by the Lorentz curves in Fig. 1.6(b). Notice that 90% of energy is contained in about 6-7% of the components in the wavelet-transformed data set compared to nearly 50% of the components in the original (time) domain.

Example 1.2.5 Wavelets whiten data. In this example, we show another interesting property of wavelets. Orthogonal wavelet transformations map white noise to white noise, which is a consequence of orthogonality. However, signals that are correlated in the time domain become almost uncorrelated in the wavelet domain. Informally, the wavelet transformation acts as an approximation to the Karhunen-Loève transformation. To exemplify this statement, a time series of 256 components was generated from a random process with stationary increments, ARIMA(1,1,1) process. Such processes exhibit long-range dependence and their autocovariance functions [Fig. 1.7(a)] show slow decay. The autocovariance function of the wavelet-transformed time series exhibits very different behavior. Only the covariances at the first few lags are significant at a 5% significance level.

Related discussion can be found in Johnstone and Silverman [222], Mallat [277],

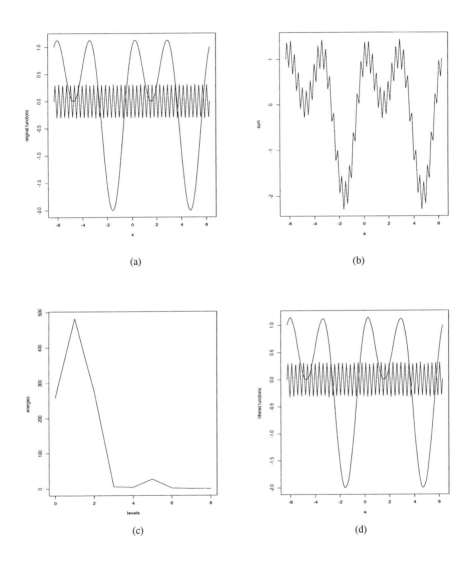

Fig. 1.5 Filtering property of wavelets. Two functions $y_1 = \sin x + \cos 2x$ and $y_2 = \frac{1}{5}\arcsin(\sin 20x)$, and their sum $y_1 + y_2$ are plotted in panels (a) and (b). Panel (c) shows the separation of "energy" to different levels in wavelet decomposition, while panel (d) shows filtered functions.

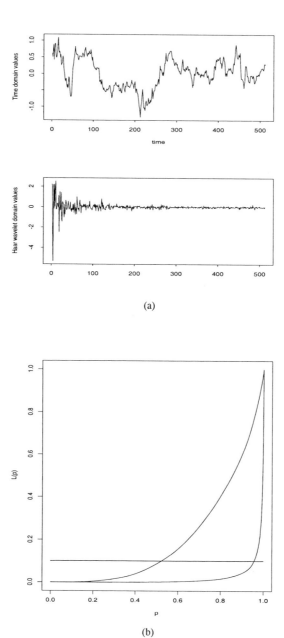

Fig. 1.6 (a) Atmospheric turbulence measurements of u velocity component (upper panel) and their wavelet transformation (lower panel). (b) Lorentz curves of the original and transformed measurements. The curve corresponding to transformed measurements has higher curvature.

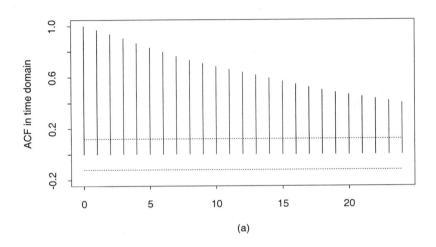

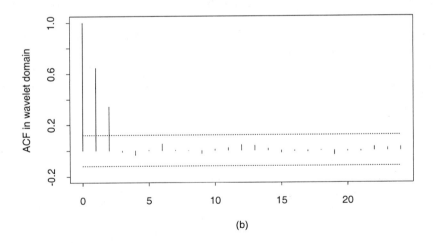

Fig. 1.7 Illustration of the whitening effect of wavelet transformations. Autocovariance function for a time series [ARIMA(1,1,1)] in the time domain [panel (a)] and the wavelet domain [panel (b)].

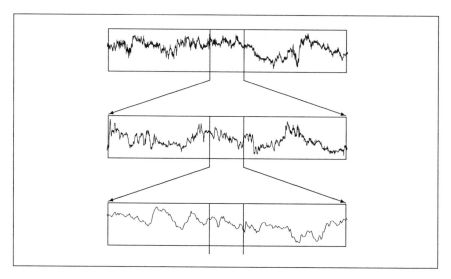

Fig. 1.8 Self-similarity of a turbulence time series.

Walter [439], and Wornell [461].

Example 1.2.6 Wavelets detect self-similar phenomena. Being self-similar themselves, wavelets are especially apt to describe phenomena exhibiting self-similarity in different scales (Fig. 1.8). Early research on wavelets was generated to address related problems in geophysics, especially in turbulence. An overview can be found in Kumar and Foufoula-Georgiou [250]. A curious phenomenon is that atmospheric turbulence measurements of different physical quantities, such as air velocities, ozone and humidity concentrations, temperature, and so on, follow identical power laws (as predicted by Kolmogorov's [242] theory). Such laws describe the energy transport in the inertial range of turbulent flows. A nice reference is a book by Frisch [160].

One of the theoretical laws is the "$-\frac{5}{3}$" law. It states that the log-power spectrum in the inertial range decreases linearly, with the slope of $-\frac{5}{3}$. Fig. 1.9 shows the wavelet-spectrum of air velocity measurements and it's near-perfect compliance with the $-\frac{5}{3}$ law.

There are problems in which wavelets should be used with caution. For instance, in the wavelet domain, the dependence structure in the transformed time series is influenced by the choice of the decomposing wavelet. In some cases, the extent of such non-robustness hinders practical generalizations. When non-robustness is of particular concern, researchers usually fix a *good* wavelet for a class of problems, as is the case with the prevalent use of the Haar and Walsh bases in processing the turbulence data.

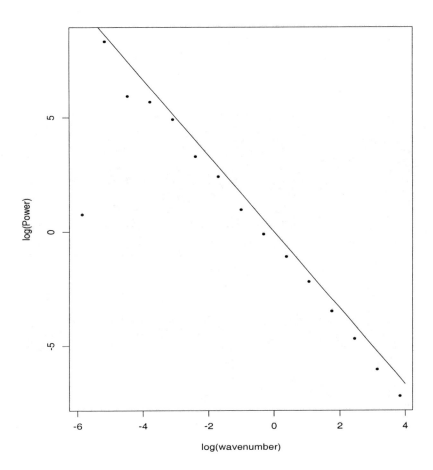

Fig. 1.9 Wavelet power-spectrum and Kolmogorov's $-\frac{5}{3}$ law. Dots represent the logarithms of the cumulative level-energies. The variable `log(wavenumber)` is linearly related to the level j.

16 INTRODUCTION

Wavelets are not replacements for the standard Fourier methods, they are alternatives. If the signal is a linear combination of harmonics, clearly wavelets are suboptimal building blocks. For instance, if f is given as a lacunary (sparse) Fourier series like $\sum_{j=0}^{\infty} 2^{-j} \sin(2\pi 2^j x)$, then wavelets will be inferior in tasks of denoising and compression as compared to the Fourier transformation.

The whitening property, discussed in Example 1.2.5, impairs the performance of wavelet-based methods in prediction problems.

Another example in which wavelets should be cautiously used comes from image processing. In this example the interpretation of an object changes when the resolution changes. The two women in the background in Salvador Dalí's picture *Mercado de esclavos con aparicion del busto invisible de Voltaire*[2] (Fig. 1.10) at a coarser resolution level can be interpreted as a bust of Voltaire. Clearly, the meaning of the object changes in different scales.

Fig. 1.10 *Mercado de esclavos con aparicion del busto invisible de Voltaire,* a 1940 painting by Salvador Dalí.

1.3 WAVELETS AND STATISTICS

Statistical multiscale modeling has, in recent years, become a burgeoning area in both theoretical and applied statistics, and is beginning to impact developments

---

[2] *Slave Market with the Disappearing Bust of Voltaire (1940), Oil on canvas.* $18\frac{1}{4} \times 25\frac{3}{8}$ *in.* Collection of The Salvador Dalí Museum, St. Petersburg, Florida. ©1998 Salvador Dalí Museum, Inc.

in statistical methodology as well as in various applied scientific fields. Wavelet-based methods are developing in statistics in areas such as regression, density and function estimation, factor analysis, modeling and forecasting in time series analysis, and spatial statistics. Emerging connections of Bayesian statistical modeling and wavelets are generating exciting new directions for the interface of the two research areas, with significant potential for future impact on applied work.

The attention of the statistical community was attracted when Mallat established a connection between wavelets and signal processing and Donoho and Johnstone showed that wavelet thresholding had desirable statistical optimality properties. Since then, wavelets have proved useful in many statistical disciplines, notably in nonparametric statistics and time series analysis. Bayesian concepts and modeling approaches have, more recently, been identified as providing promising contexts for wavelet-based denoising applications.

In addition to replacing traditional orthonormal bases in a variety statistical problems, wavelets brought novel techniques and invigorated some of the existing ones. Even in the cases in which the traditional orthogonal series are simply replaced by wavelet bases, wavelets often offer better localization and parsimony. For example, Čencov's [66] linear density estimator in the form of a Fourier series uses traditional orthonormal bases (Hermite, Fourier) to express its empirical Fourier coefficients. Wavelets achieve the same convergence rates and at the same time provide efficient non-linear approximations and adaptivity to unknown smoothness (via wavelet shrinkage). Wavelet shrinkage is achieved via explicit or implicit use of statistical models in the wavelet domain.

We elaborate further on the modeling in the wavelet domain and formalize some of the concepts already mentioned.

Low Entropy Modeling Environment. As we mentioned before, wavelet transformations tend to disbalance the data on input. Even though the transformations preserve the ℓ_2-norm of the data, the energy of the transformed data (an engineering term for the ℓ_2-norm) is concentrated in only a few wavelet coefficients. This concentration narrows the class of plausible statistical models and facilitates the thresholding. Different formalizations of this disbalancing property can yield a variety of criteria for the best basis selection. For more discussion, see Coifman and Wickerhauser [94], Donoho [123], and Mallat [277], among others.

Ockham's Razor Principle. Wavelets, as building blocks of models, are well localized in both time and scale (frequency). Signals with rapid local changes (signals with discontinuities, cusps, sharp spikes, etc.) can be precisely represented with just a few wavelet coefficients. Generally, this statement does not apply to other standard orthonormal bases that may require many "compensating" coefficients to describe discontinuity artifacts or to suppress Gibbs' effects. The latest "generation" of wavelets form over-complete dictionaries and provide parsimonious representations of real phenomena with complicated time and frequency behaviors.

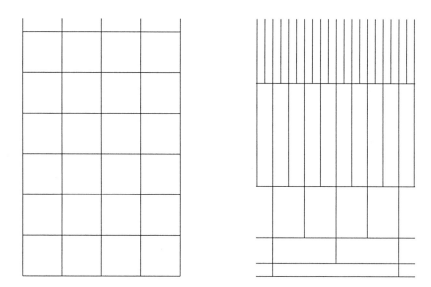

Fig. 1.11 Localized Fourier and wavelet paving of time–scale space.

By-Passing the Curse of Heisenberg. Heisenberg's principle states that in modeling time-frequency phenomena, we cannot be precise in the time domain and in the frequency domain simultaneously. In other words, squares and rectangles in the pavement of the time–scale plane (as given in Fig. 1.11) have areas bounded from below by a universal constant.

Wavelets automatically trade-off the time-frequency precision by their innate nature. The parsimony mentioned above can be ascribed to the ability of wavelets to cope with limitations of Heisenberg's principle in a data-dependent manner.

Whitening Property. There is ample theoretical and empirical evidence that wavelet transformations tend to simplify the dependence structure in the original data. It is even possible to construct a biorthogonal basis that will decorrelate a given stationary time series (a wavelet-counterpart of the Karhunen-Loève transformation). For a discussion and examples, see Walter [439].

Smoothness Control. Under mild conditions wavelets provide unconditional bases for many important smoothness spaces ($\mathbb{L}_p, p > 1$; Besov Spaces $\mathbb{B}_{pq}^\sigma$; Hölder Spaces $\mathbb{C}^\alpha$). Using simpler terminology, this means that by controlling the magnitude of the coefficients in the wavelet domain one controls the smoothness of the decomposed function. This connection provides the theoretical framework for wavelet smoothing and wavelet function and density estimation.

1.4 AN APPETIZER: CALIFORNIA EARTHQUAKES

We conclude this introductory chapter with a real-life example. The example we provide emphasizes basic differences between wavelet-based and standard denoising methods. It shows the ability of wavelets to "zoom-in" and adapt their space-scale "descriptors" to the data at hand.

A researcher from the geology department at Duke University was interested in the possibility of predicting earthquakes by monitoring water–levels in the nearby wells. To do this, he obtained water level measurements from six wells located in California that were taken every hour for approximately six years. The goal was to smooth the data, eliminate the noise, and inspect the signal at pre-earthquake time. Here is some background (provided by Dr. Stuart Rojstaczer, Duke University).

> The ability of water wells to act as strain meters has been observed for centuries. The Chinese, for example, have records of water flowing from wells prior to earthquakes. Lab studies indicate that a seismic slip occurs along a fault prior to rupture. Recent work has attempted to quantify this response, in an effort to use water wells as sensitive indicators of volumetric strain. If this is possible, water wells could aid in earthquake prediction by sensing precursory earthquake strain. Water level records from six wells in southern California are collected over a six year time span. At least 13 moderate size earthquakes (Magnitude 4.0 - 6.0) occurred in close proximity to the wells during this time interval. There is a significant amount of noise in the water level record which must first be filtered out. Environmental factors such as earth tides and atmospheric pressure create noise with frequencies ranging from seasonal to semidiurnal. The amount of rainfall also affects the water level, as do surface loading, pumping, recharge (such as an increase in water level due to irrigation), and sonic booms, to name a few.

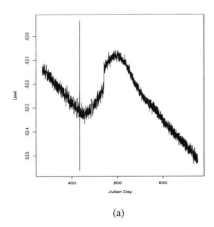
(a)

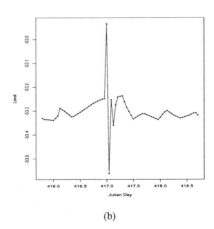
(b)

Fig. 1.12 (a) California water-level data set. (b) Water-level oscillation at the earthquake time.

Once the noise is subtracted from the signal, the record can be analyzed for changes in water level, either an increase or a decrease depending upon whether the aquifer is experiencing a tensile or compressional volume strain, just prior to an earthquake.

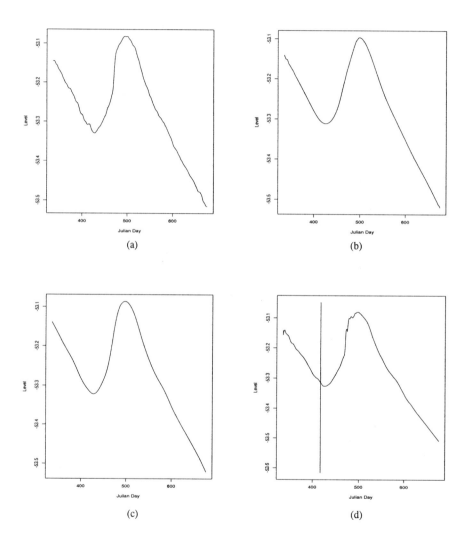

Fig. 1.13 Comparison of several smoothing methods. (a) Data smoothed by kernel method (normal window, $k = 5$); (b) Data smoothed by loess method; (c) Data smoothed by supsmu method; and (d) Wavelet smoothed data (Daubechies' wavelet with four vanishing moments).

A plot of the raw data for hourly measurements collected over one year (8192 =

2^{13} observations), is given in Fig. 1.12(a). The line-like artifact [enlarged in Fig. 1.12(b)] represents a line connecting two extreme level values at the earthquake time (Julian day of 417).

The measurements were smoothed by three traditional methods (kernel, lowess, and supsmu) and by wavelet shrinkage. Fig. 1.13(a), (b), and (c) are processed data, smoothed by the kernel method (normal kernel, bandwidth = 5), by the locally weighted regression smoother (implemented in S-Plus as `lowess`), and by the local cross-validation smoother (implemented in S-Plus as `supsmu`) method.

Rather than discussing whether the filtering indicates that the earthquake could have been predicted, we emphasize differences in the outputs of traditional and wavelet smoothing methods. Notice that in all traditional methods the artifact of interest (earthquake jump) is lost. The application of nonlinear wavelet shrinkage to the data, results in a smooth signal with the jump at earthquake time preserved. The wavelet-smoothed data are given in Fig. 1.13(d). Only 20 of 8192 (0.244%) coefficients (those 20 with largest magnitude) were used in describing the wavelet shrinkage estimator.

2
Prerequisites

In this chapter, we introduce notation and briefly review several mathematical concepts necessary for the definition and derivation of the basic properties of wavelets. Some fundamental concepts from the theory of Hilbert spaces, Fourier analysis, linear algebra, and signal processing will be used to define the multiresolution analysis and to develop wavelet formalism.

2.1 GENERAL

For denoting the sets of natural, integer, real, and complex numbers we use notation $\mathbb{N}, \mathbb{Z}, \mathbb{R}$, and $\mathbb{C}$. The modulus of a complex number $z \in \mathbb{C}$ will be denoted by $|z|$, and the complex conjugate by $\bar{z}$. The set of positive real numbers will be denoted by $\mathbb{R}^+$. It is tacitly assumed that all functions are measurable. The support of a function f, denoted $\mathrm{supp}(f)$, is the closure in $\mathbb{R}$ of the set $\{x \in \mathbb{R} : f(x) \neq 0\}$.

The indicator of a relation ρ, $\mathbf{1}(\rho)$, is defined to be 1 if the relation ρ is satisfied and 0 otherwise. The Kronecker delta $\delta_{u,v}$ can be defined using the indicator function as $\mathbf{1}(u = v)$. We also define δ_u to be $\delta_{u,0}$. Maximum and minimum of a and b are denoted by $a \vee b$ and $a \wedge b$, respectively.

Let $f_+ = f \cdot \mathbf{1}(f \geq 0) = f \vee 0$ be a positive part of a function f and $f_- = -f \cdot \mathbf{1}(f \leq 0) = -(f \wedge 0)$ be its negative part. By definition, $|f| = f_+ + f_-$ and $f = f_+ - f_-$. We will sometimes use O-notation; $a_n = O(b_n)$ would mean that a_n/b_n is asymptotically bounded away from 0 and ∞; $a_n = o(b_n)$ would mean that $a_n/b_n \to 0$.

A Lebesgue point of a function f is any point x such that

$$\lim_{r \to 0} \frac{1}{2r} \int_{-r}^{r} |f(x+t) - f(x)|\, dt = 0.$$

The Dirac function $\delta(x)$ (not to be confused with the Kronecker symbol $\delta_{k,l}$) is defined as

$$\delta(x) = \lim_{a \to 0} \frac{1}{a} \mathbf{1}[0 \le x \le a]. \tag{2.1}$$

The Dirac function (2.1) satisfies the following relations

$$\int_{\mathbb{R}} \delta(x)\, dx = 1,$$

$$\int_{\mathbb{R}} f(x)\delta(x - x_0)\, dx = f(x_0).$$

It can be thought of as a generalized derivative of a Heaviside step function $H(x) = \mathbf{1}(x \ge 0)$.

2.2 HILBERT SPACES

Hilbert spaces are natural generalizations of finite dimensional Euclidean spaces $\mathbb{R}^n$.

Working with abstract Hilbert spaces is beneficial in several respects. Our geometric intuition, based on properties of Euclidean $\mathbb{R}^2$ or $\mathbb{R}^3$ spaces, can in part be easily extended to an arbitrary Hilbert space. An example is the *projection theorem* (Theorem 2.2.1). The norm in the Hilbert space is connected with a quadratic expression and the process of norm-minimization falls in the class of linear problems. All separable Hilbert spaces are (abstractly) equivalent to one another.

The Inner Product Space. A complex vector space $\mathcal{H}$ is said to be an inner product space if for any two elements $x, y \in \mathcal{H}$ there exists a complex number $\langle x, y \rangle$ (called the inner product of x and y) that satisfies

(i) $\langle x, y \rangle = \overline{\langle y, x \rangle}$
(ii) $\langle x + y, z \rangle = \langle x, z \rangle + \langle y, z \rangle$, for all x, y, and $z \in \mathcal{H}$.
(iii) $\langle \alpha x, y \rangle = \alpha \langle x, y \rangle$ for $x, y \in \mathcal{H}$ and $\alpha \in \mathbb{C}$.
(iv) $\langle x, x \rangle \ge 0$, for all $x \in \mathcal{H}$.
(v) $\langle x, x \rangle = 0$, if and only if $x = 0$.

The norm $||x||$ of an element $x \in \mathcal{H}$ is defined via inner product, $||x|| = \sqrt{\langle x, x \rangle}$.

Example 2.2.1 Euclidean space $\mathbb{R}^n$.
$x = (x_1, \ldots, x_n)$
$y = (y_1, \ldots, y_n)$
$\langle x, y \rangle = \sum_{i=1}^{n} x_i y_i, \quad ||x|| = \sqrt{\sum_{i=1}^{n} x_i^2}$

Example 2.2.2 $\mathbb{L}_2(\mathbb{R})$ space (space of all square-integrable functions).
$f \in \mathbb{L}_2(\mathbb{R})$ if $\int |f|^2 < \infty$.
$\langle f, g \rangle = \int fg, \quad ||f|| = \sqrt{\int f^2}$.
If $f, g \in \mathbb{L}_2(\mathbb{C})$, $\langle f, g \rangle = \int f\bar{g}$ and $||f|| = \sqrt{\int f\bar{f}}$.

Example 2.2.3 ℓ_2 space (space of all square-summable sequences).
$\underline{x} = \{x_n\} \in \ell_2$ if $\sum_{n \in \mathbb{Z}} |x_n|^2 < \infty$.
$\langle \underline{x}, \underline{y} \rangle = \sum_{i \in \mathbb{Z}} x_i \overline{y_i}, \quad ||\underline{x}|| = \sqrt{\sum_{i \in \mathbb{Z}} |x_i|^2}$.

Example 2.2.4 A function f belongs to the *Lebesgue space* $\mathbb{L}_p(\mathbb{A})$, $1 \leq p < \infty$, if

$$||f||_p = \left(\int_{\mathbb{A}} |f(x)|^p \, dx \right)^{1/p} < \infty, \quad \text{and}$$
$$||f||_\infty = \operatorname*{ess\,sup}_{x \in \mathbb{A}} |f(x)| < \infty.$$

To "upgrade" the linear space $\mathcal{H}$ equipped with a norm to the Hilbert space one needs the *completeness property*.

Definition 2.2.1 *The sequence $\{x_n\}_{n \in \mathbb{N}}$ is called a Cauchy sequence in $\mathcal{H}$ if and only if (iff)*

$$||x_m - x_n|| \to 0,$$

whenever $m, n \to \infty$.

The space $\mathcal{H}$ is complete if any Cauchy sequence $\{x_n\}$ is convergent, i.e., $x_n \to x \in \mathcal{H}$.

2.2.1 Projection Theorem

A linear subspace $\mathcal{V}$ of a Hilbert space $\mathcal{H}$ is said to be a closed subspace of $\mathcal{H}$ if $\mathcal{V}$ contains all its limiting points, i.e., if $x_n \in \mathcal{V}$ and $||x_n - x|| \to 0$, as $n \to \infty$, then $x \in \mathcal{V}$.

The orthogonal complement of a subset $\mathcal{V}$ of $\mathcal{H}$ is defined to be the set $\mathcal{V}^\perp$ of all elements of $\mathcal{H}$ that are orthogonal to every element of $\mathcal{V}$, i.e., $x \in \mathcal{V}^\perp$ if and only if $\langle x, y \rangle = 0$, for all $y \in \mathcal{V}$.

Corollary 2.2.1 *If $\mathcal{V}$ is any subset of the Hilbert space $\mathcal{H}$, then $\mathcal{V}^\perp$ is a closed subspace of $\mathcal{H}$.*

Theorem 2.2.1 *(Projection theorem) If $\mathcal{V}$ is a closed subspace of the Hilbert space $\mathcal{H}$ and $x \in \mathcal{H}$, then*

(i) There is a unique element $\hat{x} \in \mathcal{V}$ such that

$$||x - \hat{x}|| = \inf_{y \in \mathcal{V}} ||x - y||,$$

and

(ii) $\hat{x} \in \mathcal{V}$ and $||x - \hat{x}|| = \inf_{y \in \mathcal{V}} ||x - y||$, if and only if $\hat{x} \in \mathcal{V}$, and $(x - \hat{x}) \in \mathcal{V}^\perp$.

The element $\hat{x}$ is called the orthogonal projection of x onto $\mathcal{V}$ and is denoted by $\text{Proj}_\mathcal{V} x$.

Example 2.2.5 Let $x_1 = (1, 0, 1)'$ and $x_2 = (1, 2, 0)'$ be two elements in (the Hilbert space) $\mathbb{R}^3$. Let $\mathcal{M} = \overline{span}\{x_1, x_2\}$ be the subspace spanned by x_1 and x_2. We will find the projection of the vector $y = (1, 1, 1)$ on $\mathcal{M}$.

Since $\mathcal{M}$ is spanned by x_1 and x_2, the projection $\hat{y}$ is of the form $\alpha_1 x_1 + \alpha_2 x_2$ for some constants α_1 and α_2. Since $y - \hat{y}$ belongs to $\mathcal{M}^\perp$ it is orthogonal to x_1 and x_2. This orthogonality translates to a system of two equations involving α_1 and α_2.

$$\langle y - \alpha_1 x_1 - \alpha_2 x_2, x_1 \rangle = 0,$$
$$\langle y - \alpha_1 x_1 - \alpha_2 x_2, x_2 \rangle = 0.$$

The above system is equivalent to

$$2\alpha_1 + \alpha_2 = 2$$
$$\alpha_1 + 5\alpha_2 = 3,$$

which has the solution $\alpha_1 = \frac{7}{9}$, and $\alpha_2 = \frac{4}{9}$. The projection is $\hat{y} = (\frac{11}{9}, \frac{8}{9}, \frac{7}{9})$.

2.2.2 Orthonormal Sets

The closed span $\overline{span}\{x_\lambda, \lambda \in \Lambda\}$ of any subset $\{x_\lambda, \lambda \in \Lambda\}$ of $\mathcal{H}$ is defined to be smallest closed subspace of $\mathcal{H}$ that contains each element $x_\lambda, \lambda \in \Lambda$.

Example 2.2.6 Let x_1, x_2 be two vectors in $\mathbb{R}^3$ such that $x_2 \neq \alpha x_1$, $\alpha = const$. Then $\overline{span}\{x_1, x_2\}$ is the plane containing the vectors x_1 and x_2.

Definition 2.2.2 *A set $\{e_\lambda, \lambda \in \Lambda\}$ of elements from $\mathcal{H}$ is orthonormal if*

$$\langle e_s, e_t \rangle = \delta_{s,t}, \ s, t \in \Lambda.$$

Example 2.2.7 (i) The set $\{(1,1,0), (1,-1,0), (0,0,1)\}$ is orthonormal in $\mathbb{R}^3$.
(ii) The set $\{\frac{1}{\sqrt{2\pi}}, \frac{1}{\sqrt{\pi}}\sin nx, \frac{1}{\sqrt{\pi}}\cos nx, \ n = 1, 2, \dots\}$ is orthonormal in the space of all square-integrable real functions on $[-\pi, \pi]$, $\mathbb{L}_2([-\pi, \pi])$, with an inner product defined by $\langle f, g \rangle = \int_{-\pi}^{\pi} f(x)g(x)\, dx$.

Let $\{e_1, e_2, \dots, e_n\}$ be an orthonormal subset of $\mathcal{H}$ and let $\mathcal{M} = \overline{span}\{e_1, e_2, \dots, e_n\}$. Then,
(i) For any $x \in \mathcal{H}$, $\text{Proj}_\mathcal{M} x = \sum_{i=1}^{n} \langle x, e_i \rangle e_i$,
(ii) For any $(a_1, a_2, \dots, a_n)$ and any $x \in \mathcal{H}$

$$||x - \sum_{i=1}^{n} \langle x, e_i \rangle e_i|| \leq ||x - \sum_{i=1}^{n} a_i e_i||;$$

with equality only for $a_i = \langle x, e_i \rangle$.
(iii) $\sum_{i=1}^{n} |\langle x, e_i \rangle|^2 \leq ||x||^2$ (Bessel's inequality).

Hilbert space $\mathcal{H}$ is separable if $\mathcal{H} = \overline{span}\{e_\lambda, \lambda \in \Lambda\}$ and the set $\{e_\lambda, \lambda \in \Lambda\}$ is finite or countable. Such a set is called a basis.

Theorem 2.2.2 *Let $\mathcal{H}$ be a separable Hilbert space with a basis $\{e_n, n \in \mathbb{N}\}$. Then,*
(i) For any $x \in \mathcal{H}$ and any $\epsilon > 0$, one can find N large enough and constants $a_1, a_2, \dots, a_N$ such that

$$||x - \sum_{n=1}^{N} a_n e_n|| < \epsilon.$$

(ii) $x = \sum_{n=1}^{\infty} \langle x, e_n \rangle e_n$;
(iii) (Parseval's identity) $||x||^2 = \sum_{n=1}^{\infty} |\langle x, e_n \rangle|^2$;
(iv) For any $x, y \in \mathcal{H}$, $\langle x, y \rangle = \sum_{n=1}^{\infty} \langle x, e_n \rangle \langle e_n, y \rangle$;
(v) If $x = 0$, then for all n: $\langle x, e_n \rangle = 0$.

A countable set $\{f_n| \ n \in \mathbb{N}\}$ of elements f_n from a separable Hilbert space $\mathcal{H}$, constitutes a *frame* iff there are two constants A and B ($0 < A \leq B < \infty$) such that for every $x \in \mathcal{H}$,

$$A \cdot ||x||^2 \le \sum_{n \in \mathbb{Z}} |\langle x, f_n \rangle|^2 \le B \cdot ||x||^2.$$

Constants A and B are called *frame bounds*.

If $A = B \ge 1$ the frame is called *tight*. The frame is called *exact* if it is minimal, i.e., if it ceases to be a frame whenever any single element is removed from the set.

When $\{f_n | n \in \mathbb{N}\}$ constitutes a tight frame, $x \in \mathcal{H}$ can be uniquely reconstructed from $\{a_1, a_2, \ldots, a_n, \ldots\}$ from $x = \sum_n a_n f_n$.

Example 2.2.8 [191] If $\{e_1, e_2, \ldots, e_n, \ldots\}$ is an orthonormal basis for a space $\mathcal{H}$, then

- $\{e_1, e_1, e_2, e_2, \ldots, e_n, e_n, \ldots\}$ is an inexact, tight frame with frame bounds $A = B = 2$.
- $\{e_1, \frac{e_2}{2}, \ldots, \frac{e_n}{n}, \ldots\}$ is a complete, orthogonal sequence, but not a frame.
- $\{2e_1, e_2, e_3, \ldots, e_n, \ldots\}$ is an inexact, tight frame with bounds $A = 1$ and $B = 2$.
- $\{e_1, \frac{e_2}{\sqrt{2}}, \frac{e_2}{\sqrt{2}}, \frac{e_3}{\sqrt{3}}, \frac{e_3}{\sqrt{3}}, \frac{e_3}{\sqrt{3}}, \ldots\}$ is an exact, nontight frame with bounds $A = B = 1$, but no nonredundant subsequence is a frame.

A countable set $\{f_n | n \in \mathbb{N}\}$ of elements f_n from a separable Hilbert space $\mathcal{H}$, constitutes a Riesz basis if for any $x \in \mathcal{H}$ there is a unique representation $x = \sum_n a_n f_n$ and

$$A \cdot \sum_n |a_n|^2 \le ||\sum_n a_n f_n||^2 \le B \sum_n |a_n|^2$$

for some constants A and B, $0 < A \le B < \infty$. The Riesz basis reduces to an orthonormal basis when $A = B = 1$. Every Riesz basis is a frame, but the contrary is false. However, every exact frame is a Riesz basis.

For further examples and interplay between frames, Riesz bases and developments of wavelets see Holschneider [198].

Wavelets are unconditional bases for many smoothness spaces. That property of wavelet bases will be important in function and density estimation, as we will see in Chapters 6 and 7.

Definition 2.2.3 *A series $\sum_{n \in S} a_n$ converges unconditionally if for every "1-1" and "onto" map $\pi : \mathbb{N} \mapsto S$ the series $\sum_{k=1}^{\infty} a_{\pi(k)}$ converges.*

Definition 2.2.4 *A basis $\mathcal{B} = \{e_i, i \in \mathbb{N}\}$ is called unconditional for a space V if and only if the sum in unique representation $x = \sum_{i=1}^{\infty} b_i e_i$ of an element $x \in V$ converges unconditionally.*

Any orthogonal basis or, more generally, any Riesz basis, in a Hilbert space is an unconditional basis.

2.2.3 Reproducing Kernel Hilbert Spaces

A function of two variables x and y, $\mathbb{K}(x,y)$ is called a reproducing kernel function for the function space $\mathcal{H}$ if
(i) For a fixed y, $\mathbb{K}(x,y)$ is a function in $\mathcal{H}$.
(ii) For every function $f \in \mathcal{H}$ and every y, $\mathbb{K}$ has the reproducing property,

$$f(y) = \langle f(x), \mathbb{K}(x,y) \rangle.$$

Theorem 2.2.3 *Let V be a subspace of $\mathbb{L}_2$ and let $\{e_1, e_2, \ldots\}$ be a complete orthonormal basis of V. Then V is a reproducing kernel Hilbert space with a kernel (sometimes called the Bergman kernel)*

$$\mathbb{K}(x,y) = \sum_n e_n(x) e_n(y). \quad (2.2)$$

For any function $f \in V$

$$f(y) = \int f(x) \mathbb{K}(x,y) \, dx.$$

It should be noted that the sum in (2.2) is not necessarily convergent for any Hilbert space. For instance, $\mathbb{L}_2([-\pi, \pi])$ has a kernel $\mathbb{K}(x,y) = \frac{1}{2\pi} + \frac{1}{\pi} \sum_{k=1}^{\infty} \cos kx \cos ky + \sin kx \sin ky$. This divergent kernel is a representation of the most famous kernel – the Dirac delta function, $\delta(x-y)$. The reproducing kernel expression is:

$$f(x) = \int_{-\pi}^{\pi} f(y) \delta(x-y) \, dy.$$

2.3 FOURIER TRANSFORMATION

Fourier transformation is a chief tool for exploring time-frequency phenomena in time series, signal processing, and other related fields.

Definition 2.3.1 *The Fourier transformation of a function $f \in \mathbb{L}_1(\mathbb{R})$ is defined by*

$$\hat{f}(\omega) = \mathcal{F}[f(x)] = \langle f(x), e^{i\omega x} \rangle = \int_{\mathbb{R}} f(x) \overline{e^{i\omega x}} \, dx = \int_{\mathbb{R}} f(x) e^{-i\omega x} \, dx.$$

If $\hat{f} \in \mathbb{L}_1(\mathbb{R})$ is the Fourier transformation of $f \in \mathbb{L}_1(\mathbb{R})$, then

$$f(x) = \mathcal{F}^{-1}[\hat{f}(\omega)] = \frac{1}{2\pi} \int \hat{f}(\omega) e^{i\omega x}\, d\omega,$$

at every continuity point of f.

The function $\hat{f}(\omega)$ is, in general, a complex function of the form $\hat{f}(\omega) = |\hat{f}(\omega)| e^{i\varphi(\omega)}$. The part $|\hat{f}(\omega)|$ is called the *magnitude spectrum* and the exponent $\varphi(\omega)$ is called the *phase spectrum*.

If $f(x)$ is real, then

- $\hat{f}(-\omega) = \overline{\hat{f}(\omega)}$, and
- $|\hat{f}(\omega)|$ is an even function and $\varphi(\omega)$ is an odd function of ω.

It will be clear from the context if $\hat{f}$ denotes an estimator rather than the Fourier transformation of f.

Example 2.3.1 Let

$$f_l(x) = \begin{cases} 1, & |x| \leq l/2 \\ 0, & |x| > l/2. \end{cases} \tag{2.3}$$

Then, by taking into account the representation $\sin z = \frac{e^z - e^{-z}}{2i}$ we get

$$\hat{f}_l(\omega) = \int_{-l/2}^{l/2} e^{-i\omega x}\, dx = -\frac{1}{i\omega} e^{-i\omega x} \Big|_{-l/2}^{l/2} = \frac{1}{\omega} \frac{e^{i\omega l/2} - e^{-i\omega l/2}}{i} = l \frac{\sin \omega \frac{l}{2}}{\omega \frac{l}{2}}.$$

2.3.1 Basic Properties

We provide, without proofs, a list of important properties of Fourier transformation. For a comprehensive exposition of Fourier transformations, we direct the reader to the monographs of Helson [192] and Katznelson [231].

[BOU] Boundedness. $\hat{f} \in \mathbb{L}_\infty(R)$, $\|\hat{f}\|_\infty \leq \|f\|_1$.

[UC] Uniform Continuity. $\hat{f}(\omega)$ is uniformly continuous on $-\infty < \omega < \infty$.

[DEC] Decay. For $f \in \mathbb{L}_1$, $\hat{f}(\omega) \to 0$, when $|\omega| \to \infty$, (Riemann-Lebesgue lemma).

[LIN] Linearity. $\mathcal{F}[\alpha f(x) + \beta g(x)] = \alpha \mathcal{F}[f(x)] + \beta \mathcal{F}[g(x)]$.

[DER] Derivative. $\mathcal{F}[f^{(n)}(x)] = (i\omega)^n \hat{f}(\omega)$.

[PLA] Plancherel's Identity. $\langle f, g \rangle = \frac{1}{2\pi}\langle \hat{f}, \hat{g} \rangle$; If $g = f$ one obtains Plancherel's identity: $\|f\|^2 = \frac{1}{2\pi}\|\hat{f}\|^2$.

The function $|\hat{f}(\omega)|^2$ is called the *energy spectrum*. The area below the curve $|\hat{f}(\omega)|^2$ is 2π times the energy content of the signal, $E = \int |f(x)|^2\, dx$ (Plancherel's identity in engineering terms).

[SHI] Shifting. $\mathcal{F}[f(x - x_0)] = e^{-i\omega x_0}\hat{f}(\omega)$.

[SCA] Scaling. $\mathcal{F}[f(ax)] = \frac{1}{|a|}\hat{f}(\frac{\omega}{a})$.

[SYM] Symmetry. $\mathcal{F}[\mathcal{F}[f(x)]] = 2\pi f(-x)$.

[CON] Convolution. The convolution of f and g is defined as $f \star g(x) = \int f(x - t)g(t)\, dt$. One of the most important properties of Fourier transformations is $\mathcal{F}[f \star g(x)] = \hat{f}(\omega)\hat{g}(\omega)$.

[MOD] Modulation Theorem. From the symmetry property it follows that $f(x)g(x) = \frac{1}{2\pi}F(\omega) \star G(\omega)$.

[MOM] Moment Theorem

$$\int_{\mathbb{R}} x^n f(x)\, dx = (i)^n \left.\frac{d^n \hat{f}(\omega)}{d\omega^n}\right|_{\omega=0}. \tag{2.4}$$

Example 2.3.2 By using basic properties of the Fourier transformation we find $\hat{g}(\omega)$ for $f(x)\cos\omega_0 x$.

Since $\cos\omega_0 x = (e^{\omega_0 x} + e^{-\omega_0 x})/2$, $f(x)\cos\omega_0 x = \frac{1}{2}f(x)e^{\omega_0 x} + \frac{1}{2}f(x)e^{-\omega_0 x}$. [LIN] and [SHI] properties of Fourier transformation give $\hat{g}(\omega) = \frac{1}{2}\hat{f}(\omega - \omega_0) + \frac{1}{2}\hat{f}(\omega + \omega_0)$.

For instance, $f_l(x)\cos\omega_0 x$ [windowed cosine function, f_l given by (2.3)] has the following Fourier transformation:

$$\frac{\sin(\omega - \omega_0)l/2}{\omega - \omega_0} + \frac{\sin(\omega + \omega_0)l/2}{\omega + \omega_0}.$$

Example 2.3.3 The Dirac function satisfies

- $f(x) \star \delta(x - x_0) = f(x - x_0)$,
- $\mathcal{F}[\delta(x - x_0)] = e^{-i\omega x_0}$.

2.3.2 Poisson Summation Formula and Sampling Theorem

Theorem 2.3.1 *(Poisson theorem) If function f is smooth and decays fast,*[1]

$$\sum_{n=-\infty}^{\infty} f(x - nT) = \frac{1}{T} \sum_{k=-\infty}^{\infty} \hat{f}\left(\frac{2\pi k}{T}\right) e^{i2\pi kx/T}.$$

For $T = 1$ and $x = 0$,

$$\sum_{n=-\infty}^{\infty} f(n) = \sum_{k=-\infty}^{\infty} \hat{f}(2\pi k).$$

Example 2.3.4 The identity

$$\sum_{n=-\infty}^{\infty} e^{-a|n|} = \sum_{n=-\infty}^{\infty} \frac{2a}{a^2 + (2n\pi)^2}, \quad a > 0,$$

follows from the fact that $\frac{2a}{a^2+\omega^2}$ is the Fourier transformation of $e^{-a|x|}$.

The *sampling theorem* is an important tool in communication theory. It gives the conditions when an analog (continuous) signal can be exactly recovered from its sampled values. In wavelet theory, the sampling theorem helps to understand the relationship between wavelet series and discrete wavelet transformations. We define a *bandlimited function* first.

A function f is called bandlimited on $[-\Omega, \Omega]$ if its Fourier transformation $\hat{f}$ has support contained in $[-\Omega, \Omega]$, i.e., if $\hat{f}(\omega) = 0$ for $|\omega| > \Omega$.

Theorem 2.3.2 *(Sampling theorem) Let $f(x)$ be continuous and bandlimited on $[-\Omega, \Omega]$. Then, it is uniquely determined by its sampled values at $x = \frac{n\pi}{\Omega}$.*

---

[1] A sufficient condition is, for example, $f(x) = O\left(\frac{1}{1+|x|^\alpha}\right)$ and $\hat{f}(\omega) = O\left(\frac{1}{1+|\omega|^\alpha}\right)$ for some $\alpha > 1$.

The function f can be recovered from its sampled values through the interpolation formula

$$f(x) = \sum_{n=-\infty}^{\infty} f(nT) \operatorname{sinc}_T(x - nT), \tag{2.5}$$

where $\operatorname{sinc}_T(x) = \frac{\sin(\pi x/T)}{\pi x/T}$. The maximum sampling frequency is $2T$ (Nyquist rate) and $T = \frac{\pi}{\Omega}$ is the maximum sampling period. If the sampling rate is slower than the Nyquist rate, it may happen that the same sampled values come from different continuous signals. Such an error is called *aliasing*.

2.3.3 Fourier Series

Fourier series play an important role in developing wavelet theory. Calculating wavelet filters from transfer functions and orthogonality proofs are some examples of important wavelet tasks involving Fourier series.

A periodic function $f(x) = f(x + T)$ can be expanded into a series

$$f(x) = \sum_{n=-\infty}^{\infty} F_n e^{in\frac{2\pi}{T}x},$$

where

$$F_n = \frac{1}{T} \int_{-T/2}^{T/2} f(x) e^{-in\frac{2\pi}{T}x} \, dx.$$

The set $\{e_n = \frac{1}{\sqrt{T}} e^{in\frac{2\pi}{T}x}\}$ is a complete orthonormal basis for $[-T/2, T/2]$. In terms of the trigonometric functions, sines and cosines, the Fourier series has the form

$$f(x) = \frac{a_0}{2} + \sum_{n=1}^{\infty} \left(a_n \cos \frac{n\pi x}{T} + b_n \sin \frac{n\pi x}{T} \right), \tag{2.6}$$

where

$$a_n = \frac{1}{T}\int_{-T}^{T} f(x) \cos \frac{n\pi x}{T} dx, n = 0, 1, 2, \ldots \quad \text{and}$$
$$b_n = \frac{1}{T}\int_{-T}^{T} f(x) \sin \frac{n\pi x}{T} dx, n = 1, 2, \ldots . \tag{2.7}$$

Example 2.3.5 The Fourier series for $f(x) = \text{sgn}(\cos x)$ can be found by using (2.6) and (2.7). This function is periodic with the period $T = \pi$ and even. Consequently, $a_0 = 0$ and $b_n = 0$, $n = 1, 2, \ldots$

$$\begin{aligned} a_n &= \frac{2}{\pi}\int_0^{\pi} \text{sgn}(\cos x) \cos nx \, dx \\ &= \frac{2}{\pi}\int_0^{\pi/2} \cos nx \, dx - \frac{2}{\pi}\int_{\pi/2}^{\pi} \cos nx \, dx \\ &= \frac{4}{n\pi} \sin \frac{n\pi}{2}, \; n \in \mathbb{N}. \end{aligned}$$

Therefore,

$$\text{sgn}(\cos x) = \frac{4}{\pi}\sum_{n=1}^{\infty} \sin \frac{n\pi}{2} \cos nx = \frac{4}{\pi}\sum_{k=0}^{\infty} \frac{(-1)^k}{2k+1} \cos(2k+1)x. \tag{2.8}$$

In Fig. 2.1 several partial sums of the Fourier expansion (2.8) are shown. Notice "imprecisions" next to discontinuity points. Even for extremely large values of K, these unwanted artifacts (infamous *Gibbs effect*) do not disappear.

The following property of Fourier series will be used later:
Unicity. If $f \in \mathbb{L}_1(R)$ and $a_n = 0$, $n = 0, 1, \ldots$, and $b_n = 0$, $n = 1, 2, \ldots$, then $f = 0$.

2.3.4 Discrete Fourier Transform

The discrete Fourier transformation (DFT) of a sequence $\underset{\sim}{f} = \{f_n, n = 0, 1, \ldots, N-1\}$ is defined as

$$\underset{\sim}{F} = \left\{ \sum_{n=0}^{N-1} f_n w_N^{nk}, \; k = 0, \ldots, N-1 \right\},$$

where $w_N = e^{-i2\pi/N}$. The inverse is

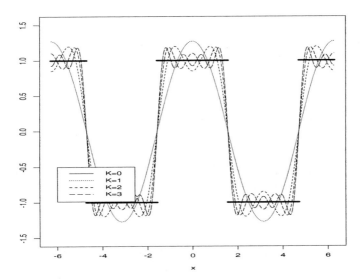

Fig. 2.1 The function sgn(cos x) and several partial sums in the Fourier series decomposition. K is the upper limit in the sum over k ($k = 0, \ldots, K$), see (2.8).

$$\underset{\sim}{f} = \left\{ \frac{1}{N} \sum_{k=0}^{N-1} F_k w_N^{-nk}, \; n = 0, \ldots, N-1 \right\}.$$

If $Q = \{Q_{nk} = e^{-i2\pi nk}\}_{N \times N}$, then $\underset{\sim}{F} = Q \cdot \underset{\sim}{f}$. The matrix Q is unitary (up to a scale factor), i.e., $Q^*Q = NI$, where I is the identity matrix and Q^* is the conjugate transpose of Q.

2.4 HEISENBERG'S UNCERTAINTY PRINCIPLE

In the introduction we mentioned Heisenberg's uncertainty principle, when discussing the time–scale adaptivity of wavelet transformations.

Heisenberg's uncertainty principle has a universal character and it states that in modeling time-frequency phenomena one cannot be arbitrarily precise in both time and frequency simultaneously. If the sides of an imaginary rectangle in the "time-frequency" plane (called Heisenberg's box) represent the time duration and the spectral bandwidth of a signal, Heisenberg's uncertainty principle claims that the area of such a rectangle is bounded from below. Here is a more formal statement of the principle.

Let $f(x) \in \mathbb{L}_2(\mathbb{R})$ be such that

$$x f(x) \in \mathbb{L}_2(\mathbb{R}). \tag{2.9}$$

The center $\bar{x}$ and the spectral bandwidth Δ_f are defined as:

$$\begin{aligned} \bar{x} &= \frac{1}{||f||^2} \int x\, |f(x)|^2\, dx, \\ (\Delta_f)^2 &= \frac{1}{||f||^2} \int (x - \bar{x})^2 |f(x)|^2\, dx. \end{aligned} \tag{2.10}$$

If f is an analog signal, then in signal processing analysis, Δ_f is called the root mean square (RMS) duration of the signal.

Similarly, in the frequency domain, we define

$$\begin{aligned} \bar{\omega} &= \frac{1}{||\hat{f}||^2} \int \omega\, |\hat{f}(\omega)|^2\, d\omega, \\ (\Delta_{\hat{f}})^2 &= \frac{1}{||\hat{f}||^2} \int \omega^2 |\hat{f}(\omega)|^2\, d\omega. \end{aligned} \tag{2.11}$$

$\Delta_{\hat{f}}$ is called the RMS duration of the bandwidth.

Heisenberg's uncertainty principle can be formally stated as Theorem 2.4.1.

Theorem 2.4.1 *Let the condition (2.9) is satisfied by both f and $\hat{f}$. Then,*

$$\Delta_f \cdot \Delta_{\hat{f}} \geq \frac{1}{2}. \tag{2.12}$$

The inequality (2.12) is sharp; the equality is achieved for $f(x) = e^{-at^2}$, $a \geq 0$.

Example 2.4.1 We will show that for $f(x) = A \cdot \mathbf{1}(x_0 - d/2 \leq x \leq x_0 + d/2)$, the RMS duration of the bandwidth $\Delta_{\hat{f}}^2$ is not finite.

Indeed, because $|\hat{f}(\omega)| = A\, d \left| \frac{\sin \omega d/2}{\omega d/2} \right|^2$ and $||\hat{f}||^2 = 2\pi A^2 d$, we obtain

$$(\Delta_{\hat{f}})^2 = \frac{2}{\pi d} \int_{\mathbb{R}} \sin^2 \frac{\omega d}{2} d\omega = \infty.$$

2.5 SOME IMPORTANT FUNCTION SPACES

We have already defined the Lebesgue space $\mathbb{L}_p$ and have given several examples of Hilbert spaces. Next, we provide definitions of several important functional spaces often encountered in formal descriptions and results concerning the regularity of functions.

As we will see later, all listed spaces can be characterized by the magnitude of wavelet coefficients of their wavelet-decomposed elements.

- $\mathbb{C}^n(\mathbb{R})$ A function belongs to the space $\mathbb{C}^n(\mathbb{R})$ if it is n-times continuously differentiable.

- $\mathbb{W}_2^s(\mathbb{R})$ Sobolev space is defined by

$$\mathbb{W}_2^s(\mathbb{R}) = \{f|\ ||(1+|\omega|^2)^{s/2}\hat{f}(\omega)||_{L_2} < \infty\}. \tag{2.13}$$

For $s = 0$, this is just $\mathbb{L}_2(\mathbb{R})$. For $s = 1, 2, 3, \ldots$, the space $\mathbb{W}_2^s(\mathbb{R})$ consists of all $\mathbb{L}_2(\mathbb{R})$ functions that are s times differentiable and whose sth derivative belongs to $\mathbb{L}_2(\mathbb{R})$. When $s = -1, -2, -3, \ldots$, the space $\mathbb{W}_2^s(\mathbb{R})$ contains all distributions with a point support of order $\leq s$, $(\delta, \delta', \ldots, \delta^{(s-1)})$.

For f and g in $\mathbb{W}_2^s(\mathbb{R})$, the inner product is defined by

$$\langle f, g \rangle_s = \frac{1}{2\pi} \int \hat{f}(\omega)\overline{\hat{g}(\omega)}(1+|\omega|^2)^s\, d\omega.$$

The space $\mathbb{W}_2^s(\mathbb{R})$ is complete with respect to this inner product and thus is a Hilbert space. Of course, the space $\mathbb{W}_p^s$ is obtained by replacing the $\mathbb{L}_2$-norm in (2.13) by the $\mathbb{L}_p$-norm.

- $\mathbb{C}^s(\mathbb{R})$ Hölder space is defined by

(i) $0 < s < 1$, $\mathbb{C}^s(\mathbb{R}) = \left\{ f \in \mathbb{L}_\infty(\mathbb{R}) : \sup_h \frac{|f(x+h) - f(x)|}{|h|^s} < \infty \right\}$.

(ii) $s = n + s'$, $0 < s' < 1$,

$$\mathbb{C}^s(\mathbb{R}) = \left\{ f \in \mathbb{L}_\infty(\mathbb{R}) \cap \mathbb{C}^n(\mathbb{R}) \mid \frac{d^n}{dx^n} f \in \mathbb{C}^{s'}(\mathbb{R}) \right\}.$$

- $\mathbb{B}_{pq}^\sigma(I)$ Besov space.

We give a definition of an inhomogeneous Besov space on $I \subset \mathbb{R}$.

Definition 2.5.1 *Let* $\Delta_h^{(0)} f(t) = f(t)$ *and*

$$\Delta_h^{(r)} = \Delta_h^{(r-1)} f(x+h) - \Delta_h^{(r-1)} f(x) = \sum_{k=0}^{r} \binom{r}{k} (-1)^k f(t+kh),$$

be the rth difference. $\Delta_h^{(r)}$ is defined for $x \in I_{rh} = \{x \in I | x + rh \in I\}$. The rth modulus of smoothness of $f \in \mathbb{L}_p(I)$ is

$$w_{r,p}(f;t) = \sup_{|h| \le t} \|\Delta_h^{(r)} f\|_{\mathbb{L}_p(I_{rh})}$$

with the supremum norm when $p = \infty$.

For selected $\sigma > 0$, $0 < p \le \infty$, and $0 < q \le \infty$ chose r so that $r - 1 \le \sigma \le r$. The Besov seminorm of index (σ, p, q) is then defined by

$$|f|_{\mathbb{B}_{p,q}^\sigma} = \left[\int_0^\infty (h^{-\sigma} w_{r,p}(f;h))^q \frac{dh}{h} \right]^{1/q},$$

if $1 \le q < \infty$, and

$$|f|_{\mathbb{B}_{p,\infty}^\sigma} = \sup_h h^{-\sigma} w_{r,p}(f;h), \qquad (2.14)$$

if $q = \infty$. The Besov norm $\|f\|_{\mathbb{B}_{p,q}^\sigma}$ is defined as $\|f\|_{\mathbb{L}_p(I)} + |f|_{\mathbb{B}_{p,q}^\sigma}$. The Besov space is the class of all functions f with a finite Besov norm. A Besov ball is a class of functions $\mathbb{B}_{p,q}(M)$ for which the norm is bounded from the above by M.

Example 2.5.1 The Besov spaces are very general spaces comprising most other spaces as special cases.
- Sobolev space $\mathbb{W}_2^s$ is $\mathbb{B}_{2,2}^s$. For $p \ne 2$, Sobolev spaces $\mathbb{W}_p^s$ do not coincide with $\mathbb{B}_{p,p}^s$.
- Hölder space $\mathbb{C}^s$ is $\mathbb{B}_{\infty,\infty}^s$.

Example 2.5.2 The following definition of Besov spaces (Meyer [294]) provides a different point of view.

The function f belongs to $\mathbb{B}_{p,q}^\sigma$, if there exist $f_0, g_0, g_1, g_2, \ldots$ in the Sobolev space $\mathbb{W}_p^m$ and a sequence $\{\epsilon_0, \epsilon_1, \epsilon_2, \ldots\} \in \ell^q$ such that

$$f = f_0 + g_0 + g_1 + g_2 + \cdots \in \mathbb{L}_p,$$
$$\|g_j\|_p \le \epsilon_j 2^{-\sigma j}, \ j = 0, 1, 2, \ldots$$
$$\|g_j^{(m)}\|_p \le C \epsilon_j 2^{(m-\sigma)j}, \ j = 0, 1, 2,$$

where $C > 0$ is a constant and m is an integer $> \sigma$.

As special cases, Besov spaces include functional spaces with high spatial irregularities such as hump algebras, Bloch spaces, and bounded variation classes; see Meyer [294] for a comprehensive discussion.

Triebel [410] and [411] are excellent monographs covering in depth the theory of function spaces. See also Meyer [294], Wojtaszczyk [459], DeVore and Popov [115], Hernández and Weiss [193], and the seminal paper of Besov [31].

2.6 FUNDAMENTALS OF SIGNAL PROCESSING

Signal processing interpretations are fundamental for understanding, constructing and implementing wavelet methods. We review some of the commonly used terms. For a more comprehensive coverage of the interplay between wavelets and signal processing we direct the reader to Strang and Nguyen [387] and Vetterli and Kovačević [419].

A *filter* H is a linear operator which maps ℓ_2 to ℓ_2. For $x \in \ell_2$ the equation $y = Hx$ has a component-wise representation $y(n) = (h \star x)(n) = \sum_k h(k)x(n-k)$, where $h(k) = h_k$, $k \in \mathbb{Z}$ are filter coefficients. The filter coefficients may be obtained when the filter H is applied to the sequence $\underline{u} = (\ldots, 0, 0, 1, 0, 0, \ldots)$ (*unit impulse at zero*),

$$\underline{h} = H\underline{u} = (\ldots, h_0, h_1, \ldots).$$

When there is no danger of confusion, the impulse response $\underline{h}$ of the filter H will be called a filter as well. Components h_i of $\underline{h}$ are called *taps* of the filter.

A *filter band* is a set of two or more filters. It is usually constructed to separate an input signal into frequency bands. When negative indices are not allowed in $\underline{h}$ the filter is called *causal*. If the number of nonzero taps is finite, the filter is called *finite impulse response* (FIR), otherwise the filter is *infinite impulse response* (IIR).

The function

$$H(\omega) = \sum_n h(n)e^{-in\omega}$$

is the *frequency response function*. When the input is $x(n) = e^{in\omega}$ and the filter taps are $(h_0, h_1, \ldots)$, the output $y(n)$ is equal to the product $x(n) \cdot H(\omega)$. Indeed,

$$\begin{aligned} y(n) &= h_0 e^{in\omega} + h_1 e^{i(n-1)\omega} + \ldots \\ &= (h_0 + h_1 e^{-i\omega} + \ldots)e^{in\omega} \\ &= H(\omega)x(n). \end{aligned}$$

The function $H(\omega)$ is a complex function, and can be represented in the form

$$H(\omega) = |H(\omega)| \, e^{-i\Phi(\omega)},$$

where $|H(\omega)|$ is the magnitude of $H(\omega)$ and $\Phi(\omega)$ is the phase angle. The filter is *low-pass* (averaging, smoothing), when $H(0) = 1$ and $H(\pi) = 0$ (the low frequencies are preserved). The filter is *high-pass* (differencing) when $H(0) = 0$ and $H(\pi) = 1$.

2.7 EXERCISES

2.1. Prove the *Schwarz inequality:* Let $x_1, x_2 \in \mathcal{H}$, then

$$|\langle x_1, x_2 \rangle| \le \langle x_1, x_1 \rangle \langle x_2, x_2 \rangle,$$

with equality when $x_2 = \alpha x_1$ for some constant α.

[Hint: Consider the inequality $\langle x_1 + \lambda x_2, x_1 + \lambda x_2 \rangle \ge 0$ and choose appropriate λ.]

2.2. *Gram-Schmidt process of orthogonalization.* Let $f_1, \ldots, f_n, \ldots$ be a sequence of linearly independent functions in $\mathbb{L}_2(\mathbb{R})$. Let

$$g_1 = f_1, \; g_2 = f_2 - \frac{\langle f_2, g_1 \rangle g_1}{\|g_1\|^2}, \ldots, g_n = f_n - \sum_{i=1}^{n-1} \frac{\langle f_n, g_i \rangle g_i}{\|g_i\|^2}, \ldots .$$

(a) Show that $\{g_1, g_2, \ldots, g_n, \ldots\}$ is an orthonormal system.

(b) Generate a set of polynomials orthonormal in $\mathbb{L}_2([-1, 1])$ from the sequence $1, x, x^2, x^3, \ldots$.

2.3. Show that the set of functions

$$f_m(x) = \sqrt{\frac{2}{\pi}} \sin mx, \quad m = 1, 2, \ldots$$

is an orthonormal system in $\mathbb{L}_2([0, \pi])$.

2.4. [459] Let $\psi_{jk}, j = 0, 1, \ldots; \; k = 0, 1, \ldots, 2^j - 1$ be the Haar basis on [0,1]. Let $f(x) = 2^N \mathbf{1}(0 \le x \le 2^{-N})$. Calculate $\sum_{j=0}^{\infty} \sum_{k=0}^{2^j - 1} \langle f, \psi_{jk} \rangle \psi_{jk}$. By considering $\sum_{j=0}^{\infty} \sum_{k=0}^{2^{j+1}-1} \langle f, \psi_{2j,k} \rangle \psi_{2j,k}$ show that $\{\psi_{jk}\}$ is not an unconditional basis in $\mathbb{L}_1[0, 1]$.

2.5. Show that $\bar{\omega}$ in (2.11) is always 0.

2.6. For $f(x) = e^{-at^2}$, $a \geq 0$, show that $(\Delta_f)^2 = \frac{1}{4a}$, and $(\Delta_{\hat{f}})^2 = a$.

2.7. Find $\hat{f}(\omega)$ for
$$f(x) = \begin{cases} e^{-\alpha x}, & x \geq 0 \\ 0, & x < 0. \end{cases}$$

[Answer: $\hat{f}(\omega) = \frac{1}{\alpha + i\omega}$.]

2.8. Find $\hat{f}(\omega)$ for $f(x) = e^{-\alpha|x|}$.

[Hint: Show first: $\int_\mathbb{R} e^{-\alpha|x|-i\omega x}\, dx = 2\int_0^\infty e^{-\alpha x} \cos \omega x\, dx$. Answer: $\frac{2\alpha}{\alpha^2+\omega^2}$.]

2.9. Find $\hat{f}(\omega)$ for $f(x) = xe^{-\alpha|x|}$.

[Hint: Show first: $\int_\mathbb{R} xe^{-\alpha|x|-i\omega x}\, dx = -2i\int_0^\infty xe^{-\alpha x} \sin \omega x\, dx$. Answer: $\frac{-4i\alpha\omega}{(\alpha^2+\omega^2)^2}$.]

2.10. *Gibbs effect.* Given the function $f(x) = \mathrm{sgn}(x)\mathbf{1}(|x| < \pi)$, show:

(a) $f(x) = \frac{4}{\pi}\sum_{n=1}^\infty \frac{1}{2n-1}\sin(2n-1)x$ is the Fourier series of $f(x)$.

(b) $S_k(x) = \frac{2}{\pi}Si[(k+1/2)x]$ is k-th partial sum of the series in (a). [$Si(y) = \int_0^y \frac{\sin x}{x}\, dx$ is the Sine-integral function].

(c) Plot $S_k(x)$ for different values of k in the neighborhood of 0.

2.11. Prove that $\{e^{icnx}\}$ is a tight frame for $c \leq 1$ with the constants $A = B = \frac{1}{c}$.

2.12. [104] Let $\mathcal{H} = \mathbb{R}^2$. Prove that the system $\{(0,1), (-\frac{\sqrt{3}}{2}, -\frac{1}{2}), (\frac{\sqrt{3}}{2}, -\frac{1}{2})\}$ is a frame. Find the frame bounds A and B. Is it the tight frame?

2.13. Show that the Heisenberg product $\Delta_f \cdot \Delta_{\hat{f}}$ [see (2.10) and (2.11)] is invariant under location, modulation and scale changes in f. That is, the product does not change if $f(x)$ is replaced by $e^{i\omega x}f((x-b)/a)$, $\omega, a, b \in \mathbb{R}$, $a \neq 0$.

2.14. (Lars Villemoes) Find the Heisenberg product for the "tent" function, $f(x) = (x+1)\cdot \mathbf{1}(-1 \leq x \leq 0) + (1-x)\cdot \mathbf{1}(0 < x \leq 1)$.

[Hint: By the derivation rule and Plancherel's formula, $\Delta_{\hat{f}}^2 = ||f'||^2/||f||^2$. Answer: $\Delta_f \cdot \Delta_{\hat{f}} = \sqrt{3/10} \approx 0.548$.]

3
Wavelets

In this chapter, we give an overview of important theoretical results in wavelet theory. We start with the continuous wavelet transformation mainly because of its historic importance. However, the emphasis will be put on critically sampled wavelets and discrete wavelet transformations.

Mallat's multiresolution analysis is the fundamental concept necessary to construct and understand the wavelet paradigm. The main mathematical tool is Fourier analysis. Theoretical results are illustrated on Haar's, Shannon's, Meyer's, and Daubechies' wavelets. We conclude the chapter with discussions of some basic properties of wavelets.

3.1 CONTINUOUS WAVELET TRANSFORMATION

The first theoretical results in wavelets are connected with continuous wavelet decompositions of $\mathbb{L}_2$ functions and go back to the early 1980s. Papers of Morlet et al. [301] and Grossmann and Morlet [180] are among the first on this subject.

Even though most of today's practical applications of wavelets involve discrete wavelet transformations and wavelet series, many researchers are reluctant to give up insightful graphical displays of continuous wavelet transformations.

By the very nature of its subject, statistics is mainly interested in discrete wavelet transformations. However, understanding continuous wavelet transformations is important since many of their properties have analogous discrete counterparts. For example, there is a significant body of research in both probability and time series where the problems are formulated in terms of continuous wavelet transformations (see Cambanis and Masry [57], Cambanis and Houdré [56], and Houdré [201]).

Let $\psi_{a,b}(x)$, $a \in \mathbb{R}\backslash\{0\}, b \in \mathbb{R}$ be a family of functions defined as translations and re-scales of a single function $\psi(x) \in \mathbb{L}_2(\mathbb{R})$,

$$\psi_{a,b}(x) = \frac{1}{\sqrt{|a|}} \psi\left(\frac{x-b}{a}\right). \tag{3.1}$$

Such normalization ensures that $||\psi_{a,b}(x)||$ is independent of a and b. The function ψ (called *the wavelet function* or *the mother wavelet*) is assumed to satisfy the *admissibility condition,*

$$C_\psi = \int_{\mathbb{R}} \frac{|\Psi(\omega)|^2}{|\omega|} d\omega < \infty, \tag{3.2}$$

where $\Psi(\omega)$ is the Fourier transformation of $\psi(x)$. The admissibility condition (3.2) implies

$$0 = \Psi(0) = \int \psi(x) dx.$$

Also, if $\int \psi(x)dx = 0$ and $\int (1+|x|^\alpha)|\psi(x)|dx < \infty$ for some $\alpha > 0$, then $C_\psi < \infty$.

This property of the function ψ, $\int \psi(x)dx = 0$, motivates the name wavelet. The diminutive comes from the fact that ψ is well localized and by appropriate scaling such localization can be made arbitrarily fine. Wavelet functions are usually normalized to "have unit energy", i.e., $||\psi_{a,b}(x)|| = 1$.

For any $\mathbb{L}_2$ function $f(x)$, the continuous wavelet transformation is defined as a function of two variables

$$CWT_f(a,b) = \langle f, \psi_{a,b} \rangle = \int f(x)\overline{\psi_{a,b}(x)}dx.$$

Here the dilation and translation parameters, a and b, respectively, vary continuously over $\mathbb{R}\backslash\{0\} \times \mathbb{R}$.

3.1.1 Basic Properties

Resolution of Identity. When the admissibility condition is satisfied, i.e., $C_\psi < \infty$, it is possible to find the inverse continuous transformation via the relation known as *resolution of identity* or *Calderón's reproducing identity,*

$$f(x) = \frac{1}{C_\psi} \int_{\mathbb{R}^2} \mathcal{CWT}_f(a,b) \psi_{a,b}(x) \frac{da\,db}{a^2}.$$

If a is restricted to $\mathbb{R}^+$, which is natural since a can be interpreted as a reciprocal of frequency, (3.2) becomes

$$C_\psi = \int_0^\infty \frac{|\Psi(\omega)|^2}{\omega} d\omega < \infty, \qquad (3.3)$$

and the *resolution of identity* relation becomes

$$f(x) = \frac{1}{C_\psi} \int_{-\infty}^\infty \int_0^\infty \mathcal{CWT}_f(a,b) \psi_{a,b}(x) \frac{1}{a^2} da\,db.$$

Example 3.1.1 In this example we provide two illustrations of continuous wavelet decompositions. Fig3.1 depicts a continuous wavelet transformation of the earthquake data set [Fig. 1.12(a)] discussed in Section 1.4. The wavelet used was the "Mexican hat," see Example 3.1.3 on page 49.

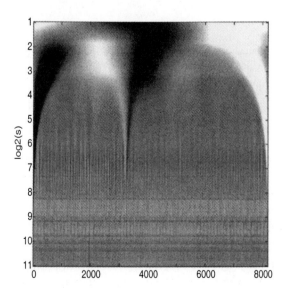

Fig. 3.1 Continuous wavelet transformation of the earthquake data set by the "Mexican hat" wavelet.

Fig. 3.2(a) gives a continuous wavelet transformation of Donoho and Johnstone's blocks test function. The wavelet used was Morlet's wavelet $[\psi(x) \propto e^{-\frac{x^2}{2\sigma^2}} e^{i\omega_0 x}]$

46 WAVELETS

which is a modulated Gaussian function. Notice the "cones of influence" in the time/frequency plane in panel (a) corresponding to points of singularity in the signal.

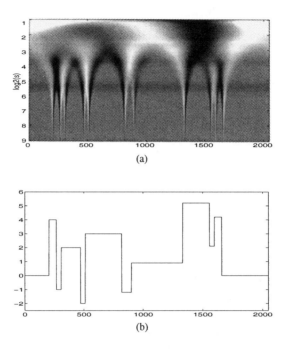

Fig. 3.2 (a) Continuous wavelet transformation of blocks by the Morlet's wavelet (b) Original data set provided for comparison.

Both decompositions in this example are obtained by WaveLab software.

Next, we list a few important properties of continuous wavelet transformations.

Shifting Property. If $f(x)$ has a continuous wavelet transformation $\mathcal{CWT}_f(a,b)$, then $g(x) = f(x - \beta)$ has the continuous wavelet transformation $\mathcal{CWT}_g(a,b) = \mathcal{CWT}_f(a, b - \beta)$.

Scaling Property. If $f(x)$ has a continuous wavelet transformation $\mathcal{CWT}_f(a,b)$, then $g(x) = \frac{1}{\sqrt{s}} f\left(\frac{x}{s}\right)$ has the continuous wavelet transformation $\mathcal{CWT}_g(a,b) = \mathcal{CWT}_f\left(\frac{a}{s}, \frac{b}{s}\right)$.

Both the shifting property and the scaling property are simple consequences of changing variables under the integral sign.

Energy Conservation.

$$\int_{-\infty}^{\infty} |f(x)|^2 dx = \frac{1}{C_\psi} \int_{-\infty}^{\infty} \int_{-\infty}^{\infty} |\mathcal{CWT}_f(a,b)|^2 \frac{1}{a^2} da\, db.$$

Localization. Let $f(x) = \delta(x - x_0)$ be the Dirac pulse at the point x_0. Then, $\mathcal{CWT}_f(a,b) = \frac{1}{\sqrt{a}} \psi(\frac{x_0 - b}{a})$.

Reproducing Kernel Property. Define $\mathbb{K}(u,v;a,b) = \langle \psi_{u,v}, \psi_{a,b} \rangle$. Then, if $F(u,v)$ is a continuous wavelet transformation of $f(x)$,

$$F(u,v) = \frac{1}{C_\psi} \int_{-\infty}^{\infty} \int_{0}^{\infty} \mathbb{K}(u,v;a,b) F(a,b) \frac{1}{a^2} da\, db,$$

i.e., $\mathbb{K}$ is a reproducing kernel. The associated reproducing kernel Hilbert space (RKHS) is defined as a $\mathcal{CWT}$ image of $\mathbb{L}_2(\mathbb{R})$ – the space of all complex-valued functions F on $\mathbb{R}^2$ for which $\frac{1}{C_\psi} \int_{-\infty}^{\infty} \int_0^\infty |F(a,b)|^2 \frac{da\, db}{a^2}$ is finite.

Characterization of Regularity. Let $\int (1 + |x|) |\psi(x)| dx < \infty$ and let $\Psi(0) = 0$. If $f \in \mathbb{C}^\alpha$ (Hölder space with exponent α), then

$$|\mathcal{CWT}_f(a,b)| \leq C|a|^{\alpha+1/2}. \tag{3.4}$$

Conversely, if a continuous and bounded function f satisfies (3.4), then $f \in \mathbb{C}^\alpha$.

For the proof and results concerning local regularity, see Jaffard [212, 211], and Holschneider and Tchamitchian [200].

Example 3.1.2 Continuous Haar transformation. Continuous wavelet transformation can be simple. Let

$$\psi_{a,b}^{\text{HAAR}}(x) = \frac{1}{\sqrt{a}}[\mathbf{1}(b \leq x < \frac{a}{2} + b) - \mathbf{1}(\frac{a}{2} + b \leq x \leq a + b)],$$
$$a \in \mathbb{R}^+, b \in \mathbb{R}. \tag{3.5}$$

Let F be a primitive for f, i.e., $F' = f$. Then,

$$\mathcal{CWT}_f(a,b) = \langle f, \psi_{a,b}^{\text{HAAR}} \rangle = \frac{2}{\sqrt{a}} \left[F(\frac{a}{2} + b) - \frac{F(b) + F(a+b)}{2} \right].$$

Fig. 3.3 gives a contour plot of $\mathcal{CWT}_f(a,b)$ for the doppler test function defined

48 WAVELETS

by (1.4). The horizontal axis in the plot is scale a, $(0 < a \leq 2)$ and vertical axis is

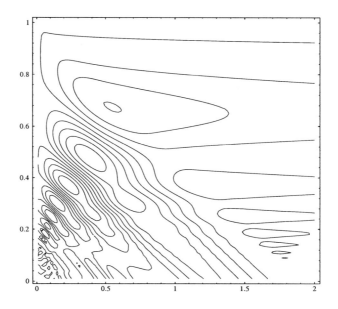

Fig. 3.3 Level plot of a continuous Haar wavelet transformation of the `doppler` function. The horizontal axis is scale a, $(0 < a \leq 2)$ and the vertical axis is shift b $(0 < b < 1)$.

shift b $(0 < b < 1)$.

3.1.2 Wavelets for Continuous Transformations

How can one construct wavelet functions? We already saw that Haar's function given in (3.5) is one choice. The following result gives an effective way of selecting a wavelet function.

Theorem 3.1.1 *Let $\xi(x)$ and $\xi^{(n)}(x)$, $n \geq 1$, be $\mathbb{L}_2(\mathbb{R})$ functions and let $\xi^{(n)}(x) \neq 0$. Then, $\psi(x) = \xi^{(n)}(x)$ is a wavelet.*

Proof: By the [DER] property of Fourier transformations: $|\Psi(\omega)| = |\omega|^n |\hat{\xi}(\omega)|$. We need to show that the admissibility constant C_ψ is finite. Indeed,

CONTINUOUS WAVELET TRANSFORMATION   49

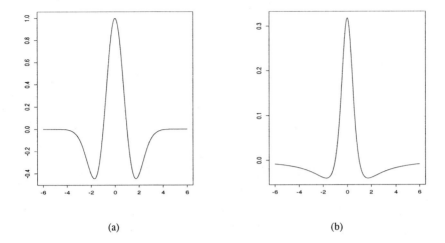

Fig. 3.4 (a) "Mexican hat" or Marr's wavelet. (b) Poisson wavelet.

$$\begin{aligned} C_\psi &= \int \frac{|\Psi(\omega)|^2}{|\omega|} d\omega \\ &= \int_{|\omega|\le 1} |\omega|^{2n-1} |\hat{\xi}(\omega)|^2 d\omega + \int_{|\omega|>1} \frac{|\omega|^{2n}|\hat{\xi}(\omega)|^2}{|\omega|} d\omega \\ &\le \int_{|\omega|\le 1} |\hat{\xi}(\omega)|^2 d\omega + \int_{|\omega|>1} |\omega^n \hat{\xi}(\omega)|^2 d\omega \\ &\le \|\xi(x)\|_{L_2}^2 + \|\xi^{(n)}\|_{L_2}^2 < \infty. \end{aligned}$$

Example 3.1.3 Mexican hat or Marr's wavelet. The function

$$\psi(x) = \frac{d^2}{dx^2}[-e^{-x^2/2}] = (1-x^2)e^{-x^2/2}$$

is a wavelet [known as the "Mexican hat" or Marr's wavelet, Fig. 3.4(a)].

By Theorem 3.1.1, it follows that C_ψ is finite. By direct calculation one may obtain $C_\psi = 2\pi$.

Example 3.1.4 Poisson wavelet. The function $\psi(x) = -(1+\frac{d}{dx})\frac{1}{\pi}\frac{1}{1+x^2}$ is a wavelet [known as the Poisson wavelet, Fig. 3.4(b)]. The analysis of functions with respect to this wavelet is related to the boundary value problem of the Laplace operator. See Exercise 3.4.

The following result can be proved by using Theorem 3.1.1:

Result: If $\psi \in \mathbb{L}_1(\mathbb{R}) \cap \mathbb{L}_2(\mathbb{R})$ is not identically equal to zero and satisfies (i) $\int_{\mathbb{R}} \psi(x)dx = 0$ and (ii) $\int_{\mathbb{R}} |x|^\alpha |\psi(x)|dx < \infty$ for some $\alpha > \frac{1}{2}$, then ψ is a wavelet (see Exercise 3.1).

For more information on continuous wavelet transformations, we direct the reader to the monographs of Daubechies [104], Holschneider [198], and Meyer [293]. See also Louis, Maaß, and Reider [267] and Vetterli and Kovačević [419].

3.2 DISCRETIZATION OF THE CONTINUOUS WAVELET TRANSFORM

The continuous wavelet transformation of a function of one variable is a function of two variables. Clearly, the transformation is redundant. To "minimize" the transformation one can select discrete values of a and b and still have a transformation that is invertible. However, sampling that preserves all information about the decomposed function cannot be coarser than the *critical sampling*.

The critical sampling (Fig. 3.5) defined by

$$a = 2^{-j}, \ b = k2^{-j}, \ j, k \in \mathbb{Z}, \tag{3.6}$$

will produce the minimal basis. Any coarser sampling will not give a unique inverse transformation; that is, the original function will not be uniquely recoverable. Moreover under mild conditions on the wavelet function ψ, such sampling produces an orthogonal basis $\{\psi_{jk}(x) = 2^{j/2}\psi(2^j x - k), \ j, k \in \mathbb{Z}\}$.

There are other discretization choices. For example, selecting $a = 2^{-j}$, $b = k$ will lead to non-decimated (or stationary) wavelet transformation, that will be discussed in Chapter 5. For more general sampling, given by

$$a = a_0^{-j}, \ b = k\,b_0\,a_0^{-j}, \ j, k \in \mathbb{Z}, \ a_0 > 1, b_0 > 0, \tag{3.7}$$

numerically stable reconstructions are possible if the system $\{\psi_{jk}, \ j, k \in \mathbb{Z}\}$ constitutes a frame. Here

$$\psi_{jk}(x) = a_0^{j/2} \psi\left(\frac{x - k\,b_0\,a_0^{-j}}{a_0^{-j}}\right) = a_0^{j/2} \psi(a_0^j x - k\,b_0),$$

is (3.1) evaluated at (3.7). For a comprehensive discussion see Daubechies [104] and Heil and Walnut [191].

Next, we consider wavelet transformations (wavelet series expansions) for values of a and b given by (3.6). An elegant theoretical framework for critically sampled

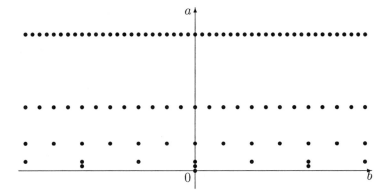

Fig. 3.5 Critical Sampling in $\mathbb{R} \times \mathbb{R}^+$ half-plane ($a = 2^{-j}$ and $b = k\, 2^{-j}$).

wavelet transformation is *Mallat's Multiresolution Analysis* [275, 276].

3.3 MULTIRESOLUTION ANALYSIS

A multiresolution analysis (MRA) is a sequence of closed subspaces $V_n, n \in \mathbb{Z}$ in $\mathbb{L}_2(\mathbb{R})$ such that they lie in a containment hierarchy

$$\cdots \subset V_{-2} \subset V_{-1} \subset V_0 \subset V_1 \subset V_2 \subset \cdots . \qquad (3.8)$$

The nested spaces have an intersection that is trivial and a union that is dense in $\mathbb{L}_2(\mathbb{R})$,

$$\cap_n V_j = \{\mathbf{0}\}, \quad \overline{\cup_j V_j} = \mathbb{L}_2(\mathbb{R}).$$

[With $\overline{A}$ we denoted the closure of a set A]. The hierarchy (3.8) is constructed such that (i) V-spaces are self-similar,

$$f(2^j x) \in V_j \text{ iff } f(x) \in V_0. \qquad (3.9)$$

and (ii) there exists *a scaling function* $\phi \in V_0$ whose integer-translates span the space V_0,

$$V_0 = \left\{ f \in \mathbb{L}_2(\mathbb{R}) \mid f(x) = \sum_k c_k \phi(x - k) \right\},$$

and for which the set $\{\phi(\bullet - k),\ k \in \mathbb{Z}\}$ is an orthonormal basis.[1]

A few technical conditions on ϕ are necessary for future developments. First, we assume $\int \phi(x)dx \neq 0$. Since $V_0 \subset V_1$, the function $\phi(x) \in V_0$ can be represented as a linear combination of functions from V_1, i.e.,

$$\phi(x) = \sum_{k \in \mathbb{Z}} h_k \sqrt{2}\phi(2x - k), \tag{3.10}$$

for some coefficients $h_k,\ k \in \mathbb{Z}$. This equation is known as the *scaling equation* (or two-scale equation) and, as we will see later, it is fundamental in constructing and exploring wavelets.

In the wavelet literature, the reader may encounter an indexing of the multiresolution subspaces, which is the reverse of that in (3.8),

$$\cdots \subset V_2 \subset V_1 \subset V_0 \subset V_{-1} \subset V_{-2} \subset \cdots. \tag{3.11}$$

This convention, sometimes called Daubechies' convention, as opposed to Mallat's convention in (3.8), is almost equally often used. For instance, by the self-similarity property (3.9), the family $\{\phi_{jk}(x) = 2^{j/2}\phi(2^j x - k),\ j \text{ fixed},\ k \in \mathbb{Z}\}$ is a basis of V_j according to Mallat's indexing, while $\{\phi_{jk}(x) = 2^{-j/2}\phi(2^{-j}x - k),\ j \text{ fixed},\ k \in \mathbb{Z}\}$ is a basis of V_j according to Daubechies' indexing.

We will use Mallat's indexing since the summation expressions over multiresolution subspaces, used throughout the book, look more natural than their counterparts from the Daubechies' indexing. However, for software implementation of wavelet transformations the Daubechies' notation seems to be more natural since "the next step" in the transformation algorithm corresponds to the "next index" in the multiresolution indexing.

The coefficients h_n in (3.10) are important in connecting the MRA to the theory of signal processing. The (possibly infinite) vector $\underline{h} = \{h_n,\ n \in \mathbb{Z}\}$ will be called a *wavelet filter*. It is a low-pass (averaging) filter as will become clear later by considerations in the Fourier domain.

To further explore properties of multiresolution analysis subspaces and their bases, we will often work in the Fourier domain. Define the function m_0 as follows:

$$m_0(\omega) = \frac{1}{\sqrt{2}} \sum_{k \in \mathbb{Z}} h_k e^{-ik\omega} = \frac{1}{\sqrt{2}} H(\omega). \tag{3.12}$$

The function in (3.12) is sometimes called the *transfer function* and it describes the

---

[1] It is possible to relax the orthogonality requirement. It is sufficient to assume that the system of functions $\{\phi(\bullet - k),\ k \in \mathbb{Z}\}$ constitutes a Riesz basis for V_0.

behavior of the associated filter $\underline{h}$ in the Fourier domain. Notice that the function m_0 is periodic with the period 2π and that the filter taps $\{h_n,\ n \in \mathbb{Z}\}$ are the Fourier coefficients of the function $H(\omega) = \sqrt{2}\, m_0(\omega)$.

In the Fourier domain, the relation (3.10) becomes

$$\Phi(\omega) = m_0\left(\frac{\omega}{2}\right) \Phi\left(\frac{\omega}{2}\right), \qquad (3.13)$$

where $\Phi(\omega)$ is the Fourier transformation of $\phi(x)$. Indeed,

$$\begin{aligned}
\Phi(\omega) &= \int_{-\infty}^{\infty} \phi(x) e^{-i\omega x}\, dx \\
&= \sum_k \sqrt{2}\, h_k \int_{-\infty}^{\infty} \phi(2x - k) e^{-i\omega x}\, dx \\
&= \sum_k \frac{h_k}{\sqrt{2}} e^{-ik\omega/2} \int_{-\infty}^{\infty} \phi(2x - k) e^{-i(2x-k)\omega/2}\, d(2x - k) \\
&= \sum_k \frac{h_k}{\sqrt{2}} e^{-ik\omega/2}\, \Phi\left(\frac{\omega}{2}\right) \\
&= m_0\left(\frac{\omega}{2}\right) \Phi\left(\frac{\omega}{2}\right). \qquad (3.14)
\end{aligned}$$

By iterating (3.13), one gets

$$\Phi(\omega) = \prod_{n=1}^{\infty} m_0\left(\frac{\omega}{2^n}\right), \qquad (3.15)$$

which is convergent under very mild conditions on rates of decay of the scaling function ϕ. There are several sufficient conditions for convergence of the product in (3.15). For instance, the uniform convergence on compact sets is assured if (i) $m_0(\omega) = 1$ and (ii) $|m_0(\omega) - 1| < C|\omega|^\epsilon$, for some positive C and ϵ. See also Theorem 3.5.1.

Next, we prove two important properties of wavelet filters associated with an orthogonal MRA, *normalization* and *orthogonality*.

Normalization.

$$\sum_{k \in \mathbb{Z}} h_k = \sqrt{2}. \qquad (3.16)$$

Proof:

$$\int \phi(x)dx = \sqrt{2} \sum_k h_k \int \phi(2x-k)dx$$
$$= \sqrt{2} \sum_k h_k \frac{1}{2} \int \phi(2x-k)d(2x-k)$$
$$= \frac{\sqrt{2}}{2} \sum_k h_k \int \phi(x)dx.$$

Since $\int \phi(x)dx \neq 0$ by assumption, (3.16) follows.

This result also follows from $m_0(0) = 1$, since $\int \phi(x)dx \neq 0$ and $\phi \in L_1(\mathbb{R})$ in the time domain translate to $\Phi(0) \neq 0$ and $\Phi(\omega) \in L_\infty$ in the Fourier domain.

Orthogonality. For any $l \in \mathbb{Z}$,

$$\sum_k h_k h_{k-2l} = \delta_l. \qquad (3.17)$$

Proof: Notice first that from the scaling equation (3.10) it follows that

$$\phi(x)\phi(x-l) = \sqrt{2} \sum_k h_k \phi(2x-k)\phi(x-l) \qquad (3.18)$$
$$= \sqrt{2} \sum_k h_k \phi(2x-k) \sqrt{2} \sum_m h_m \phi(2(x-l)-m).$$

By integrating the both sides in (3.18) we obtain

$$\delta_l = 2 \sum_k h_k \left[\sum_m h_m \frac{1}{2} \int \phi(2x-k)\phi(2x-2l-m)\, d(2x) \right]$$
$$= \sum_k \sum_m h_k h_m \delta_{k,2l+m}$$
$$= \sum_k h_k h_{k-2l}.$$

The last line is obtained by taking $k = 2l + m$.

An important special case is $l = 0$ for which (3.17) becomes

$$\sum_k h_k^2 = 1. \tag{3.19}$$

One consequence of the orthogonality condition (3.17) is the following: the convolution of filter $\underline{h}$ with itself, $\underline{f} = \underline{h} \star \underline{h}$, is an *à trous*.[2]

The fact that the system $\{\phi(\bullet - k),\ k \in \mathbb{Z}\}$ constitutes an orthonormal basis for V_0 can be expressed in the Fourier domain in terms of either $\Phi(\omega)$ or $m_0(\omega)$.

(a) In terms of $\Phi(\omega)$:

$$\sum_{l=-\infty}^{\infty} |\Phi(\omega + 2\pi l)|^2 = 1. \tag{3.20}$$

By the [PAR] property of the Fourier transformation and the 2π-periodicity of $e^{i\omega k}$ one has

$$\begin{aligned}
\delta_k &= \int_{\mathbb{R}} \phi(x)\overline{\phi(x-k)}dx \\
&= \frac{1}{2\pi} \int_{\mathbb{R}} \Phi(\omega)\overline{\Phi(\omega)} e^{i\omega k} d\omega \\
&= \frac{1}{2\pi} \int_0^{2\pi} \sum_{l=-\infty}^{\infty} |\Phi(\omega + 2\pi l)|^2 e^{i\omega k} d\omega.
\end{aligned} \tag{3.21}$$

The last line in (3.21) is the Fourier coefficient a_k in the Fourier series decomposition of

$$f(\omega) = \sum_{l=-\infty}^{\infty} |\Phi(\omega + 2\pi l)|^2.$$

Due to the uniqueness of Fourier representation, $f(\omega) = 1$.

Remark 3.3.1 By identity (3.20), any set of independent functions spanning V_0, $\{\phi(x-k), k \in \mathbb{Z}\}$, can be orthogonalized in the Fourier domain. The orthonormal basis is generated by integer-shifts of the function

---

[2] The attribute *à trous* (Fr.) ($\equiv$ with holes) comes from the property $f_{2n} = \delta_n$, i.e., each tap on even position in $\underline{f}$ is 0, except the tap f_0. Such filters are also called half-band filters, see [199].

$$\mathcal{F}^{-1}\left[\frac{\Phi(\omega)}{\sqrt{\sum_{l=-\infty}^{\infty}|\Phi(\omega+2\pi l)|^2}}\right]. \tag{3.22}$$

This normalization in the Fourier domain is sometimes used in constructing wavelet bases; see Exercise 3.14.

Remark 3.3.2 The system $\{\phi(\bullet - k),\ k \in \mathbb{Z}\}$ is a frame for V_0 iff

$$A \le \sum_{l=-\infty}^{\infty} |\Phi(\omega + 2\pi l)|^2 \le B,$$

where A and B are frame bounds.

(b) In terms of m_0 :

$$|m_0(\omega)|^2 + |m_0(\omega + \pi)|^2 = 1. \tag{3.23}$$

Since $\sum_{l=-\infty}^{\infty} |\Phi(2\omega + 2l\pi)|^2 = 1$, then by (3.13)

$$\sum_{l=-\infty}^{\infty} |m_0(\omega + l\pi)|^2 |\Phi(\omega + l\pi)|^2 = 1. \tag{3.24}$$

Now split the sum in (3.24) into two sums – one with odd and the other with even indices, i.e.,

$$1 = \sum_{k=-\infty}^{\infty} |m_0(\omega + 2k\pi)|^2 |\Phi(\omega + 2k\pi)|^2 +$$
$$\sum_{k=-\infty}^{\infty} |m_0(\omega + (2k+1)\pi)|^2 |\Phi(\omega + (2k+1)\pi)|^2.$$

To simplify the above expression, we use relation (3.20) and the 2π-periodicity of $m_0(\omega)$.

$$1 = |m_0(\omega)|^2 \sum_{k=-\infty}^{\infty} |\Phi(\omega + 2k\pi)|^2 + |m_0(\omega + \pi)|^2 \sum_{k=-\infty}^{\infty} |\Phi((\omega + \pi) + 2k\pi)|^2$$
$$= |m_0(\omega)|^2 + |m_0(\omega + \pi)|^2.$$

Remark 3.3.3 Conditions $|m_0(\omega)|^2 + |m_0(\omega + \pi)|^2 = 1$ and $\sum_{k=-\infty}^{\infty} |\Phi(2\omega + 2k\pi)|^2 = 1$ are not equivalent. The first is a necessary and the second is a sufficient condition for orthogonality. For example (see [104]), $\phi(x) = \frac{1}{3} \mathbf{1}(0 \leq x \leq 3)$ has $m_0(\omega) = \frac{1+e^{-3i\omega}}{2}$; it satisfies $m_0(0) = 1$ and $|\frac{1+e^{-3i\omega}}{2}|^2 + |\frac{1-e^{-3i\omega}}{2}|^2 = 1$. However, since $\Phi(\omega) = e^{-3i\omega/2} \frac{\sin 3\omega/2}{3\omega/2}$,

$$\sum_{k=-\infty}^{\infty} |\Phi(2\omega + 2k\pi)|^2 = \frac{1}{3} + \frac{4}{9} \cos \omega + \frac{2}{9} \cos 2\omega \neq 1.$$

3.3.1 Derivation of a Wavelet Function

Whenever a sequence of subspaces satisfies MRA properties, there exists (though not unique) an orthonormal basis for $\mathbb{L}_2(\mathbb{R})$,

$$\{\psi_{jk}(x) = 2^{j/2} \psi(2^j x - k), \ j, k \in \mathbb{Z}\} \tag{3.25}$$

such that $\{\psi_{jk}(x), \ j\text{-fixed}, \ k \in \mathbb{Z}\}$ is an orthonormal basis of the "difference space" $W_j = V_{j+1} \ominus V_j$. The function $\psi(x) = \psi_{00}(x)$ is called a *wavelet function* or informally *the mother wavelet*.

Next, we detail the derivation of a wavelet function from the scaling function. Since $\psi(x) \in V_1$ (because of the containment $W_0 \subset V_1$), it can be represented as

$$\psi(x) = \sum_{k \in \mathbb{Z}} g_k \sqrt{2} \phi(2x - k), \tag{3.26}$$

for some coefficients $g_k, \ k \in \mathbb{Z}$.

Define

$$m_1(\omega) = \frac{1}{\sqrt{2}} \sum_k g_k e^{-ik\omega}. \tag{3.27}$$

By a derivation similar to that in (3.14), we obtain the Fourier counterpart of (3.26),

$$\Psi(\omega) = m_1(\frac{\omega}{2})\Phi(\frac{\omega}{2}). \tag{3.28}$$

The spaces W_0 and V_0 are orthogonal by construction. Therefore,

$$\begin{aligned} 0 = \int \psi(x)\phi(x-k)dx &= \frac{1}{2\pi}\int \Psi(\omega)\overline{\Phi(\omega)}e^{i\omega k}d\omega \\ &= \frac{1}{2\pi}\int_0^{2\pi} \sum_{l=-\infty}^{\infty} \Psi(\omega+2l\pi)\overline{\Phi(\omega+2l\pi)}e^{i\omega k}d\omega. \end{aligned}$$

By repeating the Fourier series argument, as in (3.20), we conclude

$$\sum_{l=-\infty}^{\infty} \Psi(\omega+2l\pi)\overline{\Phi(\omega+2l\pi)} = 0.$$

By taking into account the definitions of m_0 and m_1, and by mimicking the derivation of (3.23), we find

$$m_1(\omega)\overline{m_0(\omega)} + m_1(\omega+\pi)\overline{m_0(\omega+\pi)} = 0. \tag{3.29}$$

From (3.29), we conclude that there exists a function $\lambda(\omega)$ such that

$$(m_1(\omega), m_1(\omega+\pi)) = \lambda(\omega)\left(\overline{m_0(\omega+\pi)}, -\overline{m_0(\omega)}\right). \tag{3.30}$$

By substituting $\xi = \omega + \pi$ and by using the 2π-periodicity of m_0 and m_1, we conclude that

$$\begin{aligned} \lambda(\omega) &= -\lambda(\omega+\pi), \text{ and} \\ \lambda(\omega) &\quad \text{is } 2\pi\text{-periodic.} \end{aligned} \tag{3.31}$$

Any function $\lambda(\omega)$ of the form $e^{\pm i\omega}S(2\omega)$, where S is an $\mathbb{L}_2([0, 2\pi])$, 2π-periodic function, will satisfy (3.29); however, only the functions for which $|\lambda(\omega)| = 1$ will define an orthogonal basis ψ_{jk} of $\mathbb{L}_2(\mathbb{R})$. For technical details the reader is directed to Hernández and Weiss [193].

To summarize, we choose $\lambda(\omega)$ such that

(i) $\lambda(\omega)$ is 2π-periodic,

(ii) $\lambda(\omega) = -\lambda(\omega + \pi)$, and

(iii) $|\lambda(\omega)|^2 = 1$.

Standard choices for $\lambda(\omega)$ are $-e^{-i\omega}$, $e^{-i\omega}$, and $e^{i\omega}$; however, any other function satisfying (i)-(iii) will generate a valid m_1. We choose to define $m_1(\omega)$ as

$$m_1(\omega) = -e^{-i\omega}\overline{m_0(\omega + \pi)}. \tag{3.32}$$

since it leads to a convenient and standard connection between the filters $\underline{h}$ and $\underline{g}$.

The form of m_1 and the equation (3.20) imply that $\{\psi(\bullet - k), k \in \mathbb{Z}\}$ is an orthonormal basis for W_0.

Since $|m_1(\omega)| = |m_0(\omega + \pi)|$, the orthogonality condition (3.23) can be rewritten as

$$|m_0(\omega)|^2 + |m_1(\omega)|^2 = 1. \tag{3.33}$$

By comparing the definition of m_1 in (3.27) with

$$\begin{aligned}
m_1(\omega) &= -e^{-i\omega}\frac{1}{\sqrt{2}}\sum_k \overline{h_k} e^{i(\omega+\pi)k} \\
&= \frac{1}{\sqrt{2}}\sum_k (-1)^{1-k}\overline{h_k} e^{-i\omega(1-k)} \\
&= \frac{1}{\sqrt{2}}\sum_n (-1)^n \overline{h_{1-n}} e^{-i\omega n},
\end{aligned}$$

we relate g_n and h_n as

$$g_n = (-1)^n\, \overline{h_{1-n}}. \tag{3.34}$$

In signal processing literature, the relation (3.34) is known as the *quadrature mirror relation* and the filters $\underline{h}$ and $\underline{g}$ as *quadrature mirror filters*.

Remark 3.3.4 Choosing $\lambda(\omega) = e^{i\omega}$ leads to the rarely used high-pass filter $g_n = (-1)^{n-1}\, \overline{h_{-1-n}}$. It is sometimes convenient to define g_n as $(-1)^n \overline{h_{1-n+M}}$, where M is a "shift constant." Such re-indexing of $\underline{g}$ affects only the shift-location of the wavelet function.

60   WAVELETS

3.4 SOME IMPORTANT WAVELET BASES

In this section we discuss in detail five important families of wavelets: Haar's, Shannon's, Meyer's, Franklin's, and Daubechies'.

3.4.1 Haar's Wavelets

Haar wavelets have limited practical value, however, their educational value is tremendous. Here we illustrate some of the results derived in the Section 3.3 using the Haar wavelet. We start with $\phi(x) = \mathbf{1}(0 \leq x \leq 1)$ and pretend that ψ is unknown.

The scaling equation (3.10) is very simple for the Haar case. By inspection, we conclude that the scaling equation is

$$\begin{aligned} \phi(x) &= \phi(2x) + \phi(2x - 1) \\ &= \frac{1}{\sqrt{2}}\sqrt{2}\phi(2x) + \frac{1}{\sqrt{2}}\sqrt{2}\phi(2x - 1), \end{aligned} \quad (3.35)$$

which gives the wavelet filter coefficients:

$$h_0 = h_1 = \frac{1}{\sqrt{2}}.$$

For the Haar wavelet, the transfer function becomes

$$m_0(\omega) = \frac{1}{\sqrt{2}}\left(\frac{1}{\sqrt{2}}e^{-i\omega 0}\right) + \frac{1}{\sqrt{2}}\left(\frac{1}{\sqrt{2}}e^{-i\omega 1}\right) = \frac{1 + e^{-i\omega}}{2}.$$

Now,

$$m_1(\omega) = -e^{-i\omega}\,\overline{m_0(\omega + \pi)} = -e^{-i\omega}\left(\frac{1}{2} - \frac{1}{2}e^{i\omega}\right) = \frac{1 - e^{-i\omega}}{2}.$$

Notice that $m_0(\omega) = |m_0(\omega)|e^{i\varphi(\omega)} = \cos\frac{\omega}{2}\cdot e^{-i\omega/2}$ (after $\cos x = \frac{e^{ix} + e^{-ix}}{2}$). Since $\varphi(\omega) = -\frac{\omega}{2}$, Haar's wavelet has *linear phase*, i.e., the scaling function is symmetric in the time domain. Also, the orthogonality condition $|m_0(\omega)|^2 + |m_0(\omega)|^2 = 1$ is easily verified, see Fig. 3.6.

Relation (3.28) becomes

$$\Psi(\omega) = \frac{1 - e^{-i\omega/2}}{2}\Phi\left(\frac{\omega}{2}\right) = \frac{1}{2}\Phi\left(\frac{\omega}{2}\right) - \frac{1}{2}\Phi\left(\frac{\omega}{2}\right)e^{-i\omega/2},$$

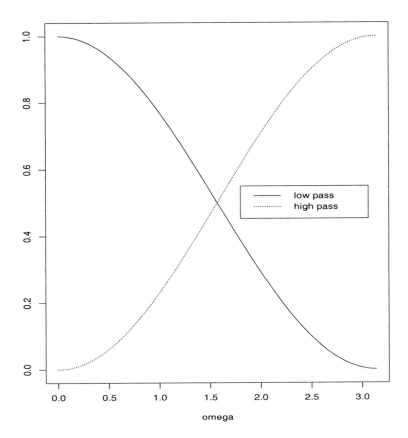

Fig. 3.6 Functions $|m_0(\omega)|^2 = \cos^2 \frac{\omega}{2}$ and $|m_1(\omega)|^2 = \sin^2 \frac{\omega}{2}$ for the Haar case. The orthogonality condition (3.33) becomes the fundamental trigonometric identity, $\sin^2 \frac{\omega}{2} + \cos^2 \frac{\omega}{2} = 1$.

62 WAVELETS

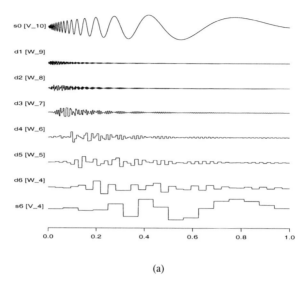

(a)

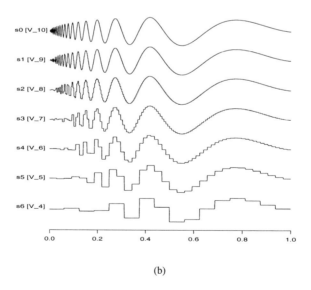

(b)

Fig. 3.7 (a) Multiresolution analysis of the doppler function. The original function in V_{10} is a direct sum of projections on V_4 and $W_4, W_5, \ldots, W_9$ subspaces. (b) Coarsening the doppler function by projecting it to V- subspaces.

and by applying the inverse Fourier transformation we obtain

$$\psi(x) = \phi(2x) - \phi(2x-1)$$

in the time-domain. Therefore we "found" the Haar wavelet function ψ. From the expression for m_1 we "conclude" that $g_0 = -g_{-1} = \frac{1}{\sqrt{2}}$.

The Haar basis is not an appropriate basis for all applications for several reasons. The building blocks in Haar's decomposition are discontinuous functions that obviously are not effective in approximating smooth functions. Although the Haar wavelets are well localized in the time domain, in the frequency domain they decay at the slow rate of $O(\frac{1}{n})$. Fig. 3.7 suggests that many levels of detail are needed to describe the signal.

3.4.2 Shannon's Wavelets

The other extreme is the *Shannon wavelet*, sometimes called the Littlewood-Paley wavelet. It is a time–scale mirror image of the Haar wavelet. The scaling function in Shannon's basis is defined in the Fourier domain as

$$\Phi(\omega) = \mathbf{1}(-\pi \leq \omega < \pi).$$

It is straightforward to derive its time-domain representation,

$$\phi(x) = \frac{1}{2\pi} \int_{-\pi}^{\pi} e^{i\omega x} d\omega = \frac{\sin(\pi x)}{\pi x}. \tag{3.36}$$

The family $\{\phi(\bullet - k), k \in \mathbb{Z}\}$ is an orthonormal system. Indeed, $\sum_{l=-\infty}^{\infty} |\Phi(\omega + 2\pi l)|^2 = 1$ [see (3.20)] is trivially satisfied. Equivalently, in the time-domain,

$$\begin{aligned}
\int \phi(x)\phi(x-k)dx &\stackrel{[\text{PAR}]}{=} \frac{1}{2\pi} \int \Phi(\omega)\overline{\Phi(\omega)} d\omega \\
&= \frac{1}{2\pi} \int_{-\pi}^{\pi} e^{i\omega k} d\omega \\
&= \frac{\sin(\pi k)}{\pi k} \\
&= \delta_k.
\end{aligned}$$

By inspecting (3.13), we find that for the Shannon wavelet $m_0(\frac{\omega}{2}) = 1$ for $\omega \in [-\pi, \pi]$ and $m_0(\frac{\omega}{2}) = 0$ for $\omega \in [-2\pi, -\pi) \cup (\pi, 2\pi]$. By taking into account 2π-periodicity, we obtain

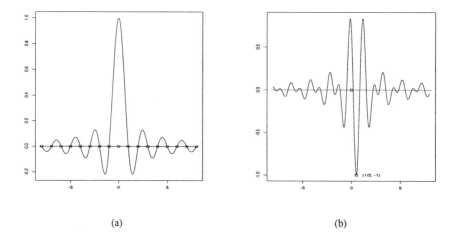

Fig. 3.8 Shannon's scaling [panel (a)] and wavelet [panel (b)] functions.

$$m_0(\omega) = \sum_{k \in \mathbb{Z}} \mathbf{1}\left(-\frac{\pi}{2} + 2k\pi \leq \omega \leq \frac{\pi}{2} + 2k\pi\right).$$

From (3.32) and (3.28),

$$\Psi(\omega) = -e^{-i\omega/2}\mathbf{1}(\pi \leq |\omega| \leq 2\pi) = -e^{-i\omega/2} \cdot \left[\Phi\left(\frac{\omega}{2}\right) - \Phi(\omega)\right].$$

The corresponding time-domain function $\psi(x)$ is

$$\psi(x) = \phi\left(x - \frac{1}{2}\right) - 2\phi(2x - 1) = \operatorname{sinc}\pi\left(x - \frac{1}{2}\right) - 2\operatorname{sinc}2\pi\left(x - \frac{1}{2}\right).$$

Graphs of ϕ and ψ are given in Fig. 3.8. The squares on the x-axis in Fig. 3.8(a) emphasize the interpolating property $\phi(n) = \delta_n$, which will be discussed in Section 3.5.6.

The support of Φ is $[-2\pi, -\pi] \cup [\pi, 2\pi]$. Since the supports of Φ and Ψ are exclusive, the functions ϕ and ψ are orthogonal (see Exercise 3.13).

As we noted before, the components of the vector h are the coefficients in the Fourier series expansion of $H(\omega) = \sqrt{2}\, m_0(\omega)$. Thus,

$$h_k = \frac{\sqrt{2}}{2\pi} \int_{-\pi/2}^{\pi/2} e^{ik\omega} = \frac{\sqrt{2}}{k\pi} \cdot \frac{e^{ik\pi/2} - e^{-ik\pi/2}}{2i}$$

$$= \frac{\sqrt{2}}{k\pi} \sin k\frac{\pi}{2} = \frac{1}{\sqrt{2}} \operatorname{sinc} \frac{\pi}{2k}, \; k \in \mathbb{Z}.$$

Although Shannon's filter is ideal in the frequency domain since it cuts the frequency band in half, it has poor time localization properties. It corresponds to an IIR filter with slowly decaying coefficients that make it non-localized in time.

3.4.3 Meyer's Wavelets

As we argued before, the Shannon wavelet exhibits ideal frequency behavior but its bad time localization hinders its practical use. Meyer [291] suggested modifying Shannon's $\Phi(\omega)$ function by smoothing sharp edges at $\omega = \pm\pi$ while preserving the orthogonality condition (3.20). He proposed a scaling function with the following equation in the Fourier domain

$$\Phi(\omega) = \begin{cases} 1, & |\omega| \leq \frac{2\pi}{3} \\ \cos[\frac{\pi}{2} \nu(\frac{3}{2\pi}|\omega| - 1)], & \frac{2\pi}{3} \leq |\omega| \leq \frac{4\pi}{3} \\ 0, & \text{otherwise} \end{cases} \quad (3.37)$$

where a smooth "taper" function ν satisfies:

$$\nu(x) + \nu(1-x) = 1$$
$$\nu(x) = 0, \; x \leq 0$$
$$\nu(x) = 1, \; x \geq 1.$$

Many different ν-functions are possible (see Walter [439] for the most general definition) but we will confine ourselves to the class of polynomials originally proposed in [291]. Meyer's ν-polynomials can be introduced in a probabilistic way.

Let $\nu_\theta(x)$ be the cumulative distribution function of a symmetric (about $\frac{1}{2}$) Beta random variable X, $X \sim Beta(\theta, \theta)$. Then for $0 \leq x \leq 1$,

$$\nu_\theta(x) = \frac{1}{B(\theta, \theta)} \int_0^x t^{\theta-1}(1-t)^{\theta-1} dt$$
$$= x^\theta F(1-\theta, \theta, 1+\theta; x)/B(\theta, \theta) \quad (3.38)$$

where $F(\alpha, \beta, \gamma; z) = \sum_k \frac{(\alpha)_k (\beta)_k}{(\gamma)_k \, k!} z^k$, $|z| < 1$ is the hypergeometric function,

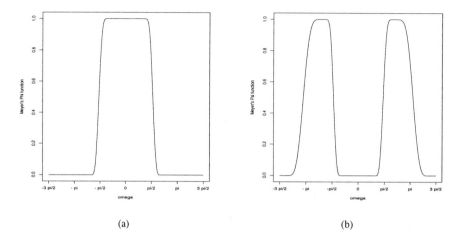

Fig. 3.9 Meyer's $\Phi(\omega)$ [panel (a)] and $\Psi(\omega)$ [panel (b)] for $\nu(x) = x^4(35 - 84x + 70x^2 - 20x^3)$.

Table 3.1 The ν-polynomials for $\theta = 1, \ldots, 5$. The corresponding $\Phi(\omega)$ have $0, 1, \ldots, 4$ continuous derivatives.

θ	$\nu_\theta(x)$
1	x
2	$x^2(2 - 3x)$
3	$x^3(10 - 15x + 6x^2)$
4	$x^4(35 - 84x + 70x^2 - 20x^3)$
5	$x^5(126 - 420x + 540x^2 - 315x^3 + 70x^4)$

$B(\theta, \theta)$ is Euler's beta function, and $(\alpha)_k$ is the product $\alpha(\alpha + 1) \ldots (\alpha + k - 1)$. For $\theta = 4$ the expression (3.38) becomes $\nu(x) = x^4(35 - 84x + 70x^2 - 20x^3)$. Corresponding $\Phi(\omega)$ and $\Psi(\omega)$ are shown in Fig. 3.9(a) and (b), respectively. Table 3.1 gives the ν-polynomials for some integer values of θ.

Now let us see how Meyer's construction works. First, note that $m_0(\frac{\omega}{2}) = \Phi(\omega)$, $-\pi \leq \omega \leq \pi$. This follows from the fact that $\Phi(\frac{\omega}{2}) = 1$ in the domain of $\Phi(\omega)$ by construction [see Fig. 3.10(a)]. Since $m_0(\omega)$ is a 2π-periodic function it follows that

$$m_0(\omega) = \sum_k \Phi[2(\omega + 2k\pi)].$$

Next, we demonstrate that, as expected, $\Phi(\omega) = m_0(\omega/2)\Phi(\omega/2)$. Indeed,

$$m_0(\omega/2)\Phi(\omega/2) = \sum_k \Phi(\omega + 4k\pi)\Phi(\omega/2)$$
$$= \Phi(\omega)\Phi(\omega/2)$$
[since $\Phi(\omega + 4k\pi)$ and $\Phi(\omega/2)$ overlap only for $k = 0$]
$$= \Phi(\omega) \quad \text{[since } \Phi(\omega/2) = 1 \text{ in the support of } \Phi(\omega)\text{]}.$$

The class of ν-polynomials provides miraculous cancellations in proving the orthogonality condition $\sum_k |\Phi(\omega + 2k\pi)|^2 = 1$. Functions $\Phi(\omega)$ and $\Phi(\omega - 2k\pi)$ overlap in the region $[\frac{2\pi}{3}, \frac{4\pi}{3}]$ only. To prove the orthogonality condition we need to show that $|\Phi(\omega)|^2 + |\Phi(\omega - 2k\pi)|^2 = 1$ for $\omega \in [\frac{2\pi}{3}, \frac{4\pi}{3}]$.

From (3.37),

$$|\Phi(\omega)|^2 + |\Phi(\omega - 2k\pi)|^2$$
$$= \cos^2\left[\frac{\pi}{2}\nu\left(\frac{3}{2\pi}\omega - 1\right)\right] + \cos^2\left[\frac{\pi}{2}\nu\left(\frac{3}{2\pi}|\omega - 2\pi| - 1\right)\right]$$
$$= \cos^2\left[\frac{\pi}{2}\nu\left(\frac{3}{2\pi}\omega - 1\right)\right] + \cos^2\left[\frac{\pi}{2}\nu\left(\frac{3}{2\pi}(-\omega + 2\pi) - 1\right)\right]$$
[since $\omega - 2\pi < 0$]
$$= \cos^2\left[\frac{\pi}{2}\nu\left(\frac{3}{2\pi}\omega - 1\right)\right] + \cos^2\left[\frac{\pi}{2} - \frac{\pi}{2}\nu\left(\frac{3}{2\pi}\omega - 1\right)\right]$$
[by $\nu(x) + \nu(1 - x) = 1$]
$$= \cos^2\left[\frac{\pi}{2}\nu\left(\frac{3}{2\pi}\omega - 1\right)\right] + \sin^2\left[\frac{\pi}{2}\nu\left(\frac{3}{2\pi}\omega - 1\right)\right] = 1.$$

Next, we find an expression for the wavelet function in the Fourier domain.

$$\Psi(\omega) = -e^{-i\omega/2}\overline{m_0(\omega/2 + \pi)}\Phi(\omega/2)$$
$$= -e^{-i\omega/2}\sum_l \Phi(\omega + 2\pi(2l + 1))\Phi(\omega/2)$$
$$= -e^{-i\omega/2}[\Phi(\omega - 2\pi) + \Phi(\omega + 2\pi)]\Phi(\omega/2)$$

since for l different than 0 and -1 functions $\Phi(\omega + 2\pi(2l + 1))$ and $\Phi(\omega/2)$ do not overlap [see Fig. 3.10(b)].

From (3.37),

$$\Psi(\omega) = \begin{cases} -e^{-i\omega/2}\sin[\frac{\pi}{2}\nu(\frac{3}{2\pi}|\omega| - 1)], & \frac{2\pi}{3} \leq |\omega| \leq \frac{4\pi}{3} \\ -e^{-i\omega/2}\cos[\frac{\pi}{2}\nu(\frac{3}{2\pi}|\omega| - 1)], & \frac{4\pi}{3} \leq |\omega| \leq \frac{8\pi}{3} \\ 0, & \text{otherwise} \end{cases} \quad (3.39)$$

68 WAVELETS

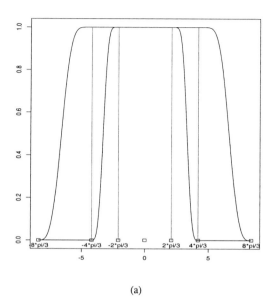

(a)

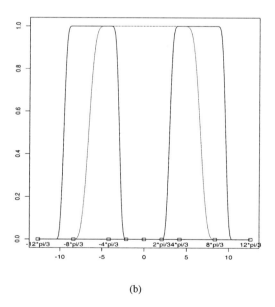

(b)

Fig. 3.10 Meyer wavelet in the Fourier domain. (a) Superimposed graphs of $\Phi(\omega)$ and $\Phi(\frac{\omega}{2})$. Notice that $\Phi(\frac{\omega}{2}) = 1$ for $|\omega| \leq \frac{2\pi}{3}$. (b) Functions $\Phi(\omega + 2\pi)$ and $\Phi(\omega - 2\pi)$ overlap with $\Phi(\frac{\omega}{2})$.

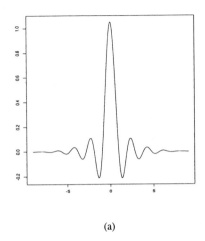
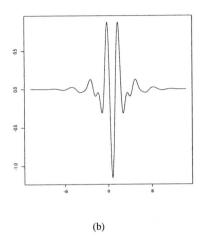

Fig. 3.11 Meyer's $\phi(x)$ [panel (a)] and $\psi(x)$ [panel (b)] for taper $\nu(x) = x^4(35 - 84x + 70x^2 - 20x^3)$.

Since $\Phi(\omega)$ is an even function,

$$\phi(x) = \frac{1}{2}\Phi(\omega)e^{i\omega x}\,d\omega = \frac{1}{\pi}\int_0^{4\pi/3} \Phi(\omega)\cos\omega x\,d\omega,$$

and

$$\psi(x) = -\frac{1}{\pi}\int_{2\pi/3}^{4\pi/3} \Phi(\omega/2)\Phi(\omega - 2\pi)\cos\omega(x - 1/2)\,d\omega.$$

Graphs of the functions $\phi(x)$ and $\psi(x)$ are shown in Fig. 3.11(a) and (b), respectively.

Filter coefficients for a Meyer wavelet are the coefficients in the Fourier expansion of $\sqrt{2}\,m_0(\omega)$ [see (3.12)],

$$h_n = \frac{\sqrt{2}}{\pi}\left(\int_0^{\pi/2} \cos n\omega\,d\omega + \int_{\pi/3}^{2\pi/3} \cos\left[\frac{\pi}{2}\nu\left(\frac{3}{\pi}\omega - 1\right)\right]\,d\omega\right).$$

Coefficients h_n are determined by numerical integration, except for the case $\nu(x) = x$ when a closed form of the above integral is available. It this case, $h_n = \sqrt{2}\pi(\frac{\sin n\pi/3}{n} + \frac{4n\sin n\pi/3}{9-4n^2} + \frac{6\cos 2n\pi/3}{9-4n^2})$.

Remark 3.4.1 Walter [439] introduced Meyer-type wavelets. The scaling and wavelet functions are defined in the Fourier domain as

$$\Phi(\omega) = \left(\int_{\omega-\pi}^{\omega+\pi} dP\right)^{1/2} \quad \text{and} \quad \Psi(\omega) = e^{-i\omega/2} \left(\int_{|\omega|/2-\pi}^{|\omega|/2+\pi} dP\right)^{1/2},$$

where P is any probability measure supported on $[-\frac{\pi}{3}, \frac{\pi}{3}]$. The orthogonality of scaling functions verifies readily,

$$\sum_k |\Phi(\omega + 2k\pi)|^2 = \sum_k \left(\int_{\omega+2k\pi-\pi}^{\omega+2k\pi+\pi} dP\right)$$
$$= \sum_k \int_{\omega-(2k-1)\pi}^{\omega+(2k+1)\pi} dP = \int_\mathbb{R} dP = 1.$$

Meyer-type (bandlimited) wavelets are used in deconvolution problems in statistics since only bandlimited wavelets allow wavelet estimators in the case of "supersmooth" error distributions. See also Walter and Zayed [446].

3.4.4 Franklin's Wavelets

The method of constructing wavelet bases described in this section is very intuitive and have been used to generate some popular wavelet bases.

Let V_0 be the space of all $\mathbb{L}_2(\mathbb{R})$ functions linear on intervals $[k, k+1]$, $k \in \mathbb{Z}$. The shifts of the Schauder basis on $[0, 1]$ (discussed on page 3)

$$\Delta_0(x) = 1,$$
$$\Delta_1(x) = x,$$
$$\Delta_2(x) = 2x\mathbf{1}(0 \le x \le \frac{1}{2}) + 2(1-2x)\mathbf{1}(\frac{1}{2} \le x \le 1),$$
$$\ldots$$

span self-similar spaces V_j.

Though the space V_0 can be spanned by translations of $\Delta_1(x) = x$, we chose the related function, $\Delta(x) = \Delta_1(x) - \Delta_1(x-1)$. The function $\Delta(x)$ is a "tent" function [Fig. 3.12(a)] given as

$$\Delta(x) = (1 - |x-1|)\,\mathbf{1}(0 \le x < 2) = \begin{cases} x & 0 \le x \le 1 \\ 2-x & 1 \le x < 2 \\ 0 & \text{otherwise} \end{cases}. \quad (3.40)$$

The family of functions

$$\{\Delta(x - k),\ k \in \mathbb{Z}\} \tag{3.41}$$

spans the space V_0. However, the functions in (3.41) are not orthogonal scaling functions. According to (3.22) in Remark 3.3.1, it is possible to orthogonalize the family (3.41). This orthogonalization was proposed by Battle and Lemarié [255, 23] and the resulting wavelet bases are known in the literature as Battle-Lemarié wavelets. The Franklin wavelet is a special case of Battle-Lemarié wavelets for which the starting family is (3.41). For a historical development, see Franklin [157].

The function (3.40) has the following form in the Fourier domain:

$$\hat{\Delta}(\omega) = e^{-i\omega} \left(\frac{\sin \frac{\omega}{2}}{\frac{\omega}{2}} \right)^2. \tag{3.42}$$

The argument can be probabilistic. The density of the sum of two i.i.d. uniform on [0,1] random variables is a tent function. Since $e^{-i\omega/2} \frac{\sin(\omega/2)}{\omega/2}$ is the Fourier transformation of the uniform density, the [CON] property implies (3.42).

To orthogonalize the system $\{\Delta(x - k),\ k \in \mathbb{Z}\}$, we have to find the sum

$$\begin{aligned} S(\omega) &= \sum_l |\hat{\Delta}(\omega + 2l\pi)|^2 \\ &= \sum_l \left(\frac{\sin(\frac{\omega}{2} + l\pi)}{\frac{\omega}{2} + l\pi} \right)^4 \end{aligned}$$

and divide $\hat{\Delta}(\omega)$ by $\sqrt{S(\omega)}$.

The following lemma gives the desired sum.

Lemma 3.4.1

$$S(\omega) = 1 - \frac{2}{3} \sin^2 \frac{\omega}{2}.$$

Proof: Notice first that $\sum_l \left(\frac{\sin(\frac{\omega}{2} + l\pi)}{\frac{\omega}{2} + l\pi} \right)^2 = 1$. This identity is a consequence of the orthogonality condition applied on the Haar scaling function:

$$\sum_l |\Phi(\omega + 2l\pi)|^2 = 1 \quad \text{for } \Phi(\omega) = e^{-i\omega/2} \frac{\sin \omega/2}{\omega/2}.$$

Since

$$1 = \sum_l \frac{\sin^2(\frac{\omega}{2} + l\pi)}{(\frac{\omega+2l\pi}{2})^2}$$

$$= \sin^2 \frac{\omega}{2} \sum_l \frac{4}{(\omega + 2l\pi)^2} \quad [\text{since } \sin(\frac{\omega}{2} + l\pi) = (-1)^l \sin(\frac{\omega}{2})]$$

it follows

$$\sum_l \frac{1}{(\omega + 2l\pi)^2} = \frac{1}{4 \sin^2 \frac{\omega}{2}}.$$

By taking the derivative of both sides twice, we obtain

$$\sum_l \frac{1}{(\omega + 2l\pi)^3} = \frac{\cos \frac{\omega}{2}}{8 \sin^3 \frac{\omega}{2}}$$

and

$$\sum_l \frac{1}{(\omega + 2l\pi)^4} = \frac{1 - \frac{2}{3} \cdot \sin^2 \frac{\omega}{2}}{16 \sin^4 \frac{\omega}{2}}.$$

Finally,

$$S(\omega) = 16 \sum_l \frac{\sin^4(\frac{\omega}{2} + l\pi)}{(\omega + 2l\pi)^4}$$

$$= 16 \cdot \sin^4(\frac{\omega}{2}) \sum_l \frac{1}{(\omega + 2l\pi)^4}$$

$$= 1 - \frac{2}{3} \sin^2 \frac{\omega}{2}.$$

The orthogonalized system is

$$\Phi(\omega) = \left(\frac{\sin \frac{\omega}{2}}{\frac{\omega}{2}}\right)^2 \cdot [S(\omega)]^{-1/2}$$

$$= \left(\frac{\sin \frac{\omega}{2}}{\frac{\omega}{2}}\right)^2 \cdot \left(1 - \frac{2}{3} \sin^2 \frac{\omega}{2}\right)^{-1/2}.$$

Since $1 - \frac{2}{3} \sin^2 \frac{\omega}{2} = \frac{2 + \cos \omega}{3}$, it follows that

$$m_0(\omega) = \frac{\Phi(2\omega)}{\Phi(\omega)} = \cos^2\frac{\omega}{2}\sqrt{\frac{2+\cos\omega}{2+\cos 2\omega}}.$$

The functions $m_1(\omega)$ and $\Psi(\omega)$ are obtained as $m_1(\omega) = -e^{-i\omega}\overline{m_0(\omega+\pi)}$ and $\Psi(2\omega) = -e^{-i\omega}\overline{m_0(\omega+\pi)}\Phi(\omega)$.

The graphs of the "tent" function (3.40), $\Phi(\omega), \Psi(\omega), m_0(\omega), \phi(x)$ and $\psi(x)$ are given in Fig. 3.12. Notice that both scaling and wavelet functions have a simple, piecewise linear form. However, in the process of orthogonalization we cannot control the support of wavelet functions. Franklin wavelets are of unbounded support and correspond to IIR filters. However, the decay of scaling and wavelet functions is exponential.

The wavelet function $\psi(x)$ is symmetric about the point $x = \frac{1}{2}$, $\psi(x) = \psi(1-x)$. Indeed,

$$\begin{aligned}\psi(1-x) &= \frac{1}{2\pi}\int \Psi(\omega)e^{i\omega(1-x)}d\omega \\ &= \frac{1}{2\pi}\int -e^{-i\omega}F(\omega)e^{i\omega}e^{-i\omega x}d\omega \\ &= \frac{1}{2\pi}\int F(-\omega)e^{i(-\omega)x}d(-\omega) = \psi(x).\end{aligned}$$

It is possible to orthogonalize the family of shifts of B-spline wavelets of an arbitrary order n. The formula for general case is

$$\sum_l \frac{1}{(\omega+2l\pi)^n} = \frac{(-1)^{n-2}}{(n-1)!}\frac{d^{n-2}}{d\omega^{n-2}}\sum_l \frac{1}{(\omega+2l\pi)^2}.$$

3.4.5 Daubechies' Compactly Supported Wavelets

Daubechies was first to construct compactly supported orthogonal wavelets with a preassigned degree of smoothness. Some scaling and wavelet functions from Daubechies' family are given in Fig. 3.13. Here we present the idea of Daubechies, omitting some technical details. Detailed treatment of this topic can be found in Daubechies' monograph [104], Chapters 6 and 7.

Suppose that ψ has N (≥ 2) vanishing moments, i.e., $\int x^n\psi(x)dx = 0$, $n = 0, 1, \ldots, N-1$. Then by Theorem 3.5.1, $m_0(\omega)$ has the form:

$$m_0(\omega) = \left(\frac{1+e^{-i\omega}}{2}\right)^N \mathcal{L}(\omega), \qquad (3.43)$$

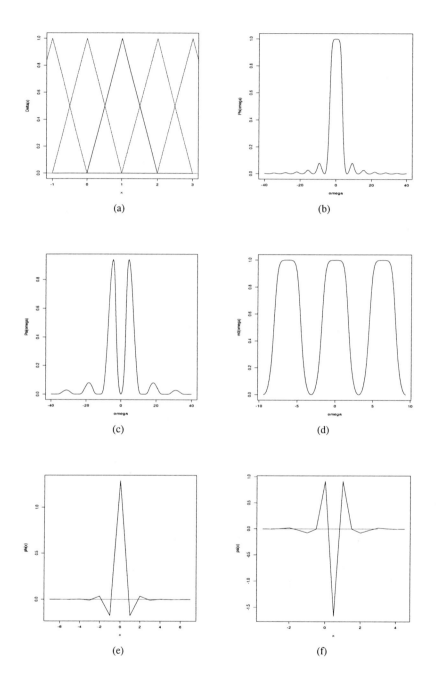

Fig. 3.12 Franklin wavelet related functions: (a) translations of the "tent" function, (b) $\Phi(\omega)$, (c) $\Psi(\omega)$, (d) $m_0(\omega)$, (e) $\phi(x)$, and (f) $\psi(x)$.

SOME IMPORTANT WAVELET BASES   75

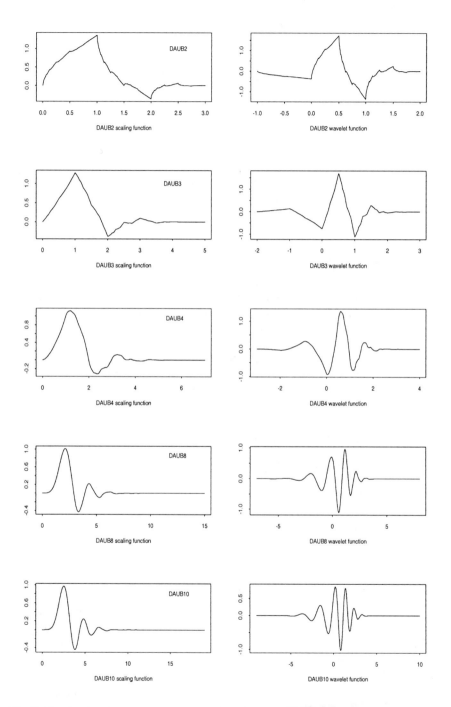

Fig. 3.13 Graphs of scaling and wavelets functions from Daubechies' family, $N = 2, 3, 4, 8$, and 10.

where $\mathcal{L}(\omega)$ is a trigonometric polynomial. When

$$M_0(\omega) = |m_0(\omega)|^2 = \left(\cos^2\frac{\omega}{2}\right)^N \cdot |\mathcal{L}(\omega)|^2,$$

the orthogonality condition (3.23) becomes

$$M_0(\omega) + M_0(\omega + \pi) = 1. \tag{3.44}$$

$|\mathcal{L}(\omega)|^2$ is a polynomial in $\cos\omega$. It can be re-expressed as a polynomial in $\sin^2\frac{\omega}{2}$ since $\cos\omega = 1 - 2\sin^2\frac{\omega}{2}$. Denote this polynomial by $P(\sin^2\frac{\omega}{2})$. In terms of P the orthogonality condition (3.44) becomes

$$(1-y)^N P(y) + y^N P(1-y) = 1, \quad (y = \sin^2\frac{\omega}{2}). \tag{3.45}$$

By Bezout's result (outlined below), there exists a unique solution of the functional equation (3.45). It can be found by the Euclidean algorithm since the polynomials $(1-y)^N$ and y^N are relatively prime.

Lemma 3.4.2 *(Bezout) If p_1 and p_2 are two polynomials of degree n_1 and n_2, respectively, with no common zeroes, then there exist unique polynomials q_1 and q_2 of degree $n_2 - 1$ and $n_1 - 1$, respectively, so that*

$$p_1(x)q_1(x) + p_2(x)q_2(x) = 1.$$

For the proof of the lemma, we direct the reader to Daubechies ([104], 169-170). The unique solution of (3.45) with degree $deg(P(y)) \leq N - 1$ is

$$\sum_{k=0}^{N-1} \binom{N+k-1}{k} y^k, \quad y = \sin^2\frac{\omega}{2}, \tag{3.46}$$

and since it is positive for $y \in [0, 1]$, it does not contradict the positivity of $|\mathcal{L}(\omega)|^2$.

Remark: If the degree of a solution is not required to be minimal then any other polynomial $Q(y) = P(y) + y^N R(\frac{1}{2} - y)$ where R is an odd polynomial preserving the positivity of Q, will lead to a different solution for $m_0(\omega)$. By choosing $R \neq 0$, one can generalize the standard Daubechies family. As we will see in Section 5.1, an example is the construction of *coiflets*.

The function $|m_0(\omega)|^2$ is now completely determined. To finish the construction we have to find its square root. A result of Riesz, known as the *spectral factorization lemma*, makes this possible.

SOME IMPORTANT WAVELET BASES 77

Lemma 3.4.3 *(Riesz) Let A be a positive trigonometric polynomial with the property $A(-x) = A(x)$. Then, A is necessarily of the form*

$$A(x) = \sum_{m=1}^{M} u_m \cos mx.$$

In addition, there exists a polynomial B of the same order $B(x) = \sum_{m=1}^{M} v_m e^{imx}$ such that $|B(x)|^2 = A(x)$. If the coefficients u_m are real, then B can be chosen so that the coefficients v_m are also real.

We first represent $|\mathcal{L}(\omega)|^2$ as the polynomial

$$\frac{a_0}{2} + \sum_{k=1}^{N-1} a_k \cos^k \omega,$$

by replacing $\sin^2 \frac{\omega}{2}$ in (3.46) by $\frac{1-\cos \omega}{2}$.

An auxiliary polynomial P_A, such that $|\mathcal{L}(e^{-i\omega})|^2 = |P_A(e^{-i\omega})|$, is formed. If $z = e^{-i\omega}$, then $\cos \omega = \frac{z+z^{-1}}{2}$ and one such auxiliary polynomial is

$$P_A(z) = \frac{1}{2} \sum_{k=1-N}^{N-1} a_{|k|} z^{N-1+k}. \tag{3.47}$$

Since $P_A(z) = z^{2N-2} P_A(\frac{1}{z})$, the zeroes of $P_A(z)$ appear in reciprocal pairs if real, and quadruples $(z_i, \bar{z}_i, z_i^{-1}, \bar{z}_i^{-1})$ if complex. Fig. 3.14 gives locations of zeroes of $P_A(z)$ for $N = 4$ [panel (a)] and $N = 20$ [panel (b)]. Without loss of generality we assume that z_j, $\bar{z}_j$ and r_j lie outside the unit circle in the complex plane. Of course, then z_j^{-1}, $\bar{z}_j^{-1}$ and r_j^{-1} lie inside the unit circle. The factorized polynomial P_A can be written as

$$P_A(z) = \frac{1}{2} a_{N-1} \left[\prod_{i=1}^{I} (z - r_i)(z - \frac{1}{r_i}) \right]$$
$$\left[\prod_{j=1}^{J} (z - z_j)(z - \bar{z}_j)(z - z_j^{-1})(z - \bar{z}_j^{-1}) \right]. \tag{3.48}$$

Here $r_1, r_2, \ldots, r_I$ are real and non zero, and $z_1, \ldots, z_J$ are complex; $I + 2J = N - 1$.

Our goal is to take a square root from $|P_A(z)|$ and the following simple substitution puts $|P_A(z)|$ in a convenient form.

Since $z = e^{-i\omega}$, we replace $|(z - z_j)(z - \bar{z}_j^{-1})|$ by $|z_j|^{-1}|z - z_j|^2$ (see Exercise 3.25), and the polynomial $|P_A|$ becomes

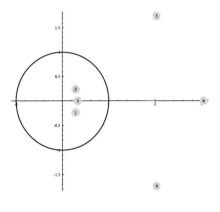

(a)

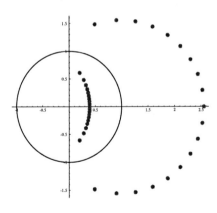

(b)

Fig. 3.14 Locations of zeroes for the polynomial $P_A(z)$ for (a) $N = 4$ and (b) $N = 20$. Notice a curious distribution of zeroes in panel (b). The outer zeroes concentrate along the arc of $|z - 1| = \sqrt{2}$, while the inner zeroes concentrate along the arc of $|z + 1| = \sqrt{2}$. For more details about the interesting properties of zeroes of Daubechies polynomials, see Strang and Nguyen [387] and Shen and Strang [374].

$$\frac{1}{2}|a_{N-1}|\prod_{i=1}^{I}|r_i^{-1}|\prod_{j=1}^{J}|z_j|^{-2}\cdot|\prod_{i=1}^{I}(z-r_i)\prod_{j=1}^{J}(z-z_j)(z-\bar{z}_j)|^2.$$

Now, $\mathcal{L}(\omega)$ is

$$\pm(\frac{1}{2}|a_{N-1}|\prod_{i=1}^{I}|r_i^{-1}|\prod_{j=1}^{J}|z_j|^{-2})^{\frac{1}{2}}$$
$$\cdot|\prod_{i=1}^{I}(z-r_i)\prod_{j=1}^{J}(z-z_j)(z-\bar{z}_j)|,\ z=e^{-i\omega}, \tag{3.49}$$

where the sign is chosen so that $m_0(0) = \mathcal{L}(0) = 1$. Note that $deg[P_A(z)] = deg[|\mathcal{L}(z)|^2] = N - 1$.

Finally, the coefficients $h_0, h_1, \ldots, h_{2N-1}$ in the polynomial $\sqrt{2}\,m_0(\omega)$ are the desired wavelet filter coefficients.

Example 3.4.1 We will find m_0 for $N = 2$.
$|\mathcal{L}(\omega)|^2 = \sum_{k=0}^{2-1}\binom{2+k-1}{k}\sin^2\frac{\omega}{2} = 1 + 2\frac{1-\cos\omega}{2} = \frac{1}{2}4 - 1\cdot\cos\omega$ gives $a_0 = 4$ and $a_1 = -1$.

The auxiliary polynomial P_A is

$$\begin{aligned} P_A(z) &= \frac{1}{2}\sum_{k=-1}^{1}a_{|k|}z^{1+k} \\ &= \frac{1}{2}(-1 + 4z - z^2) \\ &= -\frac{1}{2}\left(z - (2+\sqrt{3})\right)\left(z - (2-\sqrt{3})\right). \end{aligned}$$

One square root from the above polynomial is

$$\begin{aligned} \sqrt{\frac{1}{2}(|-1|)}\frac{1}{2+\sqrt{3}}\left(z-(2+\sqrt{3})\right) &= \frac{1}{\sqrt{2}}\sqrt{2-\sqrt{3}}\left(z-(2+\sqrt{3})\right) \\ &= \frac{1}{2}\left((\sqrt{3}-1)z - (1+\sqrt{3})\right). \end{aligned}$$

The change in sign in the expression above is necessary, since the expression should have the value of 1 at $z = 1$ or equivalently at $\omega = 0$. Finally,

$$m_0(\omega) = \left(\frac{1+e^{-i\omega}}{2}\right)^2 \frac{1}{2}\left((1-\sqrt{3})e^{-i\omega} + (1+\sqrt{3})\right)$$
$$= \frac{1}{\sqrt{2}}\left(\frac{1+\sqrt{3}}{4\sqrt{2}} + \frac{3+\sqrt{3}}{4\sqrt{2}}e^{-i\omega} + \frac{3-\sqrt{3}}{4\sqrt{2}}e^{-2i\omega} + \frac{1-\sqrt{3}}{4\sqrt{2}}e^{-3i\omega}\right).$$

Table 3.2 gives $\underline{h}$-filters for DAUB2 - DAUB10 wavelets.

3.5 SOME EXTENSIONS

3.5.1 Regularity of Wavelets

An appealing property of wavelet bases is their diversity. One can construct wavelets with different smoothness, symmetry, and support properties. Sometimes the requirements can be conflicting since some of the properties are exclusive. For example, there exists no symmetric real-valued wavelet with a compact support. Similarly, there is no $\mathbb{C}^\infty$-wavelet function with an exponential decay.

Scaling functions and wavelets can be constructed with desired degree of smoothness. The regularity (smoothness) of wavelets is connected with the rate of decay of scaling functions and ultimately with the number of vanishing moments of scaling and wavelet functions. For instance, the Haar wavelet has only the "zeroth" vanishing moment (as a consequence of the admissibility condition) resulting in a discontinuous wavelet function.

Theorem 3.5.1 is important in connecting the regularity of wavelets, the number of vanishing moments, and the form of the transfer function $m_0(\omega)$. The proof is based on the Taylor series argument and the scaling properties of wavelet functions. For details, we direct the reader to Daubechies ([104], 153-155). Let

$$\mathcal{M}_k = \int x^k \phi(x)dx \text{ and } \mathcal{N}_k = \int x^k \psi(x)dx,$$

be the kth moments of the scaling and wavelet functions, respectively.

Theorem 3.5.1 *Let $\psi_{jk}(x) = 2^{j/2}\psi(2^j x - k)$, $j, k \in \mathbb{Z}$ be an orthonormal system of functions in $\mathbb{L}_2(\mathbb{R})$,*

$$|\psi(x)| \leq \frac{C_1}{(1+|x|)^\alpha}, \ \alpha > N,$$

and $\psi \in \mathbb{C}^{N-1}(\mathbb{R})$, where the derivatives $\psi^{(k)}(x)$ are bounded for $k \leq N - 1$.

Table 3.2 The $\underline{h}$ filters for Daubechies' wavelets for $N = 2, \ldots, 10$ vanishing moments.

k	DAUB2	DAUB3	DAUB4
0	0.4829629131445342	0.3326705529500827	0.2303778133088966
1	0.8365163037378080	0.8068915093110930	0.7148465705529161
2	0.2241438680420134	0.4598775021184915	0.6308807679298592
3	-0.1294095225512604	-0.1350110200102548	-0.0279837694168604
4		-0.0854412738820267	-0.1870348117190935
5		0.0352262918857096	0.0308413818355607
6			0.0328830116668852
7			-0.0105974017850690

k	DAUB5	DAUB6	DAUB7
0	0.1601023979741926	0.1115407433501095	0.0778520540850092
1	0.6038292697971887	0.4946238903984531	0.3965393194819173
2	0.7243085284377723	0.7511339080210954	0.7291320908462351
3	0.1384281459013216	0.3152503517091976	0.4697822874051931
4	-0.2422948870663808	-0.2262646939654398	-0.1439060039285650
5	-0.0322448695846383	-0.1297668675672619	-0.2240361849938750
6	0.0775714938400454	0.0975016055873230	0.0713092192668303
7	-0.0062414902127983	0.0275228655303057	0.0806126091510831
8	-0.0125807519990819	-0.0315820393174860	-0.0380299369350144
9	0.0033357252854738	0.0005538422011615	-0.0165745416306669
10		0.0047772575109455	0.0125509985560998
11		-0.0010773010853085	0.0004295779729214
12			-0.0018016407040475
13			0.0003537137999745

k	DAUB8	DAUB9	DAUB10
0	0.0544158422431070	0.0380779473638881	0.0266700579005487
1	0.3128715909143165	0.2438346746126514	0.1881768000776480
2	0.6756307362973218	0.6048231236902548	0.5272011889316280
3	0.5853546836542239	0.6572880780514298	0.6884590394535462
4	-0.0158291052563724	0.1331973858249681	0.2811723436606982
5	-0.2840155429615815	-0.2932737832793372	-0.2498464243271048
6	0.0004724845739030	-0.0968407832230689	-0.1959462743773243
7	0.1287474266204823	0.1485407493381040	0.1273693403356940
8	-0.0173693010018109	0.0307256814793158	0.0930573646035142
9	-0.0440882539307979	-0.0676328290613591	-0.0713941471663802
10	0.0139810279173996	0.0002509471148278	-0.0294575368218849
11	0.0087460940474065	0.0223616621236844	0.0332126740593155
12	-0.0048703529934519	-0.0047232047577528	0.0036065535669515
13	-0.0003917403733769	-0.0042815036824646	-0.0107331754833277
14	0.0006754494064506	0.0018476468830567	0.0013953517470513
15	-0.0001174767841248	0.0002303857635232	0.0019924052951842
16		-0.0002519631889428	-0.0006858566949593
17		0.0000393473203163	-0.0001164668551292
18			0.0000935886703200
19			-0.0000132642028945

Then, ψ has N vanishing moments,

$$\mathcal{N}_k = 0, \quad 0 \leq k \leq N-1.$$

If, in addition,

$$|\phi(x)| \leq \frac{C_2}{(1+|x|)^\alpha}, \quad \alpha > N$$

then, the associated function $m_0(\omega)$ is necessarily of the form

$$m_0(\omega) = \left(\frac{1+e^{-i\omega}}{2}\right)^N \cdot \mathcal{L}(\omega), \qquad (3.50)$$

where $\mathcal{L}$ is a 2π-periodic, $\mathbb{C}^{N-1}$-function.

The following definition of regularity is often used,

Definition 3.5.1 *The multiresolution analysis (or, the scaling function) is said to be r-regular if, for any $\alpha \in \mathbb{Z}$,*

$$|\phi^{(k)}(x)| \leq \frac{C}{(1+|x|)^\alpha},$$

for $k = 0, 1, \ldots, r$.

The requirement that ψ possesses N vanishing moments can be expressed in terms of Ψ, m_0, or equivalently, in terms of the filter $\underline{h}$.

Assume that a wavelet function $\psi(x)$ has N vanishing moments, i.e.,

$$\mathcal{N}_k = 0, \quad k = 0, 1, \ldots, N-1. \qquad (3.51)$$

By the property [MOM] of Fourier transformations, the requirement (3.51) corresponds to

$$\left.\frac{d^k \Psi(\omega)}{d\omega^k}\right|_{\omega=0} = 0, \quad k = 0, 1, \ldots, N-1,$$

which implies

$$m_1^{(k)}(\omega)|_{\omega=0} = m_1^{(k)}(0) = 0, \quad k = 0, 1, \ldots, N-1. \qquad (3.52)$$

It is easy to check that in terms of m_0, relation (3.52) becomes

$$m_0^{(k)}(\omega)|_{\omega=\pi} = m_0^{(k)}(\pi) = 0, \ k = 0, 1, \ldots, N-1. \tag{3.53}$$

The argument is inductive. The case $k = 0$ follows from $\Psi(0) = m_1(0)\Phi(0)$ [(3.28) evaluated at $\omega = 0$] and the fact that $\Phi(0) = 1$. Since $\Psi'(0) = \frac{1}{2}m_1'(0)\Psi(0) + \frac{1}{2}m_1(0)\Psi'(0)$ it follows that $m_1'(0) = 0$, as well. Then, $m_1^{(N-1)}(0) = 0$ follows by induction.

The condition $m_1^{(k)}(0) = 0, \ k = 0, 1, \ldots, N-1$ translates to a constraint on the wavelet-filter coefficients

$$\sum_{n \in \mathbb{Z}} n^k g_n = \sum_{n \in \mathbb{Z}} (-1)^n n^k h_n = 0, \ k = 0, 1, \ldots, N-1. \tag{3.54}$$

Example 3.5.1 (Strang-Fix condition [150]). Let ψ have N vanishing moments. Since $m_0(\omega)$ has a root $\omega = \pi$ of multiplicity N,

$$i^k \Phi^{(k)}(2n\pi) = \delta_n \mathcal{M}_k, \ 0 \leq k < N,$$

and by Poisson summation formula

$$\sum_n (x-n)^k \phi(x-n) = \mathcal{M}_k. \tag{3.55}$$

In Fig. 3.15 we illustrate equation (3.55) – the ability of wavelets to exactly represent polynomials. For example, the DAUB2 scaling function approximates exactly polynomials of degree $n = 0$ and $n = 1$ [relations $\sum_k \phi(x-k) = 1$ and $\sum_k [(3-\sqrt{3})/2 + k]\phi(x-k) = x$ in Fig. 3.15(a) and (b)]. Moreover, the approximations to polynomials of higher order are very good. Fig. 3.15(c) and (d) show approximations to quadratic and cubic polynomials.

How smooth are the wavelets from the Daubechies family? There is an apparent trade-off between the length of support and the regularity index of scaling functions. Daubechies [102], Daubechies and Lagarias [106], and Volkmer [428], among others, obtained regularity exponents for wavelets in the Daubechies family.

Let ϕ be the DAUBN scaling function. There are two popular measures of regularity of ϕ: Sobolev and Hölder regularity exponents. Let α_N^* be the supremum of β such that

84 WAVELETS

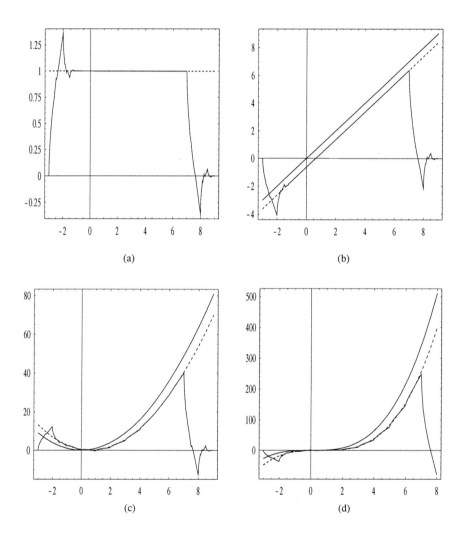

Fig. 3.15 Illustration of the Strang–Fix condition, (3.55).

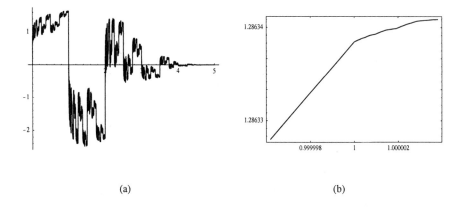

Fig. 3.16 (a) The first derivative of the DAUB3 scaling function. (b) The DAUB3 scaling function ϕ enlarged near $x = 1$. Notice that ϕ is smooth at 1, which is in accordance with the precise bound on its Hölder index, $\alpha_3 = 1.0878$.

$$\int (1 + |\omega|)^\beta |\Phi(\omega)| d\omega < \infty,$$

and let α_N be the exponent of the Hölder space $\mathbb{C}^{\alpha_N}$ to which the scaling function ϕ belongs.

Table 3.3 Sobolev α_N^* and Hölder α_N regularity exponents of Daubechies' scaling functions.

N	1	2	3	4	5	6	7	8	9	10
α_N^*	0.5	1	1.415	1.775	2.096	2.388	2.658	2.914	3.161	3.402
α_N		0.550	0.915	1.275	1.596	1.888	2.158	2.415	2.661	2.902

The following result describes the limiting behavior of α_N.

Theorem 3.5.2 *[428]*

$$\lim_{N \to \infty} \frac{\alpha_N}{N} = 1 - \frac{\log 3}{2 \log 2} \approx 0.2075.$$

From Table 3.3, we see that DAUB4 is the first differentiable wavelet, since $\alpha > 1$. More precise bounds on α_N yield that ϕ from the DAUB3 family is, in fact, the first differentiable scaling function ($\alpha_3 = 1.0878$), even though it seams to have a peak at 1, see Fig. 3.16. See also Daubechies [104], page 239, for the discussion. The two panels in Fig. 3.16 have been produced using Mathematica's Wavelet Explorer software.

Remark 3.5.1 As pointed out by Burrus, Gopinath, and Guo [49], the Sobolev and Hölder regularities are related because of inclusions

$$\mathbb{W}^{\alpha_N^* + 1/2} \subset \mathbb{C}^{\alpha_N} \subset \mathbb{W}^{\alpha_N^*}.$$

Therefore, Theorem 3.5.2 holds for the exponent α_N^*, as well.

3.5.2 The Least Asymmetric Daubechies' Wavelets: Symmlets

We referred to the Daubechies family of wavelets as *the extremal phase* family. As we pointed out before, the compactly supported wavelets cannot be of linear phase, i.e., they cannot be symmetric. However, it is possible to construct wavelets with compact support that are "more symmetric".

Equation (3.49) implies that $m_0(\omega)$ is proportional to

$$\left(\frac{1 + e^{-i\omega}}{2}\right)^N \prod_{l=1}^{L} (e^{-i\omega} - z_l)(e^{-i\omega} - \bar{z}_l) \cdot \prod_{i=1}^{I} (e^{-i\omega} - r_i),$$

and we had some freedom in selection of the zeroes z_l and r_i. Instead of taking all zeroes [see (3.48)] outside the unit circle in the complex plane, we could have selected any real zero from the pair (r_i, r_i^{-1}) and any conjugate pair from the quadruple $(z_l, \bar{z}_l, z_l^{-1}, \bar{z}_l^{-1})$. Thus, we could obtain several different "square roots" of $|\mathcal{L}(\omega)|$.

Taking all zeroes from outside (or inside) the unit circle lead to extreme deviation from the linear phase in the function $m_0(\omega)$.

In Fig. 3.17(a) all the roots of the polynomial $P_A(z)$ [see (3.47)] for $N = 4$ are shown. If, for $N = 4$, [real, (a conjugate-complex pair)] this is the nature of roots leading to a "square root" of $|\mathcal{L}(\omega)|^2$, then possible choices are

$$[3, (1, 2)], \ [6, (4, 5)], \ [3, (4, 5)], \ \text{and} \ [6, (1, 2)],$$

where the numbers 1-6 correspond to the zeroes depicted in Fig. 3.14(a).

The first two triples correspond to the extremal phase case. The resulting wavelets are not identical – they are symmetric to each other. The last two triples correspond to the minimum phase case. Deviations from the linear phase are plotted in Fig. 3.17(a). The solid and dashed lines correspond to extremal and minimum phases, respectively.

Panel 3.17 (b) compares scaling functions for the minimal phase and extremal phase families ($N = 4$). Table 3.4 gives the filter components for the SYMM4,

Table 3.4 Symmlet coefficients for $N = 4, 5$, and 6 vanishing moments.

k	SYMM4	SYMM5	SYMM6
0	0.032223100604052	0.019538882735250	0.015404109327045
1	-0.012603967262031	-0.021101834024689	0.003490712084222
2	-0.099219543576634	-0.175328089908056	-0.117990111148520
3	0.297857795605306	0.016602105764511	-0.048311742585698
4	0.803738751805133	0.633978963456791	0.491055941927974
5	0.497618667632775	0.723407690404040	0.787641141028651
6	-0.029635527646003	0.199397533976856	0.337929421728166
7	-0.075765714789502	-0.039134249302313	-0.072637522786377
8		0.029519490925706	-0.021060292512371
9		0.027333068344999	0.044724901770781
10			0.001767711864254
11			-0.007800708325032

SYMM5, and SYMM6 families.

3.5.3 Approximations and Characterizations of Functional Spaces

Any $\mathbb{L}_2(\mathbb{R})$ function f can be represented as

$$f(x) = \sum_{j,k} d_{jk} \psi_{jk}(x),$$

and this unique representation corresponds to a multiresolution decomposition $\mathbb{L}_2(\mathbb{R}) = \bigoplus_{j=-\infty}^{\infty} W_j$. Also, for any fixed j_0 the decomposition $\mathbb{L}_2(\mathbb{R}) = V_{j_0} \oplus \bigoplus_{j=j_0}^{\infty} W_j$ corresponds to the representation

$$f(x) = \sum_{k} c_{j_0,k} \phi_{j_0,k}(x) + \sum_{j \geq j_0} \sum_{k} d_{jk} \psi_{j,k}(x). \tag{3.56}$$

The first sum in (3.56) is an orthogonal projection $\mathbf{P}_{j_0}$ of f on V_{j_0}.

In general, it is possible to bound $||\mathbf{P}_{j_0} f - f|| = ||(\mathbf{I} - \mathbf{P}_{j_0}) f||$ if the regularities of functions f and ϕ are known.

When both f and ϕ have n continuous derivatives, Meyer [293] proved that there exists a constant C such that

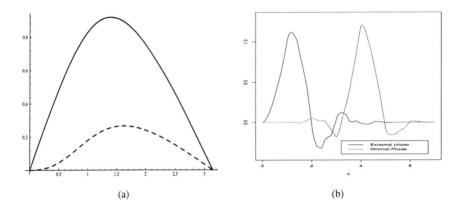

Fig. 3.17 (a) Deviation of linear phase for DAUB4 wavelet (choice $[3, (1, 2)]$, solid line) and SYMM4 wavelet (choice $[3, (4, 5)]$, dashed line). (b) Corresponding scaling functions.

$$\|(\mathbf{I} - \mathbf{P}_{j_0})f\|_{L_2} \leq C \cdot 2^{-nj_0} \|f\|_{L_2}.$$

A range of important function spaces can be fully characterized by wavelets. We list a few characterizations. For example, a function f belongs to the Hölder space $\mathbb{C}^s$ if and only if there is a constant C such that in an r-regular MRA ($r > s$) the wavelet coefficients satisfy

$$\begin{aligned}(i) \quad & |c_{j_0,k}| \leq C, \\ (ii) \quad & |d_{j,k}| \leq C \cdot 2^{-j(s+\frac{1}{2})}, \ j \geq j_0, k \in \mathbb{Z}.\end{aligned} \quad (3.57)$$

A function f belongs to the Sobolev $\mathbb{W}_2^s(\mathbb{R})$ space if and only if

$$\sum_{j,k} |d_{jk}|^2 \cdot (1 + 2^{2js}) < \infty.$$

Even the general (non-homogeneous) Besov spaces, (see page 36) can be characterized by moduli of the wavelet coefficients of its elements. For a given r-regular MRA with $r > \max\{\sigma, 1\}$, the following result (see Meyer [294], page 200) holds

Theorem 3.5.3 *Let I_j be a set of indices so that $\{\psi_i, i \in I_j\}$ constitutes an o.n. basis of the detail space W_j. There exist two constants $C' \geq C > 0$ such that, for every exponent $p \in [1, \infty]$, for each $j \in \mathbb{Z}$ and for every element $f(x) = \sum_{i \in I_j} d_i \psi_i(x)$ in W_j,*

$$C\|f\|_p \leq 2^{j/2}2^{-j/p}\left(\sum_{i\in I_j}|d_i|^p\right)^{1/p} \leq C'\|f\|_p.$$

The following characterization of Besov $\mathbb{B}_{p,q}^\sigma$ spaces can be obtained directly from this result. If the MRA has regularity $r > s$, then wavelet bases are Riesz bases for all $1 \leq p, q \leq \infty$, $0 < \sigma < r$.

The function $f = \sum_k c_{j_0 k}\phi_{j_0 k}(x) + \sum_{j \geq j_0}\sum_k d_{jk}\psi_{jk}(x)$ belongs to $\mathbb{B}_{p,q}^\sigma$ space if its wavelet coefficients satisfy

$$\left(\sum_k |c_{j_0,k}|^p\right)^{1/p} < \infty,$$

and

$$\left\{\left(\sum_{i\in I_j} 2^{j(\sigma+1/2-1/p)}|d_i|^p\right)^{1/p}, j \geq j_0\right\}$$

is an ℓ_q sequence, i.e, $\left[\sum_{j\geq j_0}\left(2^{j(\sigma+1/2-1/p)}(\sum_k |d_{j,k}|^p)^{1/p}\right)^q\right]^{1/q} < \infty.$

The results listed are concerned with global regularity. The local regularity of functions can also be studied by inspecting the magnitudes of their wavelet coefficients. For more details, we direct the reader to the work of Jaffard [212, 211] and Jaffard and Laurencot [213].

3.5.4 Daubechies-Lagarias Algorithm

In this section, we describe an algorithm for fast numerical calculation of wavelet values at a given point, based on the Daubechies-Lagarias *local pyramidal algorithm* [106, 107]. The scaling function and wavelet function in Daubechies' families have no explicit representations (except for the Haar wavelet). However, it is often necessary to find their values at arbitrary points; examples include calculation of coefficients in density estimation and non-equally spaced regression.

The Daubechies-Lagarias algorithm enables us to evaluate ϕ and ψ at a point with preassigned precision. We will illustrate the algorithm on wavelets from the Daubechies family; however, the algorithm works for all FIR quadrature mirror filters.

90   WAVELETS

Let ϕ be the scaling function of the DAUBN wavelet. The support of ϕ is $[0, 2N - 1]$. Let $x \in (0, 1)$, and let $dyad(x) = \{d_1, d_2, \ldots d_n, \ldots\}$ be the set of 0-1 digits in the dyadic representation of x ($x = \sum_{j=1}^{\infty} d_j 2^{-j}$). By $dyad(x, n)$, we denote the subset of the first n digits from $dyad(x)$, i.e., $dyad(x, n) = \{d_1, d_2, \ldots d_n\}$.

Let $\underline{h} = (h_0, h_1, \ldots, h_{2N-1})$ be the wavelet filter coefficients. Define two $(2N - 1) \times (2N - 1)$ matrices as:

$$T_0 = (\sqrt{2} \cdot h_{2i-j-1})_{1 \leq i,j \leq 2N-1} \text{ and } T_1 = (\sqrt{2} \cdot h_{2i-j})_{1 \leq i,j \leq 2N-1}. \quad (3.58)$$

Then the local pyramidal algorithm can be constructed based on Theorem 3.5.4.

Theorem 3.5.4 *[106]*

$$\lim_{n \to \infty} T_{d_1} \cdot T_{d_2} \cdots T_{d_n} \quad (3.59)$$

$$= \begin{bmatrix} \phi(x) & \phi(x) & \cdots & \phi(x) \\ \phi(x+1) & \phi(x+1) & \cdots & \phi(x+1) \\ \vdots & & & \\ \phi(x+2N-2) & \phi(x+2N-2) & \cdots & \phi(x+2N-2) \end{bmatrix}.$$

The convergence of $\|T_{d_1} \cdot T_{d_2} \cdots T_{d_n} - T_{d_1} \cdot T_{d_2} \cdots T_{d_{n+m}}\|$ to zero, for fixed m, is exponential and constructive, i.e., effective decreasing bounds on the error can be established.

Example 3.5.2 Consider the DAUB2 scaling function ($N = 2$). The corresponding filter is $\underline{h} = \left(\frac{1+\sqrt{3}}{4\sqrt{2}}, \frac{3+\sqrt{3}}{4\sqrt{2}}, \frac{3-\sqrt{3}}{4\sqrt{2}}, \frac{1-\sqrt{3}}{4\sqrt{2}} \right)$. According to (3.58) the matrices T_0 and T_1 are given as

$$T_0 = \begin{bmatrix} \frac{1+\sqrt{3}}{4} & 0 & 0 \\ \frac{3-\sqrt{3}}{4} & \frac{3+\sqrt{3}}{4} & \frac{1+\sqrt{3}}{4} \\ 0 & \frac{1-\sqrt{3}}{4} & \frac{3-\sqrt{3}}{4} \end{bmatrix} \text{ and } T_1 = \begin{bmatrix} \frac{3+\sqrt{3}}{4} & \frac{1+\sqrt{3}}{4} & 0 \\ \frac{1-\sqrt{3}}{4} & \frac{3-\sqrt{3}}{4} & \frac{3+\sqrt{3}}{4} \\ 0 & 0 & \frac{1-\sqrt{3}}{4} \end{bmatrix}.$$

Let us evaluate the scaling function at an arbitrary point, say $x = 0.45$. Twenty "decimals" in the dyadic representation of 0.45 are $dyad(0.45, 20) = \{$ 0, 1, 1, 1, 0, 0, 1, 1, 0, 0, 1, 1, 0, 0, 1, 1, 0, 0, 1, 1 $\}$. In addition to the value at 0.45, we get (for free) the values at 1.45 and 2.45 (the values 0.45, 1.45, and 2.45 are in the domain of ϕ, the interval [0,3]. The values $\phi(0.45)$, $\phi(1.45)$, and $\phi(2.45)$ may be approximated as averages of the first, second, and third row, respectively in the matrix

$$\prod_{i \in dyad(0.45, 20)} T_i = \begin{bmatrix} 0.86480582 & 0.86480459 & 0.86480336 \\ 0.08641418 & 0.08641568 & 0.08641719 \\ 0.04878000 & 0.04877973 & 0.04877945 \end{bmatrix}.$$

The Daubechies-Lagarias algorithm gives only the values of the scaling function. In applications, most of the evaluation needed involves the wavelet function. It turns out that another algorithm is unnecessary, due to the following result.

Theorem 3.5.5 *[343] Let x be an arbitrary real number, let the wavelet be given by its filter coefficients, and let $\underline{u}$ with $2N - 1$ be a vector defined as*

$$\underline{u}(x) = \{(-1)^{1-\lfloor 2x \rfloor} h_{i+1-\lfloor 2x \rfloor}, i = 0, \ldots, 2N-2\}.$$

If for some i the index $i + 1 - \lfloor 2x \rfloor$ is negative or larger than $2N - 1$, then the corresponding component of $\underline{u}$ is equal to 0.

Let the vector $\underline{v}$ be

$$\underline{v}(x, n) = \frac{1}{2N-1} \mathbf{1}' \prod_{i \in dyad(\{2x\}, n)} T_i,$$

where $\mathbf{1}' = (1, 1, \ldots, 1)$ is the row-vector of ones. Then,

$$\psi(x) = \lim_{n \to \infty} \underline{u}(x)' \underline{v}(x, n),$$

and the limit is constructive.

Proof of the theorem (Exercise 3.23) is a straightforward but tedious re-expression of (3.26).

3.5.5 Moment Conditions

In section 3.5.1 we saw that the requirement that the wavelet function possesses N-vanishing moments was expressed in terms of Φ, m_0, or $\underline{h}$.

Let us suppose that we want to design a wavelet filter $\underline{h} = \{h_0, \ldots, h_{2N-1}\}$ only by considering properties of its filter taps. Assume that

$$\mathcal{N}_k = \int_\mathbb{R} x^k \psi(x)dx = 0, \text{ for } k = 0, 1, \ldots, N-1. \tag{3.60}$$

As we discussed in Section 3.3, some relevant properties of a multiresolution analysis can be expressed as relations involving coefficients of the filter $\underset{\sim}{h}$.

For example, the normalization property gave

- $$\sum_{i=0}^{2N-1} h_i = \sqrt{2},$$

the requirement for vanishing moments of ψ led to

- $$\sum_{i=0}^{2N-1} (-1)^i i^k h_i = 0, \ k = 0, 1, \ldots, N-1,$$

and, finally, the orthogonality property was expressed as

- $$\sum_{i=0}^{2N-1} h_i h_{i+2k} = \delta_k \ k = 0, 1, \ldots, N-1.$$

We obtained $2N+1$ equations with $2N$ unknowns; however the system is solvable since the equations are not linearly independent. For example, the equation

$$h_0 - h_1 + h_2 - \cdots - h_{2N-1} = 0,$$

can be expressed as a linear combination of the others.

Example 3.5.3 For $N = 2$, we obtain the system:

$$\begin{cases} h_0 + h_1 + h_2 + h_3 = \sqrt{2} \\ h_0^2 + h_1^2 + h_2^2 + h_3^2 = 1 \\ -h_1 + 2h_2 - 3h_3 = 0 \\ h_0 h_2 + h_1 h_3 = 0 \end{cases},$$

which has the familiar solution $h_0 = \frac{1+\sqrt{3}}{4\sqrt{2}}, h_1 = \frac{3+\sqrt{3}}{4\sqrt{2}}, h_2 = \frac{3-\sqrt{3}}{4\sqrt{2}}$, and $h_3 = \frac{1-\sqrt{3}}{4\sqrt{2}}$.

For $N = 4$, the system is

$$\begin{cases} h_0 + h_1 + h_2 + h_3 + h_4 + h_5 + h_6 + h_7 = \sqrt{2} \\ h_0^2 + h_1^2 + h_2^2 + h_3^2 + h_4^2 + h_5^2 + h_6^2 + h_7^2 = 1 \\ h_0 - h_1 + h_2 - h_3 + h_4 - h_5 + h_6 - h_7 = 0 \\ h_0 h_2 + h_1 h_3 + h_2 h_4 + h_3 h_5 + h_4 h_6 + h_5 h_7 == 0 \\ h_0 h_4 + h_1 h_5 + h_2 h_6 + h_3 h_7 = 0 \\ h_0 h_6 + h_1 h_7 = 0 \\ 0 h_0 - 1 h_1 + 2 h_2 - 3 h_3 + 4 h_4 - 5 h_5 + 6 h_6 - 7 h_7 = 0 \\ 0 h_0 - 1 h_1 + 4 h_2 - 9 h_3 + 16 h_4 - 25 h_5 + 36 h_6 - 49 h_7 = 0 \\ 0 h_0 - 1 h_1 + 8 h_2 - 27 h_3 + 64 h_4 - 125 h_5 + 216 h_6 - 343 h_7 = 0. \end{cases}$$

3.5.6 Interpolating (Cardinal) Wavelets

In an orthogonal multiresolution analysis, the space V_0 can be represented as

$$V_0 = \bigoplus_{j_0 \le j < 0} W_j \oplus V_{j_0},$$

for some $j_0 < 0$. It is customary to assume that the level corresponding to V_0 space contains coefficients $c_{0,n}$ which are close to $f(n)$ – the finest sampling available. In practice such closeness is assumed and for $\underline{c}^{(0)} = \{c_{0,n}, \; n \in \mathbb{Z}\}$ one simply takes the sampled values $f(n)$, $n \in \mathbb{Z}$. However $c_{0,n} = f(n)$ is exact only if the multiresolution analysis is generated by an *interpolating (cardinal)* scaling function ϕ.

Definition 3.5.2 *A scaling function ϕ is interpolating if $\phi(n) = \delta_n$.*

For non-interpolating wavelet families the aliasing error [caused by assuming $c_{0,n} = f(n)$] is given by Xia and Zhang [468].

An example of a family with an interpolating scaling function is Shannon's family [see Theorem 2.3.2 and (2.5)] but shortcomings of Shannon's wavelets have already been discussed. There are several constructions of interpolating wavelets in the literature. Beylkin and Saito [33] define a class of interpolating wavelets via autocorrelations of compactly supported wavelets. Xia and Zhang [468] prove that a cardinal orthogonal scaling function corresponds to an *à trous* filter. For example, they demonstrate that the wavelet basis corresponding to

$$m_0(\omega) = \left(\frac{1 + e^{i\omega}}{2}\right)^3 \frac{4(1 - 3e^{i\omega} + 2^{2i\omega})}{1 - 5e^{2i\omega}}$$

is interpolating. An excellent overview of interpolating wavelets with statistical flavor can be found in Walter [439]. See also Sweldens and Piessens [393] and Sweldens

[391] for nice reviews.

Wavelets from Daubechies' family are not interpolating. However, they are almost shift-interpolating wavelets.

Definition 3.5.3 *A scaling function is shift-interpolating if*

$$\phi(n+\tau) = \delta_n,$$

for some constant τ.

From the fact that $\sum_n (x-n)\phi(x-n) = \mathcal{M}_1$, it follows that $\tau = \mathcal{M}_1$.

Example 3.5.4 (Bock and Pliego [35]) The DAUB2 wavelet is almost shift-interpolating.

$$\begin{aligned}
\mathcal{M}_1 &= \frac{3-\sqrt{3}}{2} \\
\phi(\mathcal{M}_1) &= 1 \\
\phi(\mathcal{M}_1 + 1) &\approx 0 \\
\phi(\mathcal{M}_1 + 2) &\approx 0 \ .
\end{aligned}$$

3.5.7 Pollen-Type Parameterization of Wavelets

In this section, we discuss the construction of a library that contains uncountably many different wavelet bases.

Let $\underline{h}$ be a wavelet filter of length $2N$. Pollen [344, 345] showed that there is a continuous mapping from $[0, 2\pi]^{N-1}$ to a set of "wavelet solutions" in the form of a sequence $\underline{h} = \{h_0, h_1, \ldots, h_{2N-1}\}$.

Pollen representations of all wavelet solutions of lengths 4 ($N = 2$) and 6 ($N = 3$) are given in Tables 3.5 and 3.6.

Table 3.5 Pollen parameterization for $N = 2$ (four-tap filters). [$s = 2\sqrt{2}$]

n	h_n for $N = 2$
0	$(1 + \cos\varphi - \sin\varphi)/s$
1	$(1 + \cos\varphi + \sin\varphi)/s$
2	$(1 - \cos\varphi + \sin\varphi)/s$
3	$(1 - \cos\varphi - \sin\varphi)/s$

Example 3.5.5 A special case of Pollen's representation for $\varphi = \frac{\pi}{6}$ gives the DAUB2 filter.

Table 3.6 Pollen parameterization for $N = 3$ (six-tap filters) [$s = 2\sqrt{2}$].

n	h_n for $N = 3$
0	$(1 + \cos\varphi_1 - \cos\varphi_2 - \cos\varphi_1 \cos\varphi_2 + \sin\varphi_1 - \cos\varphi_2 \sin\varphi_1 - \sin\varphi_2 + \cos\varphi_1 \sin\varphi_2 - \sin\varphi_1 \sin\varphi_2)/(2s)$
1	$(1 - \cos\varphi_1 + \cos\varphi_2 - \cos\varphi_1 \cos\varphi_2 + \sin\varphi_1 + \cos\varphi_2 \sin\varphi_1 - \sin\varphi_2 - \cos\varphi_1 \sin\varphi_2 - \sin\varphi_1 \sin\varphi_2)/(2s)$
2	$(1 + \cos\varphi_1 \cos\varphi_2 + \cos\varphi_2 \sin\varphi_1 - \cos\varphi_1 \sin\varphi_2 + \sin\varphi_1 \sin\varphi_2)/s$
3	$(1 + \cos\varphi_1 \cos\varphi_2 - \cos\varphi_2 \sin\varphi_1 + \cos\varphi_1 \sin\varphi_2 + \sin\varphi_1 \sin\varphi_2)/s$
4	$(1 - \cos\varphi_1 + \cos\varphi_2 - \cos\varphi_1 \cos\varphi_2 - \sin\varphi_1 - \cos\varphi_2 \sin\varphi_1 + \sin\varphi_2 + \cos\varphi_1 \sin\varphi_2 - \sin\varphi_1 \sin\varphi_2)/(2s)$
5	$(1 + \cos\varphi_1 - \cos\varphi_2 - \cos\varphi_1 \cos\varphi_2 - \sin\varphi_1 + \cos\varphi_2 \sin\varphi_1 + \sin\varphi_2 - \cos\varphi_1 \sin\varphi_2 - \sin\varphi_1 \sin\varphi_2)/(2s)$

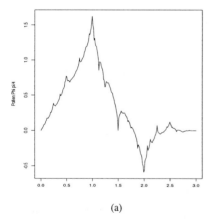
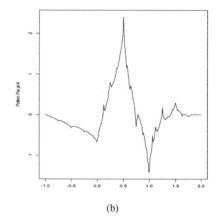

(a) (b)

Fig. 3.18 Pollen parameterization, $\varphi = \frac{\pi}{4}$. Scaling [panel (a)] and wavelet [panel (b)] functions.

Pollen wavelets provide an interesting library of wavelets that comprises uncountably many different wavelet bases continuously indexed by $\varphi \in [0, 2\pi]$.

Fig. 3.18 depicts wavelet and scaling functions for $\varphi = \frac{\pi}{4}$. Its low-pass coefficients are $h_0 = \frac{\sqrt{2}}{4}$, $h_1 = \frac{2+\sqrt{2}}{4}$, $h_2 = \frac{\sqrt{2}}{4}$, and $h_3 = \frac{-2+\sqrt{2}}{4}$, and the plots in Fig. 3.18 are obtained by point-to-point application of the Daubechies-Lagarias algorithm. For further details on Pollen representations see Pollen [344] and Tewfik, Sinha, and

Jorgensen [401].

3.6 EXERCISES

3.1. If $\psi \in \mathbb{L}_1(\mathbb{R}) \cap \mathbb{L}_2(\mathbb{R})$ is not identically equal to zero and satisfies (i) $\int_\mathbb{R} \psi(x)dx = 0$ and (ii) $\int_\mathbb{R} |x|^\alpha |\psi(x)|dx < \infty$ for some $\alpha > 1/2$, prove that ψ is a wavelet. (Hint: Define $F(x) = \int_{-\infty}^x \psi(t)dt$ and prove that $F(x)$ is in $\mathbb{L}_2$ by utilizing (ii). Apply Theorem 3.1.1).

3.2. Why does $\Psi(0) \neq 0$ imply $C_\psi = \infty$ in (3.2)?

3.3. Prove that $\mathcal{F}(\phi_{a,b}(x)) = \sqrt{a} e^{-ib\omega} \Phi(a\omega)$, if Φ is the Fourier transformation of $\phi(x)$.

3.4. Find and plot the Fourier transformation for the Poisson wavelet (Example 3.1.4). Calculate the admissibility constant C_ψ.

3.5. Prove that $\Phi(0) = 1$.

3.6. *Franklin wavelet.* Find coefficients $\{h_n, n \in \mathbb{Z}\}$ if

$$m_0(\omega) = \cos^2\frac{\omega}{2} \sqrt{\frac{2+\cos\omega}{2+\cos 2\omega}}.$$

3.7. [439] Find coefficients $\{h_n, n \in \mathbb{Z}\}$ if $m_0(\omega) = (\int_{2\omega-\pi}^{2\omega+\pi} dP)^{1/2} \mathbf{1}(|\omega| \leq \frac{2\pi}{3})$, where P is a uniform measure on $[-\epsilon, \epsilon] \subset [-\frac{\pi}{3}, \frac{\pi}{3}]$. Prove that any probability measure P on $[-\frac{\pi}{3}, \frac{\pi}{3}]$ defines a valid $m_0(\omega)$.

3.8. Prove that $\{\psi_{0k}(x), k \in \mathbb{Z}\}$ is an orthonormal system i.e., show that $\sum_{k \in \mathbb{Z}} |\psi(\omega + 2k\pi)|^2 = 1$.

3.9. Find $\Phi(\omega)$ for Haar's wavelet. [$\Phi(\omega) = e^{-i\omega/2}\text{sinc}\frac{\omega}{2}$]

3.10. The function $m_0(\omega) = \frac{1+e^{-3i\omega}}{2}$ satisfies (3.23) and $m_0(0) = 1$. Show that the corresponding system $\{\phi(x-n), n \in \mathbb{Z}\}$ is not orthonormal. Why is this the case?

3.11. Let $\Omega = [0, 1)$, $\mathcal{F} = \mathcal{B}([0, 1))$, and let P be the Lebesgue measure. Let $f \in \mathbb{L}_1$ be a Borel measurable function. Define

$$\mathcal{F}_j = \{x | \phi_{00}, \psi_{00}, \ldots, \psi_{j0} \ldots, \psi_{j,2^j-1}\}$$

to be a σ-algebra generated by the Haar basis functions up to level j.

(i) Prove that $E(f|\mathcal{F}_j)$ is a martingale adapted to the filtration $\mathcal{F}_j$.

(ii) By using Levi's martingale convergence theorem show that

$$E(f|\mathcal{F}_j) \to E(f|\mathcal{F}_\infty) = f \quad \text{a.s., and}$$

$$\int_0^1 |E(f|\mathcal{F}_j) - f| \to 0.$$

3.12. Let ϕ_{jk}^2 be the probability density of random variable X_{jk}. Calculate EX_{jk}^n, $n = 1, 2, 3$; $j = 0$, $k = 0$ for the DAUB2 and DAUB10 wavelets.

3.13. Prove that if the supports of Φ and Ψ are exclusive, the functions ϕ and ψ are orthogonal. (Hint: Use the [PAR] property of Fourier transformations.)

3.14. (*Battle-Lemarié family*) Form an orthonormal basis of $\mathbb{L}_2$ by orthogonalizing a piecewise quadratic B-spline

$$\phi(x) = \begin{cases} \frac{1}{2}(x+1)^2, & -1 \leq x \leq 0 \\ \frac{3}{4} - (x - \frac{1}{2})^2, & 0 \leq x \leq 1 \\ \frac{1}{2}(x-2)^2, & 1 \leq x \leq 2 \\ 0, & \text{otherwise} \end{cases}$$

in the Fourier domain. [Hint: $\Phi(\omega) = e^{-i\omega/2}(\frac{\sin \omega/2}{\omega/2})^3$, $\sum_k |\Phi(\omega + 2\pi k)|^2 = \frac{8}{15} + \frac{13}{30}\cos\omega + \frac{1}{30}\cos^2\omega$.]

3.15. Show that $\mathcal{M}_m = C \sum_{k=1}^n \binom{n}{k} m_k \mathcal{M}_{n-k}$, where $m_k = \sum_l l^k h_l$. Find the constant C. [Hint: Use (3.10)]

3.16. Is it true that when j increases, $\Phi_{jk}(\omega) = \mathcal{F}(\phi_{jk}(x))$ becomes more localized?

3.17. (*Jig-saw effect*) Prove that for any scaling function resulting from an orthogonal MRA,

$$1 = \sum_n \phi(x - n).$$

This relation is obvious for the Haar wavelet, but curious for other non-trivial wavelet bases. It is interesting that spikes in different integer-translates of the DAUB2 scaling function cancel each other and all translates add to 1.

3.18. Prove that Shannon's wavelet function can be expressed as

$$\psi(x) = \text{sinc}\frac{\pi x}{2} \cos \frac{3\pi x}{2}.$$

3.19. Find $\Psi(\omega)$ for Shannon's wavelet.

3.20. Show that if ϕ is an interpolating scaling function then $h_{2k} = \frac{1}{\sqrt{2}}$.

3.21. The DAUB2 scaling function is close to the shift-interpolating function. Calculate values $\phi(\mathcal{M}_1+1)$ and $\phi(\mathcal{M}_1+2)$, for $\mathcal{M}_1 = \frac{3-\sqrt{3}}{2}$ using the Daubechies-Lagarias algorithm, page 89.

3.22. [345, 459] Let $D_j = \{k2^{-j}, k \in \mathbb{Z}\}$ and $D = \cup_{j=0}^{\infty} D_j$.

Prove that for the DAUB2 scaling function $\phi(x)$ the following relations hold:

(a) $\quad 2\phi(x) + \phi(x+1) = x + \dfrac{1+\sqrt{3}}{2},$

(b) $\quad 2\phi(x+2) + \phi(x+1) = -x + \dfrac{3-\sqrt{3}}{2},$

(c) $\quad \phi(x) - \phi(x+2) = x + \dfrac{-1+\sqrt{3}}{2},$

(d) $\quad \phi\left(\dfrac{x}{2}\right) = \dfrac{1+\sqrt{3}}{4}\phi(x),$

(e) $\quad \phi\left(\dfrac{x+1}{2}\right) = \dfrac{1-\sqrt{3}}{4}\phi(x) + \dfrac{1+\sqrt{3}}{4}x + \dfrac{2+\sqrt{3}}{4},$

(f) $\quad \phi\left(\dfrac{x+2}{2}\right) = \dfrac{1+\sqrt{3}}{4}\phi(x+1) + \dfrac{1-\sqrt{3}}{4}x + \dfrac{\sqrt{3}}{4},$

(g) $\quad \phi\left(\dfrac{x+3}{2}\right) = \dfrac{1-\sqrt{3}}{4}\phi(x+1) - \dfrac{1+\sqrt{3}}{4}x + \dfrac{1}{4},$

(h) $\quad \phi\left(\dfrac{x+4}{2}\right) = \dfrac{1+\sqrt{3}}{4}\phi(x+2) - \dfrac{1-\sqrt{3}}{4}x + \dfrac{3-2\sqrt{3}}{4},$

(i) $\quad \phi\left(\dfrac{x+5}{2}\right) = \dfrac{1-\sqrt{3}}{4}\phi(x+2).$

if $x \in D \cap [0, 1]$.

3.23. Prove Theorem 3.5.5.

3.24. Find $\underline{h}$ and plot the Pollen scaling and wavelet functions corresponding to $\varphi = \frac{\pi}{3}$.

3.25. Prove the identity needed for spectral decomposition of (3.48): If $z = e^{-i\omega}$, then $|(z - z_j)(z - \bar{z}_j^{-1})| = |z_j|^{-1}|z - z_j|^2$. [Hint: Start with $(e^{-i\omega} - z_j)(e^{-i\omega} - \bar{z}_j^{-1})$ factor out $-\bar{z}_j^{-1}e^{-i\omega}$ from the second factor and then take the absolute value.]

3.26. Demonstrate that the filter with non-zero taps $h_0 = \frac{\sqrt{2}}{2}$, $h_3 = \frac{-\sqrt{2}}{10}$, and $h_{3-2k} = 12\frac{\sqrt{2}}{5^{k+1}}$, $k = 1, 2, \ldots$ is a wavelet low-pass filter. Plot an approximation to its scaling function.

3.27. An interesting parameterization of wavelets can be obtained by replacing the vanishing moment condition $\sum_n (-1)^n n h_n = 0$ with the so-called *lock* condition; for details see La Borde [251]. For four-tap filters, the lock condition is:

$$\sum (-1)^i \alpha_i h_i = 0,$$

where $\alpha_i = f(i, c), c \in \mathbb{R}$ for some continuous function $f(x, c)$. The library of wavelets obtained in this way is parameterized by c. For example, four-tap filters are obtained by solving

$$h_0^2 + h_1^2 + h_2^2 + h_3^2 = 1$$
$$h_0 h_2 + h_1 h_3 = 0$$
$$h_0 + h_1 + h_2 + h_3 = \sqrt{2}$$
$$h_3 - A h_2 + B h_1 - C h_0 = 0$$

Plot scaling and wavelet functions for some values of $A = f(1, c), B = f(2, c)$ and $C = f(3, c)$. Take $f(x, c) = e^{-x^2/c}$ and $c = 0.5, 3, 5.84, 10$, and 100. Find the value of c that gives the DAUB2 wavelet.

3.28. [Battle] The following is a version of the Balian-Low theorem for wavelets. Let ψ be an orthonormal wavelet. Prove

$$\int |e^{|t|} \psi(t)|^2 dt \cdot \int |e^{|\omega|} \Psi(\omega)|^2 d\omega = \infty.$$

4

Discrete Wavelet Transformations

Discrete wavelet transformations (DWT) are applied to discrete data sets and produce discrete outputs. Transforming signals and data vectors by DWT is a process that resembles the fast Fourier transformation (FFT), the Fourier method applied to a set of discrete measurements.

Table 4.1 The analogy between Fourier and wavelet methods

Fourier Methods	Fourier Integrals	Fourier Series	Discrete Fourier Transformations
Wavelet Methods	Continuous Wavelet Transformations	Wavelet Series	Discrete Wavelet Transformations

The analogy between Fourier and wavelet methods is even more complete (Table 4.1) when we take into account the continuous wavelet transformation and wavelet series expansions, covered in Chapter 3.

4.1 INTRODUCTION

Discrete wavelet transformations map data from the time domain (the original or input data vector) to the wavelet domain. The result is a vector of the same size. Wavelet transformations are linear and they can be defined by matrices of dimension $n \times n$ if they are applied to inputs of size n. Depending on boundary conditions, such matrices can be either orthogonal or "close" to orthogonal. When the matrix is

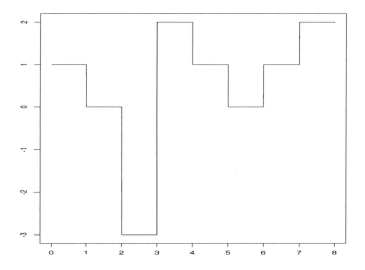

Fig. 4.1 A function interpolating $\underset{\sim}{y}$ on [0,8).

orthogonal, the corresponding transformation is a rotation in $\mathbb{R}^n$ in which the data represent coordinates of a point. The coordinates of the point in the rotated space comprise the discrete wavelet transformation of the original coordinates.

Example 4.1.1 Let the vector be $(1,2)$ and let $M(1,2)$ be the point in $\mathbb{R}^2$ with coordinates given by the data vector. The rotation of the coordinate axes by an angle of $\frac{\pi}{4}$ can be interpreted as a DWT in the Haar wavelet basis. The rotation matrix is

$$W = \begin{pmatrix} \cos\frac{\pi}{4} & \sin\frac{\pi}{4} \\ \cos\frac{\pi}{4} & -\sin\frac{\pi}{4} \end{pmatrix} = \begin{pmatrix} \frac{1}{\sqrt{2}} & \frac{1}{\sqrt{2}} \\ \frac{1}{\sqrt{2}} & -\frac{1}{\sqrt{2}} \end{pmatrix},$$

and the discrete wavelet transformation of $(1,2)'$ is $W \cdot (1,2)' = (\frac{3}{\sqrt{2}}, -\frac{1}{\sqrt{2}})'$. Notice that *the energy* (squared distance of the point from the origin) is preserved, $1^2 + 2^2 = (\frac{1}{2})^2 + (\frac{\sqrt{3}}{2})^2$, since W is a rotation.

Example 4.1.2 Let $\underset{\sim}{y} = (1, 0, -3, 2, 1, 0, 1, 2)$. The associated function f is given in Fig. 4.1. The values $f(n) = y_n$, $n = 0, 1, \ldots, 7$ are interpolated by a piecewise constant function. We assume that f belongs to Haar's multiresolution space V_0.

The following matrix equation gives the connection between $\underset{\sim}{y}$ and the wavelet coefficients (data in the wavelet domain).

$$\begin{bmatrix} 1 \\ 0 \\ -3 \\ 2 \\ 1 \\ 0 \\ 1 \\ 2 \end{bmatrix} = \begin{bmatrix} \frac{1}{2\sqrt{2}} & \frac{1}{2\sqrt{2}} & \frac{1}{2} & 0 & \frac{1}{\sqrt{2}} & 0 & 0 & 0 \\ \frac{1}{2\sqrt{2}} & \frac{1}{2\sqrt{2}} & \frac{1}{2} & 0 & -\frac{1}{\sqrt{2}} & 0 & 0 & 0 \\ \frac{1}{2\sqrt{2}} & \frac{1}{2\sqrt{2}} & -\frac{1}{2} & 0 & 0 & \frac{1}{\sqrt{2}} & 0 & 0 \\ \frac{1}{2\sqrt{2}} & \frac{1}{2\sqrt{2}} & -\frac{1}{2} & 0 & 0 & -\frac{1}{\sqrt{2}} & 0 & 0 \\ \frac{1}{2\sqrt{2}} & -\frac{1}{2\sqrt{2}} & 0 & \frac{1}{2} & 0 & 0 & \frac{1}{\sqrt{2}} & 0 \\ \frac{1}{2\sqrt{2}} & -\frac{1}{2\sqrt{2}} & 0 & \frac{1}{2} & 0 & 0 & -\frac{1}{\sqrt{2}} & 0 \\ \frac{1}{2\sqrt{2}} & -\frac{1}{2\sqrt{2}} & 0 & -\frac{1}{2} & 0 & 0 & 0 & \frac{1}{\sqrt{2}} \\ \frac{1}{2\sqrt{2}} & -\frac{1}{2\sqrt{2}} & 0 & -\frac{1}{2} & 0 & 0 & 0 & -\frac{1}{\sqrt{2}} \end{bmatrix} \cdot \begin{bmatrix} c_{00} \\ d_{00} \\ d_{10} \\ d_{11} \\ d_{20} \\ d_{21} \\ d_{22} \\ d_{23} \end{bmatrix}.$$

The solution is

$$\begin{bmatrix} c_{00} \\ d_{00} \\ d_{10} \\ d_{11} \\ d_{20} \\ d_{21} \\ d_{22} \\ d_{23} \end{bmatrix} = \begin{bmatrix} \sqrt{2} \\ -\sqrt{2} \\ 1 \\ -1 \\ \frac{1}{\sqrt{2}} \\ -\frac{5}{\sqrt{2}} \\ \frac{1}{\sqrt{2}} \\ -\frac{1}{\sqrt{2}} \end{bmatrix}.$$

Thus,

$$\begin{aligned} f &= \sqrt{2}\phi_{-3,0} - \sqrt{2}\psi_{-3,0} + \psi_{-2,0} - \psi_{-2,1} \\ &\quad + \frac{1}{\sqrt{2}}\psi_{-1,0} - \frac{5}{\sqrt{2}}\psi_{-1,1} + \frac{1}{\sqrt{2}}\psi_{-1,2} - \frac{1}{\sqrt{2}}\psi_{-1,3}. \end{aligned} \quad (4.1)$$

The solution is easy to verify. For example, when $x \in [0, 1)$,

$$f(x) = \sqrt{2} \cdot \frac{1}{2\sqrt{2}} - \sqrt{2} \cdot \frac{1}{2\sqrt{2}} + 1 \cdot \frac{1}{2} + \frac{1}{\sqrt{2}} \cdot \frac{1}{\sqrt{2}} = 1/2 + 1/2 = 1 \, (= y_0).$$

Performing wavelet transformations by multiplying the input vector with an appropriate orthogonal matrix is conceptually straightforward, but of limited practical value. Storing and manipulating transformation matrices when inputs are long (> 2000) may not even be feasible.

Example 4.1.3 Fig. 4.2 depicts a discrete wavelet transformation of the California well data [panel (a); $n = 8192$]. The decomposing wavelet is SYMM4 and the decomposition corresponds to $V_{13} = V_7 \oplus W_7 \oplus W_8 \cdots \oplus W_{12}$, in the language of multiresolution analysis. Panel (c) is a "stack plot" of wavelet coefficients. The rows correspond to levels. The relative positions of coefficients within a level reflect their time positions. Panels (b) and (d) depict, respectively, the coefficients

104   DISCRETE WAVELET TRANSFORMATIONS

from the "smooth" or "coarse" level, i.e. coefficients corresponding to V_7, and to concatenated "detail" coefficients corresponding to spaces $W_7, W_8, \ldots, W_{12}$, in that order. The plot is made with the help of the S+WAVELET package. In that software, vectors of coefficients associated with the spaces $V_7, W_7, W_8, \ldots, W_{12}$ are denoted $s6, d6, d5, \ldots, d1$, respectively.

Notice that magnitudes (energies) of detail coefficients decrease as the levels become finer (when j increases). The more regular the (sampled) function is, the faster the decay of the magnitudes. This behavior is explained by Meyer's result in (3.57). However, a few detail coefficients describing the earthquake jump are large compared to other coefficients in the same level, suggesting a local, "energetic" event.

4.2   THE CASCADE ALGORITHM

In the context of image processing, Burt and Adelson [50, 51] developed orthogonal and biorthogonal pyramid algorithms. Pyramid or cascade procedures process an image at different scales, ranging from fine to coarse, in a tree-like algorithm. The images can be denoised, enhanced or compressed by appropriate scale-wise treatments.

Mallat [275, 276] was the first to link wavelets, multiresolution analyses and cascade algorithms in a formal way. Mallat's cascade algorithm gives a constructive and efficient recipe for performing the discrete wavelet transformation. It relates the wavelet coefficients from different levels in the transformation by filtering with $\underline{h}$ and $\underline{g}$. Mallat's algorithm can be viewed as a wavelet counterpart of the Danielson-Lanczos algorithm in fast Fourier transformations.

It is convenient to link the original data with the space V_J, often V_0. Then, coarser smooth and complementing detail spaces are (V_{J-1}, W_{J-1}), (V_{J-2}, W_{J-2}), etc. Decreasing the index in V-spaces is equivalent to coarsening the approximation to the data.

By a straightforward substitution of indices in the scaling equations (3.10) and (3.26), one obtains

$$\phi_{j-1,k}(x) = \sum_{l \in \mathbb{Z}} h_{l-2k} \phi_{jk}(x) \text{ and } \psi_{j-1,k}(x) = \sum_{l \in \mathbb{Z}} g_{l-2k} \phi_{jk}(x). \qquad (4.2)$$

The relations in (4.2) are fundamental in developing the cascade algorithm.

Consider a multiresolution analysis $\cdots \subset V_{j-1} \subset V_j \subset V_{j+1} \subset \ldots$. Since $V_j = V_{j-1} \oplus W_{j-1}$, any function $v_j \in V_j$ can be represented uniquely as $v_j(x) = v_{j-1}(x) + w_{j-1}(x)$, where $v_{j-1} \in V_{j-1}$ and $w_{j-1} \in W_{j-1}$. It is customary to denote the coefficients associated with $\phi_{jk}(x)$ and $\psi_{jk}(x)$ by c_{jk} and d_{jk}, respectively.

Thus,

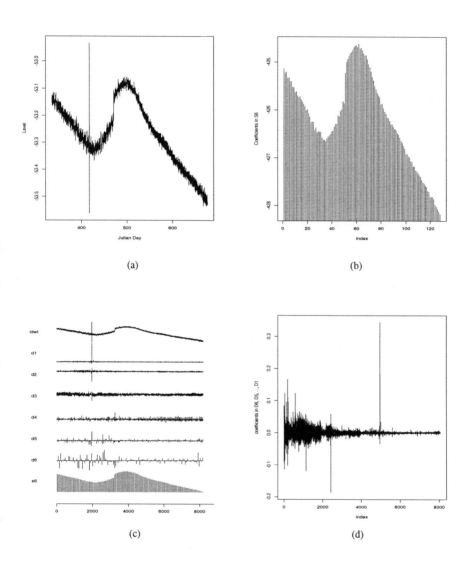

Fig. 4.2 (a) California well data set. (b) A stack-plot of wavelet coefficients. (c) The coefficients corresponding to V_7 (vector $s6$). (d) Concatenated vectors $s6, d6, d5, \ldots, d1$. The wavelet used in the decomposition was SYMM4.

$$v_j(x) = \sum_k c_{j,k}\phi_{j,k}(x)$$
$$= \sum_l c_{j-1,l}\phi_{j-1,l}(x) + \sum_l d_{j-1,l}\psi_{j-1,l}(x)$$
$$= v_{j-1}(x) + w_{j-1}(x).$$

By using the general scaling equations (4.2), orthogonality of $w_{j-1}(x)$ and $\phi_{j-1,l}(x)$ for any j and l, and additivity of inner products, we obtain

$$\begin{aligned} c_{j-1,l} &= \langle v_j, \phi_{j-1,l}\rangle \\ &= \langle v_j, \sum_k h_{k-2l}\phi_{j,k}\rangle \\ &= \sum_k h_{k-2l}\langle v_j, \phi_{j,k}\rangle \\ &= \sum_k h_{k-2l} c_{j,k}. \end{aligned} \qquad (4.3)$$

Similarly $d_{j-1,l} = \sum_k g_{k-2l} c_{j,k}$.

The cascade algorithm works in the reverse direction as well. Coefficients in the next finer scale corresponding to V_j can be obtained from the coefficients corresponding to V_{j-1} and W_{j-1}. The relation

$$\begin{aligned} c_{j,k} &= \langle v_j, \phi_{j,k}\rangle \\ &= \sum_l c_{j-1,l}\langle\phi_{j-1,l}, \phi_{j,k}\rangle + \sum_l d_{j-1,l}\langle\psi_{j-1,l}, \phi_{j,k}\rangle \\ &= \sum_l c_{j-1,l} h_{k-2l} + \sum_l d_{j-1,l} g_{k-2l}, \end{aligned} \qquad (4.4)$$

describes a single step in the reconstruction algorithm.

Example 4.2.1 Fig. 4.3 shows the cascade algorithm in action. For DAUB2, the scaling equation at integers is

$$\phi(n) = \sum_{k=0}^{3} h_k \sqrt{2}\phi(2n-k).$$

Recall that $\underline{h} = \{h_0, h_1, h_2, h_3\} = \{\frac{1+\sqrt{3}}{4\sqrt{2}}, \frac{3-\sqrt{3}}{4\sqrt{2}}, \frac{3+\sqrt{3}}{4\sqrt{2}}, \frac{1-\sqrt{3}}{4\sqrt{2}}\}$.

Since $\phi(0) = \sqrt{2}h_0\phi(0)$ and $\sqrt{2}h_0 \neq 1$, it follows that $\phi(0) = 0$. Also, $\phi(3) = 0$. For $\phi(1)$ and $\phi(2)$ we obtain the system

$$\begin{bmatrix} \phi(1) \\ \phi(2) \end{bmatrix} = \sqrt{2} \cdot \begin{bmatrix} h_1 & h_0 \\ h_3 & h_2 \end{bmatrix} \cdot \begin{bmatrix} \phi(1) \\ \phi(2) \end{bmatrix}.$$

From $\sum_k \phi(x - k) = 1$ it follows that $\phi(1) + \phi(2) = 1$. Solving for $\phi(1)$ and $\phi(2)$ we obtain

$$\phi(1) = \frac{1 + \sqrt{3}}{2} \text{ and } \phi(2) = \frac{1 - \sqrt{3}}{2}.$$

The upper left panel in Fig. 4.3 shows $\phi(1)$ and $\phi(2)$.

Now, one can refine ϕ,

$$\phi\left(\frac{1}{2}\right) = \sum_k h_k \sqrt{2} \phi(1 - k) = h_0 \sqrt{2} \phi(1) = \frac{2 + \sqrt{3}}{4},$$

$$\phi\left(\frac{3}{2}\right) = \sum_k h_k \sqrt{2} \phi(3 - k) = h_1 \sqrt{2} \phi(2) + h_2 \sqrt{2} \phi(1)$$

$$= \frac{3 + \sqrt{3}}{4} \cdot \frac{1 - \sqrt{3}}{2} + \frac{3 - \sqrt{3}}{4} \cdot \frac{1 + \sqrt{3}}{2} = 0,$$

$$\phi\left(\frac{5}{2}\right) = \sum_k h_k \sqrt{2} \phi(5 - k) = h_3 \sqrt{2} \phi(2) = \frac{2 - \sqrt{3}}{4},$$

or ψ,

$$\psi(-1) = \psi(2) = 0,$$

$$\psi\left(-\frac{1}{2}\right) = \sum_k g_k \sqrt{2} \phi(-1 - k) = h_1 \sqrt{2} \phi(1) = -\frac{1}{4}, \quad [g_n = (-1)^n h_{1-n}]$$

$$\psi(0) = \sum_k g_k \sqrt{2} \phi(0 - k) = g_{-2} \sqrt{2} \phi(2) + g_{-1} \sqrt{2} \phi(1)$$

$$= -h_2 \sqrt{2} \phi(1) = -\frac{\sqrt{3}}{4},$$

etc.

The sequential refinement for ψ shown in Fig. 4.3 is carried up to a dyadic step of 2^{-7}.

108 DISCRETE WAVELET TRANSFORMATIONS

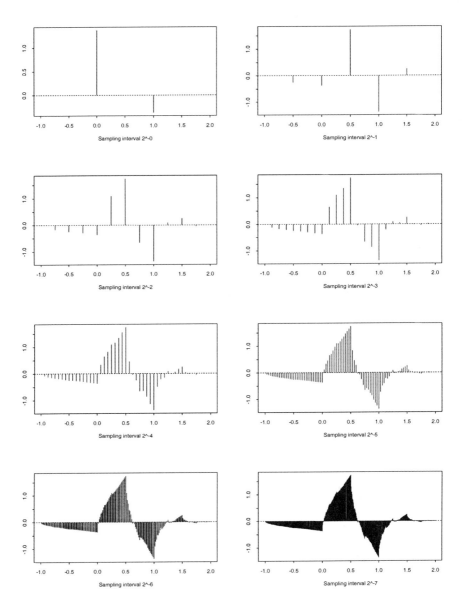

Fig. 4.3 An illustration of Mallat's cascade algorithm. By sequential refinement on dyadic points the DAUB2 wavelet function is constructed.

4.3 THE OPERATOR NOTATION OF DWT

Operator notation is an elegant way to describe the cascade algorithm. One can chose to express operators either in terms of matrices or filters. In this section, we choose the latter. For a comprehensive treatment of DWT in terms of operators, we direct the reader to the excellent monograph by Strang and Nguyen, [387].

We next overview decimation, dilation, and convolution operators acting on the ℓ_2 spaces.

Definition 4.3.1 *Decimation* $[\downarrow 2]$ *is a mapping from* $\ell(\mathbb{Z})$ *to* $\ell(2\mathbb{Z})$ *defined coordinatewise as*

$$([\downarrow 2]\, \underline{a})_k = \sum_n a_n \delta_{n-2k} = a_{2k}. \quad (4.5)$$

When applied to a sequence it retains only the values on positions with even indices. Dilation is a mapping from $\ell(2\mathbb{Z})$ *to* $\ell(\mathbb{Z})$ *defined as*

$$([\uparrow 2]\, \underline{a})_k = \sum_n a_n \delta_{k-2n};$$

it expands the sequence by inserting zeroes between the original values.

The graph below shows actions of $[\downarrow 2]$ and $[\uparrow 2]$ on the sequence $\{1, 0, -3, 2, 1, 0, 1, 2\}$ positioned so that the first element has an index of -2.

index k	-2	-1	0	1	2	3	4	5
a_k	1	0	-3	2	1	0	1	2
$[\downarrow 2]\, \underline{a}$			1		-3		1	1
$[\uparrow 2]\, \underline{a}$	0	0	-3	0	2	0	1	0

Let **H** and **G** be convolutions with filters $\underline{h}$ and $\underline{g}$,

$$\mathbf{H} : \ell(\mathbb{Z}) \mapsto \ell(\mathbb{Z}) \quad (\mathbf{H}\underline{a})_k = \sum_n h_{n-k} a_n$$

$$\mathbf{G} : \ell(\mathbb{Z}) \mapsto \ell(\mathbb{Z}) \quad (\mathbf{G}\underline{a})_k = \sum_n g_{n-k} a_n.$$

The table below is an example of filtering. It shows the application of the filter $\{1, 1/2\}$ to the sequence $\underline{a} = \{1, 0, -3, 2, 1, 0, 1, 2\}$.

110 DISCRETE WAVELET TRANSFORMATIONS

index k	-2	-1	0	1	2	3	4	5	6
a_k	1	0	-3	2	1		0	1	2
filter				1	1/2				
		1	-3/2	-2	5/2	1	1/2	2	

Let $\mathcal{H} \equiv [\downarrow 2] \, \mathbf{H}$ and $\mathcal{G} \equiv [\downarrow 2] \, \mathbf{G}$.

It is easy to verify that for an ℓ_2 sequence $a = \{a_n\}$, the operators $\mathcal{H}$ and $\mathcal{G}$ satisfy the following coordinate-wise relations:

$$(\mathcal{H}a)_k = \sum_n h_{n-2k} a_n$$

$$(\mathcal{G}a)_k = \sum_n g_{n-2k} a_n.$$

An application of operators $\mathcal{H}$ and $\mathcal{G}$ corresponds to one step in the discrete wavelet transformation.

For an illustration, we assume that the data vector is of a length that is a power of 2. Extensions to different lengths are possible, and a comprehensive discussion and algorithmic solutions are given in Wickerhauser [457].

Denote the original signal by $\underline{c}^{(J)} = \{c_k^{(J)}\}$. If the signal is of length 2^J, then $\underline{c}^{(J)}$ can be interpolated by the function $f(x) = \sum c_k^{(J)} \phi(x-k)$ from V_J. In each step of the wavelet transformation, we move to the next coarser approximation (level) $\underline{c}^{(j-1)}$ by applying the operator $\mathcal{H}$, $\underline{c}^{(j-1)} = \mathcal{H}\underline{c}^{(j)}$. The "detail information," lost by approximating $\underline{c}^{(j)}$ by the "averaged" $\underline{c}^{(j-1)}$, is given by $\underline{d}^{(j-1)} = \mathcal{G}\underline{c}^{(j)}$.

The discrete wavelet transformation of a sequence $y = \underline{c}^{(J)}$ of length 2^J can then be represented as

$$(\underline{c}^{(J-k)}, \underline{d}^{(J-k)}, \underline{d}^{(J-k+1)}, \ldots, \underline{d}^{(J-2)}, \underline{d}^{(J-1)}). \tag{4.6}$$

Notice that the lengths of y and its transformation in (4.6) coincide. Because of decimation, the length of $\underline{c}^{(j)}$ is twice the length of $\underline{c}^{(j-1)}$, and $2^J = 2^{J-k} + \sum_{i=1}^{k} 2^{J-i}$, $1 \leq k \leq J$.

For an illustration of (4.6), see Fig. 4.4. By utilizing the operator notation, it is possible to summarize the discrete wavelet transformation (curtailed at level k) in a single line:

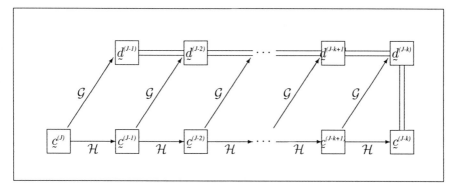

Fig. 4.4 Decomposition algorithm

$$\underline{y} \mapsto (\mathcal{H}^k \underline{y}, \mathcal{G}\mathcal{H}^{k-1}\underline{y}, \ldots, \mathcal{G}\mathcal{H}^2 \underline{y}, \mathcal{G}\mathcal{H}\underline{y}, \mathcal{G}\underline{y}).$$

The number k can be any arbitrary integer between 1 and J and it is associated with the coarsest "smooth" space, V_{J-k}, up to which the transformation was curtailed. In terms of multiresolution spaces, (4.6) corresponds to the multiresolution decomposition $V_{J-k} \oplus W_{J-k} \oplus W_{J-k+1} \oplus \cdots \oplus W_{J-1}$. When $k = J$ the vector $\underline{c}^{(0)}$ contains a single element, $c^{(0)}$.

If the filter length is greater than 2, one has to define actions of the filter beyond the boundaries of the sequence to which the filter is applied. We will discuss some standard policies of handling the boundaries later.

The reconstruction formula is also simple in terms of operators; we first define adjoint operators $\mathcal{H}^\star$ and $\mathcal{G}^\star$ as follows:

$$(\mathcal{H}^\star a)_n = \sum_k h_{n-2k} a_k$$
$$(\mathcal{G}^\star a)_n = \sum_k g_{n-2k} a_k.$$

Operators $\mathcal{H}^\star$ and $\mathcal{G}^\star$ are applied on $\underline{c}^{(j-1)}$ and $\underline{d}^{(j-1)}$, respectively, and the results are added. The vector $\underline{c}^{(j)}$ is reconstructed as

$$\underline{c}^{(j)} = \mathcal{H}^\star \underline{c}^{(j-1)} + \mathcal{G}^\star \underline{d}^{(j-1)} = \mathcal{R}^\star(\underline{c}^{(j-1)}, \underline{d}^{(j-1)}), \tag{4.7}$$

which is an operator-form re-expression of (4.4). Recursive application of (4.7) leads to

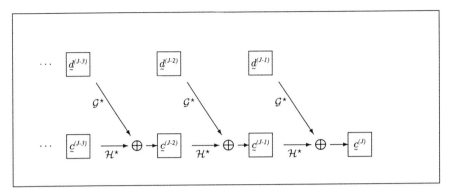

Fig. 4.5 Reconstruction algorithm

$$(\mathcal{H}^k y, \mathcal{G}\mathcal{H}^{k-1} y, \ldots, \mathcal{G}\mathcal{H}^2 y, \mathcal{G}\mathcal{H} y, \mathcal{G} y)$$
$$= (\underline{c}^{(J-k)}, \underline{d}^{(J-k)}, \underline{d}^{(J-k+1)}, \ldots, \underline{d}^{(J-2)}, \underline{d}^{(J-1)})$$
$$\mapsto \sum_{i=1}^{k-1}(\mathcal{H}^\star)^{k-1-i}\mathcal{G}^\star \underline{d}^{(J-k+i)} + (\mathcal{H}^\star)^k \underline{c}^{(J-k)} = \underline{y}.$$

Example 4.3.1 Let $y = (1, 0, -3, 2, 1, 0, 1, 2)$ be an exemplary set we want to transform by Haar's DWT. Let $k = J = 3$, i.e., the coarsest approximation and detail levels will contain a single point each. The decomposition algorithm applied on $y = (1, 0, -3, 2, 1, 0, 1, 2)$ is given schematically in Fig. 4.6.

For the Haar wavelet, the operators $\mathcal{H}$ and $\mathcal{G}$ are given by $(\mathcal{H}a)_k = \sum_n h_{n-2k} a_n = \sum_m h_m a_{m+2k} = h_0 a_{2k} + h_1 a_{2k+1} = \frac{a_{2k}+a_{2k+1}}{\sqrt{2}}$. Similarly, $(\mathcal{G}a)_k = \sum_n g_{n-2k} a_n = \sum_m g_m a_{m+2k} = g_0 a_{2k} + g_1 a_{2k+1} = \frac{a_{2k}-a_{2k+1}}{\sqrt{2}}$.

The reconstruction algorithm is given in Fig. 4.7. In the process of reconstruction, $(\mathcal{H}^\star a)_n = \sum_k h_{n-2k} a_k$, and $(\mathcal{G}^\star a)_n = \sum_k g_{n-2k} a_k$. For instance, the first line in Fig. 4.7 recovers the object $\{1, 1\}$ from $\sqrt{2}$ by applying $\mathcal{H}^\star$. Indeed, $(\mathcal{H}^\star\{a_0\})_0 = h_0\sqrt{2} = 1$ and $(\mathcal{H}^\star\{a_0\})_1 = h_1\sqrt{2} = 1$.

The careful reader might have already noticed that when the length of the filter is larger than 2, boundary problems occur since the convolving filter goes outside the range of data-indices. Due to its minimal filter length, Haar's wavelet does not exhibit problems with boundaries.

There are several approaches to resolving the boundary problem. The signal may be continued in a periodic way $(\ldots, y_{n-1}, y_n | y_1, y_2, \ldots)$, symmetric way $(\ldots, y_{n-1}, y_n | y_{n-1}, y_{n-2}, \ldots)$, padded by a constant, or extrapolated as a polynomial. Wavelet transformations can be confined to an interval (in the sense of Cohen, Daubechies and Vial [86]) and periodic and symmetric extensions can be viewed as special cases. We will discuss periodized wavelets in Chapter 5.

THE OPERATOR NOTATION OF DWT   113

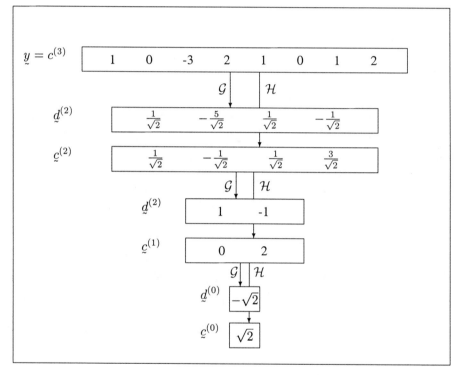

Fig. 4.6 An illustration of a decomposition procedure.

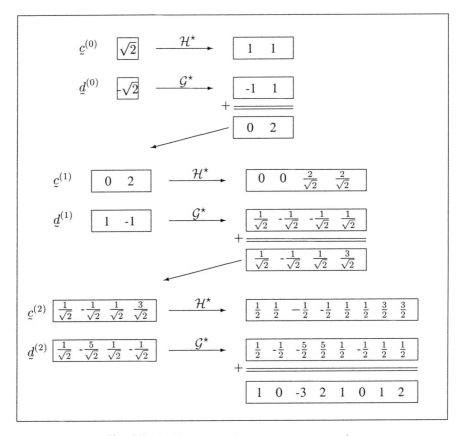

Fig. 4.7 An illustration of a reconstruction procedure

If the length of the data set is not a power of 2, but of the form $M \cdot 2^K$, for M odd and K a positive integer, then only K steps in the decomposition algorithm can be performed. For precise descriptions of conceptual and calculational hurdles caused by boundaries and data sets whose lengths are not a power of 2, we direct the reader to Bruce and Gao [42], Taswell and McGill [399], and monograph by Wickerhauser [457].

4.3.1 Discrete Wavelet Transformations as Linear Transformations

The change of basis in V_1 from $\mathcal{B}_1 = \{\phi_{1k}(x), k \in Z\}$ to $\mathcal{B}_2 = \{\phi_{0k}, k \in Z\} \cup \{\psi_{0k}, k \in Z\}$ can be performed by matrix multiplication. Therefore, it is possible to define discrete wavelet transformation by matrices. We have already seen a transformation matrix corresponding to Haar's inverse transformation in Example 4.1.2.

Let the length of the input signal be 2^J, let $\underline{h} = \{h_s,\ s \in \mathbb{Z}\}$ be the wavelet filter, and let N be an appropriately chosen constant.

Denote by H_k a matrix of size $(2^{J-k} \times 2^{J-k+1})$, $k = 1, \ldots$ with entries

$$h_s, \quad s = (N-1) + (i-1) - 2(j-1) \text{ modulo } 2^{J-k+1}, \tag{4.8}$$

at the position (i, j).

Note that H_k is a circulant matrix, its ith row is 1st row circularly shifted to the right by $2(i-1)$ units. This circularity is a consequence of using the *modulo* operator in (4.8).

By analogy, define a matrix G_k by using the filter g. A version of G_k corresponding to the already defined H_k can be obtained by changing h_i by $(-1)^i h_{N+1-i}$. The constant N is a shift parameter and affects the position of the wavelet on the time scale. For filters from the Daubechies family, a standard choice for N is the number of vanishing moments. See also Remark 3.3.4.

The matrix $\begin{bmatrix} H_k \\ G_k \end{bmatrix}$ is a basis-changing matrix in 2^{J-k+1} dimensional space; consequently, it is unitary.

Therefore,

$$I_{2^{J-k}} = [H'_k\ G'_k] \begin{bmatrix} H_k \\ G_k \end{bmatrix} = H'_k \cdot H_k + G'_k \cdot G_k.$$

and

$$I = \begin{bmatrix} H_k \\ G_k \end{bmatrix} \cdot [H'_k\ G'_k] = \begin{bmatrix} H_k \cdot H'_k & H_k \cdot G'_k \\ G_k \cdot H'_k & G_k \cdot G'_k \end{bmatrix}.$$

116 DISCRETE WAVELET TRANSFORMATIONS

This implies,

$H_k \cdot H'_k = I$, $G_k \cdot G'_k = I$, $G_k \cdot H'_k = H_k \cdot G'_k = 0$, and $H'_k \cdot H_k + G'_k \cdot G_k = I$.

Now, for a sequence y the J-step wavelet transformation is $\underline{d} = W_J \cdot \underline{y}$, where

$$W_1 = \begin{bmatrix} H_1 \\ G_1 \end{bmatrix}, \quad W_2 = \begin{bmatrix} \begin{bmatrix} H_2 \\ G_2 \end{bmatrix} \cdot H_1 \\ G_1 \end{bmatrix},$$

$$W_3 = \begin{bmatrix} \begin{bmatrix} \begin{bmatrix} H_3 \\ G_3 \end{bmatrix} \cdot H_2 \\ G_2 \end{bmatrix} \cdot H_1 \\ G_1 \end{bmatrix}, \ldots$$

Example 4.3.2 Suppose that $\underline{y} = \{1, 0, -3, 2, 1, 0, 1, 2\}$ and the filter is $\underline{h} = (h_0, h_1, h_2, h_3) = \left(\frac{1+\sqrt{3}}{4\sqrt{2}}, \frac{3+\sqrt{3}}{4\sqrt{2}}, \frac{3-\sqrt{3}}{4\sqrt{2}}, \frac{1-\sqrt{3}}{4\sqrt{2}}\right)$. Then, $J = 3$ and the matrices H_k and G_k are of dimension $2^{3-k} \times 2^{3-k+1}$.

$$H_1 = \begin{bmatrix} h_1 & h_2 & h_3 & 0 & 0 & 0 & 0 & h_0 \\ 0 & h_0 & h_1 & h_2 & h_3 & 0 & 0 & 0 \\ 0 & 0 & 0 & h_0 & h_1 & h_2 & h_3 & 0 \\ h_3 & 0 & 0 & 0 & 0 & h_0 & h_1 & h_2 \end{bmatrix}$$

$$G_1 = \begin{bmatrix} -h_2 & h_1 & -h_0 & 0 & 0 & 0 & 0 & h_3 \\ 0 & h_3 & -h_2 & h_1 & -h_0 & 0 & 0 & 0 \\ 0 & 0 & 0 & h_3 & -h_2 & h_1 & -h_0 & 0 \\ -h_0 & 0 & 0 & 0 & 0 & h_3 & -h_2 & h_1 \end{bmatrix}.$$

Since,

$$H_1 \cdot \underline{y} = \{2.19067, -2.19067, 1.67303, 1.15539\}$$
$$G_1 \cdot \underline{y} = \{0.96593, 1.86250, -0.96593, 0.96593\}.$$

$W_1 y = \{2.19067, -2.19067, 1.67303, 1.15539 \mid 0.96593, 1.86250, -0.96593, 0.96593\}$.

$$H_2 = \begin{bmatrix} h_1 & h_2 & h_3 & h_0 \\ h_3 & h_0 & h_1 & h_2 \end{bmatrix} \quad G_2 = \begin{bmatrix} -h_2 & h_1 & -h_0 & h_3 \\ -h_0 & h_3 & -h_2 & h_1 \end{bmatrix}.$$

In this example, due to the lengths of the filter and the data, we can perform the transformation for two steps only, W_1 and W_2.

The two-step DAUB2 discrete wavelet transformation of y is
$W_2 \cdot \underline{y} = \{1.68301, 0.31699 \mid -3.28109, -0.18301 \mid 0.96593, 1.86250, -0.96593, 0.96593\}$,

because

$$H_2 \cdot H_1 \cdot \underset{\sim}{y} = H_2 \cdot \{2.19067, -2.19067, 1.67303, 1.15539\}$$
$$= \{1.68301, 0.31699\}$$
$$G_2 \cdot H_1 \cdot \underset{\sim}{y} = G_1 \cdot \{2.19067, -2.19067, 1.67303, 1.15539\}$$
$$= \{-3.28109, -0.18301\}.$$

The S+WAVELETS command
```
> dwt(c(1, 0, -3, 2, 1, 0, 1, 2), wavelet="d4", n.levels=2)
```
produces the same output.

4.4 EXERCISES

4.1. Let $\underset{\sim}{h} = \{1, 2, 1\}$ and $\underset{\sim}{y} = \{1, 0, -3, 2, 1, 0, 1, 2\}$. Let the first elements in $\underset{\sim}{h}$ and $\underset{\sim}{y}$ have indices 0 and -2, respectively. Find $H\underset{\sim}{y}$ and $H^2\underset{\sim}{y}$.

4.2. What are the columns in the 8 ×8 matrix in Example 4.1.2?

4.3. Using the cascade algorithm, find $\psi(\frac{1}{2}), \psi(1)$, and $\phi(\frac{1}{4})$ in Example 4.2.1.

4.4. Show that for the Haar transformation and data $\{y_1, \ldots, y_{2^n}\}$, $c_{00} = (\sqrt{2})^n \bar{y}$.

4.5. By mimicking the procedures in Fig. 4.6 and 4.7, perform Haar's direct transformation on the sequence $\{1, 2, \ldots, 16\}$. Perform the inverse Haar transformation on $\{34, -16, -4\sqrt{2}, -4\sqrt{2}, -2, -2, -2, -2, -\sqrt{2}/2, -\sqrt{2}/2, -\sqrt{2}/2, -\sqrt{2}/2, -\sqrt{2}/2, -\sqrt{2}/2, -\sqrt{2}/2, -\sqrt{2}/2\}$.

5
Some Generalizations

In this chapter, we give an overview of several generalizations on the wavelet theme – all of them important in statistics. We start with coiflets whose vanishing moments for scaling functions minimize aliasing error introduced by replacing continuous functions with their sampled values. Another important class of wavelets, the biorthogonal wavelets, is generated by relaxing the requirement of orthogonality. We will discuss the theoretical background for their construction and give several examples of biorthogonal bases. Wavelet packets and non-decimated wavelets are related generalizations of orthogonal wavelets suitable for generating large libraries of bases. From an algorithmic point of view, wavelet packets can be obtained by permuting low- and high-pass filters in the process of wavelet decomposition, while non-decimated wavelets can be obtained by permuting the down-sampling operators with or without a unit shift. We will also discuss periodized wavelets which are important for modeling on compacts. We conclude the chapter with brief coverage of multivariate wavelet transformation and give references and pointers for additional reading. Several other developments are mentioned, but their detailed discussion is beyond the scope of this book.

5.1 COIFLETS

If the wavelet function ψ has N vanishing moments, then the so-called one-point quadrature formula $\int f(x)\psi(x)\,dx = 0$ is N-order accurate.[1] Is it possible to obtain such N-order accurate one-point quadrature formulas for $\int f(x)\phi(x)\,dx$? Such accuracy is possible if ϕ has vanishing moments as well. *Coiflets* were introduced by Daubechies in [105] with such motivation.

Coiflets were originally proposed as a family of wavelets close to interpolating wavelets. In addition to the regularity requirement $\mathcal{N}_0 = \mathcal{N}_1 = \cdots = \mathcal{N}_{L-1} = 0$, it is required that the scaling function ϕ possess vanishing moments, as well. In this case, the one-point quadrature formula $\int f(x)\phi(x)\,dx = 0$ is $(L-1)$-accurate.

Theorem 5.1.1 *[391] Let the moments of ϕ be $\mathcal{M}_1 = \mathcal{M}_2 = \cdots = \mathcal{M}_{L-1} = 0$, and let $f^{(i)}$ be bounded for $0 \leq i \leq L$. If $h = 2^{-j}$,*

$$\sum_k f(kh)\phi(2^j x - k) = \sum_k \phi(k) f(x - kh) + O(h^L).$$

Proof:

$$\begin{aligned}
\sum_k f(kh)\phi(2^j x - k) &= \sum_k \sum_{i=0}^{L-1} \frac{(kh-x)^i}{i!} f^{(i)}(x) \phi(2^j x - k) + O(h^L) \\
&= \sum_{i=0}^{L-1} f^{(i)}(x) \frac{(-h)^i}{i!} \sum_k (2^j x - k)^i \phi(2^j x - k) + O(h^L) \\
&= \sum_{i=0}^{L-1} f^{(i)}(x) \frac{(-h)^i}{i!} \sum_k k^i \phi(k) + O(h^L) \\
&= \sum_k \phi(k) \sum_{i=0}^{L-1} f^{(i)}(x) \frac{(-kh)^i}{i!} + O(h^L) \\
&= \sum_k \phi(k) f(x - kh) + O(h^L).
\end{aligned}$$

This result states that a wavelet decomposition of f with coefficients $c_{jk} = f(kh)$ results in an approximation $\tilde{f}(x) = \sum_k \phi(k) f(x - kh)$. The function $\tilde{f}(x)$ can be interpreted as a "blurred version" of $f(x)$ since the convolution with ϕ acts as a low-pass filter. For coiflets, $\tilde{f}(x) = f(x) + O(h^L)$. The closeness of coiflets to interpolating wavelets was also explored by Antoniadis [11], see Exercise 5.3.

---

[1] The degree of accuracy of a quadrature formula is N if it yields an exact result for every polynomial of degree less than or equal to N.

5.1.1 Construction of Coiflets

The condition $\mathcal{M}_1 = \mathcal{M}_2 = \cdots = \mathcal{M}_{L-1} = 0$ (of course $\mathcal{M}_0 = 1$) is equivalent to

$$\left.\frac{d^k \Phi(\omega)}{d\omega^k}\right|_{\omega=0} = 0, \quad k = 1, 2, \ldots, L-1. \tag{5.1}$$

Vanishing moments for both the scaling and wavelet functions lead to two constraints on the form of m_0. The first,

$$m_0(\omega) = \left(\frac{1 + e^{-i\omega}}{2}\right)^L \cdot \mathcal{L}_1(\omega),$$

is a consequence of Theorem 3.5.1, while the second,

$$m_0(\omega) = 1 + (1 - e^{-i\omega})^L \cdot \mathcal{L}_2(\omega),$$

comes from (5.1). $\mathcal{L}_1(\omega)$ and $\mathcal{L}_2(\omega)$ are trigonometric polynomials. By applying arguments similar to those used in the construction of Daubechies' wavelets, it is possible to find $\mathcal{L}_1$ and $\mathcal{L}_2$ so that orthogonality is satisfied.

We elaborate the case when L is an even number, $L = 2K$.
Since $(\frac{1+e^{-i\omega}}{2})^{2K} = e^{-i\omega K}(\cos^2 \frac{\omega}{2})^K$ and $(1 - e^{-i\omega})^{2K} = e^{-iK\omega}(2i \sin \frac{\omega}{2})^{2K}$, the orthogonality condition $|m_0(\omega)|^2 + |m_0(\omega + \pi)|^2 = 1$ takes the form

$$\left(\cos^2 \frac{\omega}{2}\right)^K P_1(\omega) = 1 + \left(\sin^2 \frac{\omega}{2}\right)^K P_2(\omega).$$

By utilizing the Bezout lemma, Daubechies demonstrated that P_1 has the form

$$P_1(\omega) = \sum_{k=0}^{K-1} \binom{K-1+k}{k} \left(\sin^2 \frac{\omega}{2}\right)^k + \left(\sin^2 \frac{\omega}{2}\right)^K \cdot f(\omega),$$

for an arbitrary trigonometric polynomial f. Daubechies proposed the "ansatz" function, $f(\omega) = \sum_{n=0}^{2K-1} f_n e^{-in\omega}$, and gave an effective solution for m_0. For details, consult Daubechies [105], also [104], pages 258–259.

Coiflets are less asymmetric than the wavelets from the DAUB or SYMM families. The price for this additional symmetry is paid by a larger support. The length of the support for coiflets is $3L - 1$ compared to $2L - 1$ for the standard Daubechies' family. The filter has $3L$ non-zero taps.

Table 5.1 gives filter coefficients for the COIF2, COIF4, and COIF6 filters. Notice the shifted indices; indices for $L = 2$ and $L = 3$ range from -2 to 3 and from -4

Table 5.1 COIF2 - COIF6 filters with the indexing of taps.

index	COIF2	COIF4	COIF6
-6	0	0	-0.003793512864381
-5	0	0	0.007782596425673
-4	0	0.016387336463204	0.023452696142077
-3	0	-0.041464936786872	-0.065771911281469
-2	$(\sqrt{15} - 3) \cdot \sqrt{2}/32$	-0.067372554723726	-0.061123390002973
-1	$(1 - \sqrt{15}) \cdot \sqrt{2}/32$	0.386110066822763	0.405176902409118
0	$(3 - \sqrt{15}) \cdot \sqrt{2}/16$	0.812723635449413	0.793777222626087
1	$(\sqrt{15} + 3) \cdot \sqrt{2}/16$	0.417005184423239	0.428483476377370
2	$(\sqrt{15} + 13) \cdot \sqrt{2}/32$	-0.076488599078281	-0.071799821619155
3	$(9 - \sqrt{15}) \cdot \sqrt{2}/32$	-0.050594344186464	-0.082301927106300
4	0	0.023680171946848	0.034555027573298
5	0	0.005611434819369	0.015880544863669
6	0	-0.001823208870911	-0.009007976136731
7	0	-0.000720549445520	-0.002574517688137
8	0	0	0.001117518770831
9	0	0	0.000466216959821
10	0	0	-0.000070983302506
11	0	0	-0.000034599773197

to 7, respectively. For a general L, the indices range between $-L$ and $2L - 1$. Both scaling and wavelet functions have support $[-L, 2L - 1]$.

Pairs of scaling and wavelet functions are plotted in Fig. 5.1 for $L = 2, 4, 6, 8,$ and 10. Notice that the scaling functions are almost symmetric — and yet they are compactly supported. The standard convention is to denote a coiflet by its filter length, say COIF3L. To preserve consistency with the DAUB and SYMM notations used before in which the number of vanishing moments enumerates the family, we use the notation COIFL.

5.2 BIORTHOGONAL WAVELETS

Cohen, Daubechies, and Feauveau [84] introduced biorthogonal wavelet bases.[2] The construction in [84] is similar to construction of Daubechies' compactly supported wavelets with the orthogonality condition relaxed.

Biorthogonal bases are defined as a pair of mutually orthogonal bases neither of which is an orthogonal basis. To illustrate the biorthogonality idea we give a simple

---

[2] In the signal processing world, subband filtering schemes related to biorthogonal wavelets were known prior to construction in [84], see Vetterli and Herley [417, 418].

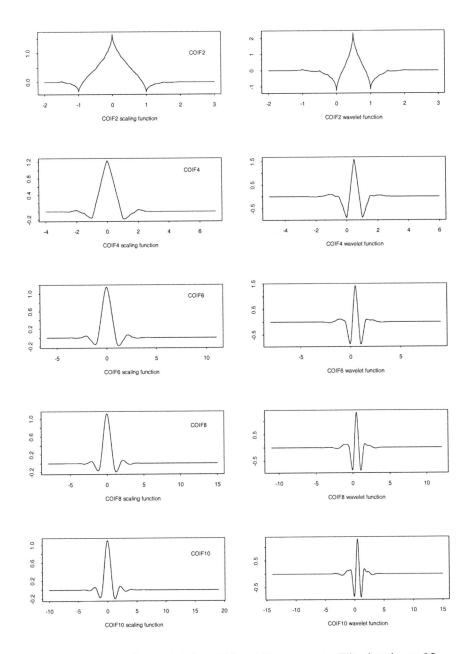

Fig. 5.1 Coiflets with $L = 2, 4, 6, 8,$ and 10 vanishing moments. Filter lengths are $3L = 6, 12, 18, 24,$ and 30. Supports are $[-L, 2L - 1]$, respectively.

example from linear algebra.

Example 5.2.1 Let $\underline{e}_1 = (1,0)$ and $\underline{e}_2 = (\frac{1}{2}, \frac{\sqrt{3}}{2})$ be a basis for $\mathbb{R}^2$. Let T be the matrix in which $\underline{e}_1$ and $\underline{e}_2$ are rows. Then a dual basis can be obtained from the columns of matrix T^{-1}. From

$$T = \begin{bmatrix} 1 & 0 \\ \frac{1}{2} & \frac{\sqrt{3}}{2} \end{bmatrix}, \quad T^{-1} = \begin{bmatrix} 1 & 0 \\ \frac{1}{\sqrt{3}} & \frac{2}{\sqrt{3}} \end{bmatrix},$$

we find vectors forming the dual basis $\tilde{e}_1 = (1, -\frac{1}{\sqrt{3}})$ and $\tilde{e}_2 = (0, \frac{2}{\sqrt{3}})$. Observe that $\underline{e}_i \tilde{e}_j = \delta_{i-j}$ (from $T \cdot T^{-1} = I$), but the bases are not orthogonal.

What is the connection between the bases $(\underline{e}_1, \underline{e}_2)$ and $(\tilde{e}_1, \tilde{e}_2)$? If we fix a vector $\underline{a} \in \mathbb{R}^2$, say, $\underline{a} = (1,1)$, then

$$(1,1) = a_1 \underline{e}_1 + a_2 \underline{e}_2 = \left(a_1 + \frac{a_2}{2}, \frac{\sqrt{3} a_2}{2} \right)$$

which has solutions $a_1 = 1 - \frac{1}{\sqrt{3}}$ and $a_2 = \frac{2}{\sqrt{3}}$. It is easy to verify that

$$\langle (1,1), \tilde{e}_1 \rangle = 1 - \frac{1}{\sqrt{3}} \text{ and } \langle (1,1), \tilde{e}_2 \rangle = \frac{2}{\sqrt{3}}.$$

Thus in representing a vector in the basis $\{\underline{e}_1, \underline{e}_2\}$, the coefficients are calculated as inner products with counterpart vectors in the dual bases $\{\tilde{e}_1, \tilde{e}_2\}$.

Note that we could represent $(1,1)$ as

$$(1,1) = b_1 \tilde{e}_1 + b_2 \tilde{e}_2,$$

with $b_1 = \langle (1,1), e_1 \rangle$ and $b_2 = \langle (1,1), e_2 \rangle$ resulting from $(T \cdot T^{-1})' = I$.

Why might one choose non-orthogonal bases when a multitude of different orthogonal wavelet bases are available? The wavelets generated by a pair of quadrature mirror filters (resulting from an orthogonal multiresolution analysis) cannot simultaneously be both FIR and of linear phase, i.e., with a finite number of real and symmetric coefficients. The only exception is the Haar wavelet, but its shortcomings have already been discussed. Biorthogonal wavelets are often constructed to be symmetric and compactly supported. In some applications, notably in image processing,[3]

---

[3] It is believed that our visual system is more tolerant of symmetric discrepancies than it is of non-symmetric ones.

the symmetry of a decomposing wavelet is desirable. In addition, the technique for the construction of wavelet filters by taking the square root of $|m_0(\omega)|^2$ (as in the construction of Daubechies' compactly supported wavelets, page 73) does not extend to the multivariate case. In a summary, the biorthogonal wavelet transformations offer more flexibility than standard orthogonal wavelet transformations at the expense of violating Parseval's identity. In statistical applications, this expense may be forbidding since the biorthogonal transformation of normal noise on input results in a more complex object — a level-wise stationary random process. An illustration is given in Fig. 5.2. Panel (a) shows the autocorrelation function of a sequence of 8192 random normal $\mathcal{N}(0, 1)$ numbers. The sequence, transformed by the biorthogonal wavelet bs2.2 exhibits a lag-correlation and heteroscedasticy. The autocorrelation function (ACF) of the first level of the output shows significant correlations at lags 1 and 2 [panel (b)]. The sample variances are 1.416718, 0.8817415, 0.778781, 0.772317, and 0.7548167, for the levels $d1 - d5$, respectively.

Biorthogonal wavelets arise naturally in several statistical applications. For example, given the covariance structure of a stationary time series it is possible to construct a wavelet basis as to fully de-correlate the input sequence. Such wavelet counterparts of Karhunen-Loève transformations are possible by utilizing biorthogonal wavelets. Wavelet-based deconvolution problems also call for biorthogonal bases, see Walter [439].

5.2.1 Construction of Biorthogonal Wavelets

In this section, we give a brief description of the construction of biorthogonal wavelets. Let

$$\cdots \subset V_{-1} \subset V_0 \subset V_1 \subset \cdots$$
$$\cdots \subset \tilde{V}_{-1} \subset \tilde{V}_0 \subset \tilde{V}_1 \subset \cdots$$

be two nested multiresolution ladders (primary and dual) such that translations of ϕ, $\{\phi(\bullet - k), k \in \mathbb{Z}\}$, span V_0 and translations of $\tilde{\phi}$, $\{\tilde{\phi}(\bullet - k), k \in \mathbb{Z}\}$, span $\tilde{V}_0$. Also let

$$\int \phi(x)\, \tilde{\phi}(x - n)\, dx = \delta_n.$$

The self-similarity of the subspaces V_i and $\tilde{V}_i$ is expressed by a pair of scaling equations

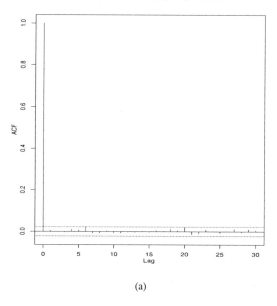

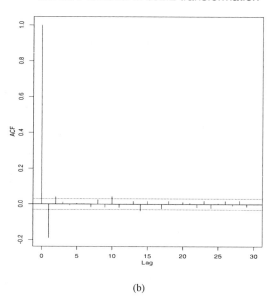

Fig. 5.2 An illustration of the violation of Parseval's identity by a biorthogonal transformation. (a) The ACF of the normal $\mathcal{N}(0, 1)$ noise on the input. The sample size was 8192. (b) The ACF of the finest level of details in a biorthogonal transformation (bs2.2). Note significant 1- and 2-lag correlations.

$$\phi(x) = \sum_k h_k \sqrt{2}\, \phi(2x - k),$$
$$\tilde{\phi}(x) = \sum_k \tilde{h}_k \sqrt{2}\, \tilde{\phi}(2x - k). \qquad (5.2)$$

The complement spaces W_0 ($V_1 = V_0 \cup W_0$) and $\tilde{W}_0$ ($\tilde{V}_1 = \tilde{V}_0 \cup \tilde{W}_0$) are spanned by families $\psi(\bullet - k)$ and $\tilde{\psi}(\bullet - k)$, respectively.

The functions ψ and $\tilde{\psi}$ also satisfy the scaling equations,

$$\psi(x) = \sum_k g_k \sqrt{2}\, \phi(2x - k),$$
$$\tilde{\psi}(x) = \sum_k \tilde{g}_k \sqrt{2}\, \tilde{\phi}(2x - k).$$

The families $\{\phi_{jk}(x) = 2^{j/2}\phi(2^j x - k),\ k \in \mathbb{Z}\}$, $\{\tilde{\phi}_{jk}(x),\ k \in \mathbb{Z}\}$, $\{\psi_{jk}(x),\ k \in \mathbb{Z}\}$, and $\{\tilde{\psi}_{jk}(x),\ k \in \mathbb{Z}\}$ are bases for the spaces $V_j, \tilde{V}_j, W_j$ and $\tilde{W}_j$, respectively. The biorthogonality condition can be expressed in terms of inner products of basis functions:

$$\int \phi_{jk}(x)\tilde{\phi}_{jk'}(x)\,dx = \delta_{k-k'}$$
$$\int \psi_{jk}(x)\tilde{\phi}_{jk'}(x)\,dx = 0 \qquad (5.3)$$
$$\int \psi_{jk}(x)\tilde{\psi}_{j'k'}(x)\,dx = \delta_{j-j'}\delta_{k-k'}.$$

The spaces V_0 and W_0 are not orthogonal; the "angle"[4] is usually acute. Analogous results hold for the tilde-spaces.

In the Fourier domain, the scaling equations (5.2) become

$$\Phi(\omega) = m_0\left(\frac{\omega}{2}\right)\Phi\left(\frac{\omega}{2}\right),$$
$$\tilde{\Phi}(\omega) = \tilde{m}_0\left(\frac{\omega}{2}\right)\tilde{\Phi}\left(\frac{\omega}{2}\right), \qquad (5.4)$$

where $m_0(\omega) = \frac{1}{\sqrt{2}}\sum_n h_n e^{-in\omega}$ and $\tilde{m}_0(\omega) = \frac{1}{\sqrt{2}}\sum_n \tilde{h}_n e^{-in\omega}$ are the primary and dual transfer functions.

---

[4] The angle between multiresolution subspaces V_0 and W_0 is defined as: $\varphi(V_0, W_0) = \inf_{v_0 \in V_0, w_0 \in W_0} \cos^{-1}\frac{|\langle v_0, w_0 \rangle|}{\|v_0\|\|w_0\|}$.

The Fourier transformations of ϕ and $\tilde{\phi}$ are $\Phi(\omega) = \prod_{n=1}^{\infty} m_0(\frac{\omega}{2^n})$ and $\tilde{\Phi}(\omega) = \prod_{n=1}^{\infty} \tilde{m}_0(\frac{\omega}{2^n})$ and these products converge under mild assumptions about the rate of tail decay of ϕ and $\tilde{\phi}$. A set of conditions necessary to assure the convergence can be found in Cohen [83].

The functions m_0 and $\tilde{m}_0$ satisfy the *biorthogonality condition*

$$\overline{m_0(\omega)}\tilde{m}_0(\omega) + \overline{m_0(\omega+\pi)}\tilde{m}_0(\omega+\pi) = 1. \tag{5.5}$$

Let $m_1(\omega) = \frac{1}{\sqrt{2}} \sum_n g_n e^{-in\omega}$ and $\tilde{m}_1(\omega) = \frac{1}{\sqrt{2}} \sum_n \tilde{g}_n e^{-in\omega}$. Then,

$$\begin{aligned} m_1(\omega) &= -e^{-i\omega}\overline{\tilde{m}_0(\omega+\pi)} \\ \tilde{m}_1(\omega) &= -e^{-i\omega}\overline{m_0(\omega+\pi)} \end{aligned} \tag{5.6}$$

are solutions (not unique) of biorthogonal equations:

$$\overline{m_1(\omega)}\tilde{m}_1(\omega) + \overline{m_1(\omega+\pi)}\tilde{m}_1(\omega+\pi) = 1 \tag{5.7}$$

and

$$\overline{m_0(\omega)}\tilde{m}_1(\omega) + \overline{m_0(\omega+\pi)}\tilde{m}_1(\omega+\pi) = 0. \tag{5.8}$$

From (5.6),

$$g_n = (-1)^n \tilde{h}_{1-n} \text{ and } \tilde{g}_n = (-1)^n h_{1-n}.$$

Any $\mathbb{L}_2(\mathbb{R})$ function can be represented in a biorthogonal basis as

$$f(x) = \sum_{jk} \langle f, \tilde{\psi}_{jk} \rangle \psi_{jk}(x) = \sum \langle f, \psi_{jk} \rangle \tilde{\psi}_{jk}(x).$$

For more details on the construction of biorthogonal wavelets, we direct the reader to Cohen, Daubechies, and Feauveau [84].

5.2.2 B-Spline Wavelets

Next, we describe the construction of an important class of biorthogonal wavelets, B-spline wavelets. Readers interested in the details of spline wavelet bases are directed to the monographs of Daubechies [104] and Chui [75].

The connection between splines and wavelets appears to be natural in the multiresolution framework. Let multiresolution spaces V_j be spanned by functions that are piecewise polynomials of degree $N-1$ on dyadic intervals $[k2^{-j},(k-1)2^{-j}]$, with $\mathbb{C}^{N-1}$ continuity at the nodes (located at the endpoints of dyadic intervals). The family of B-splines provides generators for such spaces. The B-spline of order M can be defined (probabilistically) as a density of the sum of M i.i.d. uniform on $[0,1]$ random variables $X_1, \ldots, X_M$. The choice $M = 2$ corresponds to the "tent" function $\phi_2(x) = \max\{0, 1 - |x - 1|\}$. It is an easy exercise to show that in the Fourier domain

$$\Phi_M(\omega) = \left(\frac{\sin\frac{\omega}{2}}{\frac{\omega}{2}}\right)^M. \tag{5.9}$$

It is assumed that a B-spline is a dual-scaling function of order $\tilde{N}$. We distinguish two cases for values of $\tilde{N}$.

Case 1. $\tilde{N} = 2\tilde{l}$. By using the form of $\tilde{\Phi}_{\tilde{N}}(\omega)$, as given in (5.4), and the relation between $\tilde{m}_0$ and $\tilde{\Phi}_{\tilde{N}}$, we conclude that

$$\tilde{m}_0(\omega) = \left(\cos^2\frac{\omega}{2}\right)^{\tilde{l}}.$$

We assume that m_0 has the form

$$m_0(\omega) = \left(\cos^2\frac{\omega}{2}\right)^l P(\sin^2\frac{\omega}{2})$$

for some polynomial P. The biorthogonality equation (5.5) becomes

$$(1-y)^{l+\tilde{l}}P(y) + y^{l+\tilde{l}}P(1-y) = 1 \quad \left[y = \sin^2\frac{\omega}{2}\right],$$

which we recognize as the Bezout equation (see Lemma 3.4.2). One of its solution is

$$P(y) = \sum_{m=0}^{l+\tilde{l}-1} \binom{l+\tilde{l}-1+m}{m} y^m,$$

which translates to

$$m_0(\omega) = \left(\cos\frac{\omega}{2}\right)^{2l} \sum_{m=0}^{l+\tilde{l}-1} \binom{l+\tilde{l}-1+m}{m} \left(\sin^2\frac{\omega}{2}\right)^m.$$

Case 2. $\tilde{N} = 2\tilde{l} + 1$. When $\tilde{N}$ is odd, $\tilde{m}_0(\omega)$ has the form

$$\tilde{m}_0(\omega) = \left(\cos^2\frac{\omega}{2}\right)^{\tilde{l}} \cdot e^{-i\omega/2},$$

where the factor $e^{-i\omega/2}$ accounts for the symmetry of ϕ about $\frac{1}{2}$.

By repeating the arguments from Case 1, we find that

$$m_0(\omega) = e^{-i\omega/2} \left(\cos\frac{\omega}{2}\right)^{2l+1} \sum_{m=0}^{l+\tilde{l}} \binom{l+\tilde{l}+m}{m} \left(\sin^2\frac{\omega}{2}\right)^m.$$

In both cases, the index l is arbitrary; $N = 2l$ in Case 1 and $N = 2l+1$ in Case 2. The notation $BS\tilde{N}.N$ completely describes biorthogonal wavelets generated by B-splines.

Fig. 5.3 gives a graphical representation of the scaling equations for $\tilde{\phi}_2$. Notice a geometric proof of a scaling equation for $\tilde{\phi}(x)$:

$$\tilde{\phi}(x) = \frac{1}{2\sqrt{2}} \cdot \sqrt{2}\tilde{\phi}(2x+1) + \frac{1}{\sqrt{2}} \cdot \sqrt{2}\tilde{\phi}(2x) + \frac{1}{2\sqrt{2}} \cdot \sqrt{2}\tilde{\phi}(2x-1). \quad (5.10)$$

Fig. 5.4 gives biorthogonal pairs for $\tilde{N} = 2$ and $N = 2$ (top four panels), and $\tilde{N} = 3$ and $N = 3$ (bottom four panels). It is interesting that the BS2.2 scaling and wavelet functions are very irregular. They are infinite at all dyadics. Since dyadics are of measure 0 in $\mathbb{R}$ the scaling and wavelet functions are finite almost everywhere. However, since dyadics are dense in $\mathbb{R}$, a precise plot will not "look nice". An interesting discussion on that issue can be found in Ragozin, Bruce, and Gao [348].

A related family of *semi-orthogonal wavelets*, based on B-splines, have been introduced by Chui and Wang [76, 77]. Their primary and dual scaling and wavelet functions have an explicit form. If the primary functions are compactly supported, the duals are not — but the decay of duals is exponential. In the semi-orthogonal case $V_j = \tilde{V}_j$ and $W_j = \tilde{W}_j$, however, the primary and dual bases are different. The basis functions in V_j and W_j are not orthogonal but $W_j \perp V_j$ and $W_i \perp W_j$, $i \neq j$. See Unser, Aldroubi, and Eden [412, 413] for a related construction of biorthogonal families.

Tian [402] and Tian and Wells [403] constructed coiflet-like biorthogonal wavelet bases in which the scaling functions possess vanishing moments. The family bs2.2 is a special case. Recently, Sweldens [392] demonstrated that biorthogonal wavelets

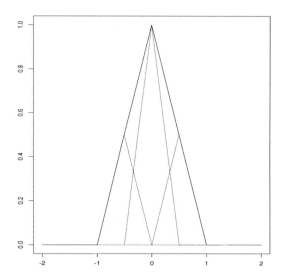

Fig. 5.3 Graphical illustration of the scaling equation for $\tilde{\phi}_2$ in (5.10).

can be efficiently constructed and implemented by the lifting scheme procedure, see Section 5.8.

It is also possible to consider two or more scaling functions whose translates span the multiresolution space V_0. This idea leads to multiwavelets. The usual scaling equation $\phi(x) = \sum_{k \in \mathbb{Z}} h_k \sqrt{2}\phi(2x - k)$, becomes a matrix equation, where h_ks are matrices and $\phi(x)$ is a vector. An increased complexity, loss of commutativity, and existence of nilpotent elements are expenses for the freedom of using more than one scaling function. Multiwavelets can be symmetric, compactly supported and have an explicit and simple expression for the scaling and wavelet functions. Example 5.2.2 gives the dilation equation in a multiwavelet case.

Example 5.2.2 Haar's Hat. Let $\phi_1(x)$ be Haar's box function and let $\phi_2(x) = 2x - 1$. Then, the dilation equation has the matrix form

$$\begin{bmatrix} \phi_1(x) \\ \phi_2(x) \end{bmatrix} = \begin{bmatrix} 1 & 0 \\ -\frac{1}{2} & \frac{1}{2} \end{bmatrix} \cdot \begin{bmatrix} \phi_1(2x) \\ \phi_2(2x) \end{bmatrix}$$
$$+ \begin{bmatrix} 1 & 0 \\ \frac{1}{2} & \frac{1}{2} \end{bmatrix} \cdot \begin{bmatrix} \phi_1(2x - 1) \\ \phi_2(2x - 1) \end{bmatrix}. \quad (5.11)$$

Geronimo, Hardin and Massopust [173] suggested construction of orthogonal and symmetric, compactly supported, continuous multiwavelets. Additional references

132 SOME GENERALIZATIONS

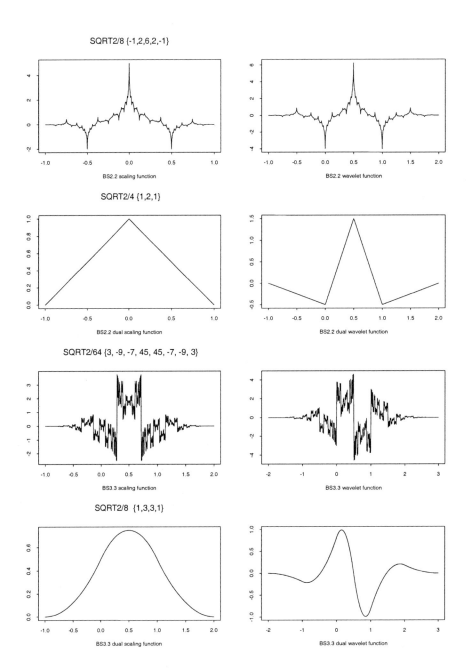

Fig. 5.4 Biorthogonal pairs of scaling and wavelet functions for the B-spline BS2.2 and BS3.3 bases. The coefficients $\underline{h}$ and $\underline{\tilde{h}}$ are given above figures of the primary and dual scaling functions.

on multiwavelets include Xia et al. [467], Strang and Nguyen [387], and Strang and Strela [388].

5.3 WAVELET PACKETS

An orthogonal multiresolution analysis leads to a multitude of orthogonal wavelet-like bases known as *wavelet packets*. They are linear combinations of wavelet functions and represent a powerful generalization of standard orthonormal wavelet bases. Wavelet packet libraries are rapidly constructible and searchable for the "best basis" for parsimonious representation of a given data set or an $\mathbb{L}_2$ function.

Wavelet packets are introduced by Coifman and Meyer [89]. Some other important references are Coifman et al. [90, 91] and Wickerhauser [457].

We start with an example:

Example 5.3.1 The Walsh basis. Consider the scaling and wavelet functions from the Haar basis, $\phi(x) = \mathbf{1}(0 \leq x < 1)$ and $\psi(x) = \mathbf{1}(0 \leq x < 1/2) - \mathbf{1}(1/2 < x < 1)$, and define a system of functions in the recursive manner

$$\mathcal{W}_0(x) = \phi(x); \quad \mathcal{W}_1(x) = \psi(x),$$
$$\cdots$$
$$\mathcal{W}_{2n}(x) = \mathcal{W}_n(2x) + \mathcal{W}_n(2x - 1),$$
$$\mathcal{W}_{2n+1}(x) = \mathcal{W}_n(2x) - \mathcal{W}_n(2x - 1), \, n \geq 1.$$

In the multiresolution analysis generated by $\phi(x)$,

$$\mathcal{W}_0(x) \in V_0, \quad \mathcal{W}_1(x) \in W_0$$
$$\mathcal{W}_2(x), \mathcal{W}_3(x) \in W_1$$
$$\cdots$$
$$\mathcal{W}_{2^m}(x), \mathcal{W}_{2^m+1}(x), \ldots, \mathcal{W}_{2^{m+1}-1}(x) \in W_m, \ldots$$

and

$$\overline{\text{span}}\{\mathcal{W}_{2^m}(x), \mathcal{W}_{2^m+1}(x), \ldots, \mathcal{W}_{2^{m+1}-1}(x), \, m \in \mathbb{Z}\} = W_m, m = 0, 1, \ldots \, .$$

Fig. 5.5 shows the first eight Walsh-basis functions at the level $j = 0$. The first two are Haar's scaling and wavelet functions. The third and fourth, $\mathcal{W}_2(x)$ and $\mathcal{W}_3(x)$, span W_1, the subsequent four span W_2, and so on. Notice that $\psi_{10}(x) = \frac{1}{\sqrt{2}}(\mathcal{W}_2(x) + \mathcal{W}_3(x))$ and $\psi_{11}(x) = \frac{1}{\sqrt{2}}(\mathcal{W}_3(x) - \mathcal{W}_2(x))$.

Walsh's functions are simple linear combinations of Haar's functions – why then

134 SOME GENERALIZATIONS

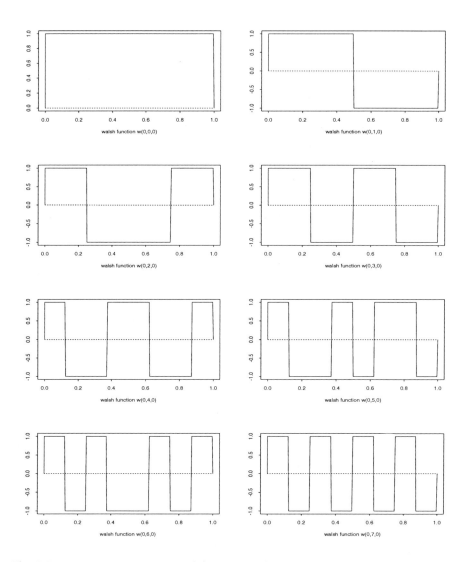

Fig. 5.5 Walsh bases functions $\mathcal{W}_n(x) = \mathcal{W}_{0,n,0}(x)$, $n = 0, 1, \ldots, 7$. Notice that the index n corresponds to the number of zeroes of the basis function. The triple subscript notation is explained in (5.13).

is the Walsh basis needed? Functions that exhibit oscillatory behavior (e.g., sound signals) can, in general, be represented more efficiently in the Walsh system than in the Haar system. Fig. 5.6 illustrates this efficiency. Panel (a) depicts a sound signal (part of the word *armadillo*).

Panel (b) gives Haar and Walsh reconstructions from the 40 best coefficients.[5] Notice that the Walsh-based approximation follows the original function more closely. Also, the local high-frequency features can be explained well by a single Walsh function. A comparable description in the Haar basis function requires several basis functions [approximations between observations 200 and 250, in Fig. 5.6(b) illustrate the point].

5.3.1 Basic Properties of Wavelet Packets

Wavelet packets generalize orthogonal wavelets the same way the Walsh basis generalizes the Haar basis. We next describe construction of wavelet packets.

Let $\underline{h}$ and $\underline{g}$ be a fixed pair of quadrature mirror filters connected with an orthogonal multiresolution analysis [see (3.34)]. Define the following sequence of functions:

$$W_{2n}(x) = \sum_k h_k \sqrt{2} W_n(2x - k),$$
$$W_{2n+1}(x) = \sum_k g_k \sqrt{2} W_n(2x - k), \quad n = 0, 1, 2, \ldots . \quad (5.12)$$

and assume that $\int W_0(x)\, dx = 1$.

The functions $W_0(x)$ and $W_1(x)$ are well-known. They are the scaling and wavelet functions, $\phi(x)$ and $\psi(x)$, respectively, corresponding to quadrature mirror filters $\underline{h}$ and $\underline{g}$.

Definition 5.3.1 *The library of all packet functions (the packet table) is the set of all functions*

$$W_{j,n,k}(x) = 2^{j/2} W_n(2^j x - k), \quad (j, n, k) \in \mathbb{Z} \times \mathbb{N} \times \mathbb{Z}. \quad (5.13)$$

A wavelet packet basis of $\mathbb{L}_2$ is any orthogonal basis selected from the packet table. Each basis is indexed by a subset of indices: j – the scaling parameter, n – the oscillation (sequency) parameter, and k – the translation parameter.

The following theorem shows that wavelet packets can form an orthogonal basis without scaling ($j = 0$).

---

[5] It is an easy exercise to check that the selection of N coefficients with the largest energy minimizes mean square error (MSE) in the class of all approximations obtained by retaining N coefficients only. So "the bigger – the better" rule applies.

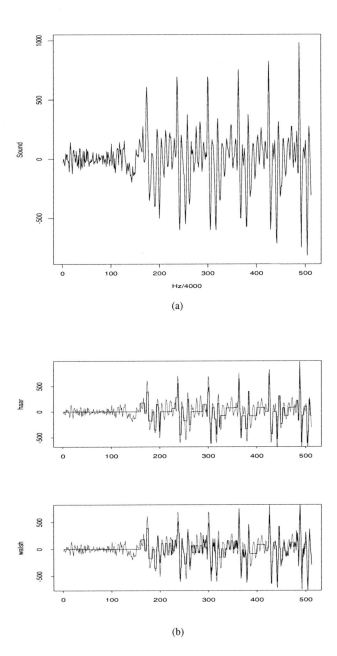

Fig. 5.6 (a) Sound data. (b) Best 40 coefficients approximation for the Haar basis (upper panel) and the Walsh basis (bottom panel).

Theorem 5.3.1 *The family of functions* $\{W_n(x-k), n \in \mathbb{N}, k \in \mathbb{Z}\}$ *is an orthonormal basis of* $\mathbb{L}_2(\mathbb{R})$.

Proof: The proof is by induction on m. When $m = 0$ and $n = 0$,

$$W_0(x) = \phi(x),$$

and $\{W_0(x-k), k \in \mathbb{Z}\}$ is an o.n. basis for V_0. For a fixed $m > 0$, assume that

$$\{W_n(x-k), 0 \leq n < 2^m, k \in \mathbb{Z}\}$$

is an orthonormal basis for V_m. Then,

$$\{\sqrt{2} W_n(2x-k), 0 \leq n < 2^m, k \in \mathbb{Z}\} \tag{5.14}$$

is an orthonormal basis for V_{m+1}. But the scaling equations (5.12) can be rewritten [as in (4.2)] as

$$W_{2n}(x-k) = \sqrt{2} \sum_l h_{2k-l} W_n(2x-l),$$

$$W_{2n+1}(x-k) = \sqrt{2} \sum_l g_{2k-l} W_n(2x-l),$$

and so (5.14) becomes

$$\{W_n(x-k), 0 \leq n < 2^{m+1}, k \in \mathbb{Z}\},$$

which completes the induction step. The statement of the theorem follows if we let $m \to \infty$.

Wavelet packets can also be introduced by using a simple algorithmic and intuitive construction in the Fourier domain. Standard wavelet transformations correspond to a particular sequence of applications of low- and high-pass filters applied to the signal on input. Applications of low- and high-pass filters in an arbitrary order will transform the function or data set on input to a wavelet packet table, an object containing a multitude of different orthogonal wavelet packet bases. We will formalize this statement next.

Let $n = \sum_j \epsilon_j 2^{j-1} = \epsilon_1 + 2\epsilon_2 + 4\epsilon_3 + \cdots + \epsilon_J 2^{J-1}$, $\epsilon_j = 0$ or 1. Let $m_0(\omega) = \frac{1}{\sqrt{2}} \sum_k h_k e^{-ik\omega}$ and $m_1(\omega) = \frac{1}{\sqrt{2}} \sum_k g_k e^{-ik\omega} = -e^{i\omega} \overline{m_0(\omega + \pi)}$.

Then, the function $W_n(x)$ has the Fourier transformation

$$\widehat{\mathcal{W}}_n(\omega) = \prod_{j=1}^{\infty} m_{\epsilon_j}\left(\frac{\omega}{2^j}\right).$$

Recall that $m_0(\omega)$ and $m_1(\omega)$ are connected with $\mathcal{H}$ and $\mathcal{G}$ operators, respectively. The order of the application of $\mathcal{H}$ and $\mathcal{G}$ is determined by n. For example, $n = 10 = 0 \cdot 2^0 + 1 \cdot 2^1 + 0 \cdot 2^2 + 1 \cdot 2^3 + 0 \cdot 2^4 + 0 \cdot 2^5 + \cdots$ corresponds to the sequence $\mathcal{HGHGHH}\ldots$

The following results (proofs can be found in Coifman, Meyer and Wickerhauser [91] or Wickerhauser [457]) describe some particular orthogonal wavelet packet bases.

Theorem 5.3.2 *The system of functions* $\{\mathcal{W}_{j,n,k}(x) = 2^{j/2}\mathcal{W}_n(2^j x - k), \; j \in \mathbb{Z}, k \in \mathbb{Z}\}$ *constitutes an orthonormal basis of* $\mathbb{L}_2(\mathbb{R})$ *for any n such that*

$$2^m \leq n < 2^{m+1}, \; m \geq 0 \text{ fixed.}$$

Theorem 5.3.3 *(General wavelet packets) Let the collection $\{j, n\} \in \mathbb{P} \subseteq \mathbb{Z} \times \mathbb{N}$ be such that the dyadic intervals $\mathcal{I}_{j,n} = \{[2^j n, 2^j(n+1)), \{j, n\} \in \mathbb{P}\}$ constitute a disjoint, countable covering of the interval $[0, \infty)$. Then, the family $\{\mathcal{W}_{j,n,k}(x), (j, n) \in \mathbb{P}, k \in \mathbb{Z}\}$ is a complete orthonormal basis of* $\mathbb{L}_2(\mathbb{R})$.

Any $\mathbb{L}_2$ function f now can be represented as

$$f(x) = \sum_{(j,n)\in\mathbb{P}} \sum_{k\in\mathbb{Z}} w_{j,n,k} \mathcal{W}_{j,n,k}(x),$$

where the set of indices $\mathbb{P}$ is as in Theorem 5.3.3.

5.3.2 Wavelet Packet Tables

The multitude of all possible wavelet packet coefficients (or equivalently, wavelet packet basis functions, since the correspondence is 1–1) can be organized in tabular form. Assume that the data vector has length 2^J, for some fixed $J > 0$. Extensions to more general lengths can be found in Wickerhauser [457]. The vectors $\mathbf{w}_{j,n} = (w_{j,n,0}, w_{j,n,1}, \ldots, w_{j,n,2^j-1})$ will be called *crystals* and the elements of crystals, $w_{j,n,k}$, will be called *atoms*. The discrete data set is conventionally assigned to $\mathbf{w}_{J,0}$.

Fig. 5.7 and Fig. 5.8 represent a wavelet packet table of depth 3 and two wavelet packet bases (functions associated with shaded rectangles). Standard wavelet bases

WAVELET PACKETS   139

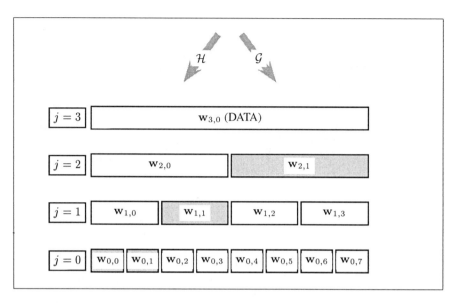

Fig. 5.7 Crystals $\mathbf{w}_{j,n} = (w_{j,n,0}, w_{j,n,1}, \ldots, w_{j,n,2^j-1})$, $0 \leq j \leq J$, $0 \leq n \leq 2^{J-j} - 1$ constitute the wavelet packet table. The data set is $\mathbf{w}_{J,0} = (w_{0,0,0}, \ldots, w_{0,0,2^J-1})$. Here $J = 3$ and shaded crystals corresponds to a standard wavelet decomposition.

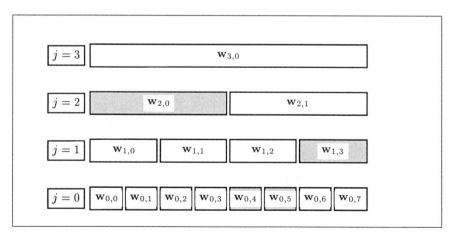

Fig. 5.8 A general wavelet packet basis of depth 3. Note that the informal "vertical projection rule" (page 140) is satisfied.

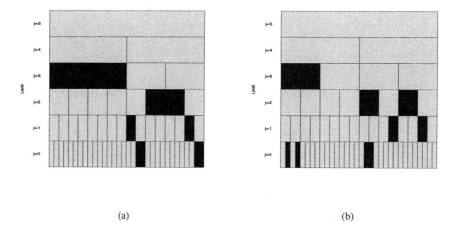

Fig. 5.9 The set of crystals in panel (a) constitutes an orthonormal basis. The set of crystals in panel (b) is not a basis.

are special cases of wavelet packet bases. One example corresponding to the choice $n = 0$, $j = 0$ and $n = 1$, $1 \leq j \leq 2$, is given in Fig. 5.7.

Fig. 5.9(a) gives another graphical representation of a wavelet packet configuration that is an orthonormal basis. The set of crystals in panel (b) contains $\mathbf{w}_{3,0} = \mathcal{HH}\mathbf{w}_{5,0}$, $\mathbf{w}_{2,4} = \mathcal{GHH}\mathbf{w}_{5,0}$, $\mathbf{w}_{2,6} = \mathcal{GGH}\mathbf{w}_{5,0}$, $\mathbf{w}_{1,11} = \mathcal{GHGG}\mathbf{w}_{5,0}$, $\mathbf{w}_{1,14} = \mathcal{GGGH}\mathbf{w}_{5,0}$, $\mathbf{w}_{0,1}$, $\mathbf{w}_{0,3}$, $\mathbf{w}_{0,17}$, and $\mathbf{w}_{0,18}$. Expressions for crystals for $j = 0$, in terms of $\mathcal{H}, \mathcal{G}$ and $\mathbf{w}_{5,0}$, are left as an exercise [see Exercise 5.11(b)].

Panel (b) corresponds to a set of functions which is not a basis. There is a simple informal "rule" to decide if a set of crystals constitutes a basis.

> **Vertical Line Rule.** *An arbitrary vertical line intersecting the packet table intersects the shaded box exactly once. More than one intersection results in redundancy while no-intersection results in non-completeness.*

5.4 BEST BASIS SELECTION

To define *the best* basis we first have to define the criteria upon which the bases will be judged. Such criteria are usually expressed in the form of a *cost functional* $\mathcal{C}$ that maps bases into $\mathbb{R}^+$.

Let $\mathcal{L}$ be a library of bases (a wavelet packet table is just one example) and let $\underset{\sim}{x}$ be the data.

Definition 5.4.1 *Given $\underset{\sim}{x}$, the cost of the basis B is $C(B\underset{\sim}{x})$, where $B\underset{\sim}{x}$ is the representation of $\underset{\sim}{x}$ in the basis B.*

The cost functions are often selected to satisfy the *additivity* requirement. Let $\underset{\sim}{x} = (x_1, x_2, \dots)$. Then,

$$C(\underset{\sim}{x}) = \sum_i \mu(|x_i|), \tag{5.15}$$

where μ is a function such that $\mu(0) = 0$.

It is assumed that the series in (5.15) converges absolutely, so that the value of $C(\underset{\sim}{x})$ is not affected by the order of summation. Additivity of the cost functional in (5.15) is crucial for fast implementation of the algorithm that searches for the best basis in the library of all bases generated by the wavelet packet table.

Example 5.4.1 For the data set $y = (1, 0, -3, 2, 1, 0, 1, 2)$ [Example 4.1.2] the Haar-based wavelet packet table of depth 3 (maximum depth since the sample size is 8) is constructed. The emphasized basis in the first table corresponds to the wavelet transformation. The basis in the second table is selected to minimize the threshold cost [(5.18), $\lambda = 1$].

1	0	-3	2	1	0	1	2
$1/\sqrt{2}$	$-1/\sqrt{2}$	$1/\sqrt{2}$	$3/\sqrt{2}$	$\boxed{1/\sqrt{2}}$	$\boxed{-5/\sqrt{2}}$	$\boxed{1/\sqrt{2}}$	$\boxed{-1/\sqrt{2}}$
0	2	$\boxed{1}$	$\boxed{-1}$	-2	0	3	1
$\boxed{\sqrt{2}}$	$\boxed{-\sqrt{2}}$	0	$\sqrt{2}$	$-\sqrt{2}$	$-\sqrt{2}$	$2\sqrt{2}$	$\sqrt{2}$

1	0	-3	2	1	0	1	2
$1/\sqrt{2}$	$-1/\sqrt{2}$	$1/\sqrt{2}$	$3/\sqrt{2}$	$\boxed{1/\sqrt{2}}$	$\boxed{-5/\sqrt{2}}$	$\boxed{1/\sqrt{2}}$	$\boxed{-1/\sqrt{2}}$
$\boxed{0}$	$\boxed{2}$	1	-1	-2	0	3	1
$\sqrt{2}$	$-\sqrt{2}$	$\boxed{0}$	$\boxed{\sqrt{2}}$	$-\sqrt{2}$	$-\sqrt{2}$	$2\sqrt{2}$	$\sqrt{2}$

5.4.1 Some Cost Measures and the Best Basis Algorithm

Next, we define some popular cost measures and give an algorithmic description of the best basis selection procedure.

Definition 5.4.2 *The entropy cost function of a vector $x = (x_1, \ldots, x_n)$ is*

$$\mathcal{E}(\underline{x}) = -\sum_{i=1}^{n} p_i \cdot \log p_i, \tag{5.16}$$

where $p_i = |x_i|^2 / ||\underline{x}||^2$.

Since $\sum_i p_i = 1$ and $p_i \geq 0$ by construction, the measure $\mathcal{E}$ is Shannon's entropy of the discrete probability distribution $\{p_1, \ldots, p_n\}$.

The functional $\mathcal{E}(\underline{x})$ is not additive but the related functional

$$C_1(\underline{x}) = -\sum_{i=1}^{n} |x_i|^2 \log |x_i|^2 \tag{5.17}$$

is. Minimizing (5.16) is equivalent to minimizing (5.17).

Some other popular cost functionals are

Threshold Cost Functional. Given the threshold $\lambda > 0$, the cost of $\underline{x}$ is the number of components x_i whose absolute value is bigger than λ.

$$C(\underline{x}) = \sum_i \mathbf{1}(|x_i| \geq \lambda). \tag{5.18}$$

ℓ_p-Norm Cost Functional ($0 < p < 2$). The ℓ_p-norm cost functional is given by

$$C(\underline{x}) = ||\underline{x}||_{\mathbb{L}_p}^p = \sum_i |x_i|^p.$$

Notice that for $p = 2$ all costs are the same because of the energy preserving property.

Sure Cost Functional. The following cost functional is based on the Stein unbiased risk estimator (SURE),

$$C(\underline{x}) = \sigma^2 [n - 2 \sum_i \mathbf{1}(|x_i| > \sigma \lambda) + \sum_i \min\{(x_i/\sigma)^2, \lambda^2\}]. \tag{5.19}$$

The Sure cost is useful in selecting a good basis for denoising tasks. The choice $\lambda = \sqrt{2 \log_e (n \log_2 n)}$ (shrinkage threshold) was proposed by Donoho and Johnstone [125]. The parameter σ in (5.19) can be replaced by an estimator in practical applications.

The entropy cost functional C_1 is predominantly used in applications because of its good discrimination properties.

A search for the best basis can be done in any family of bases. However, the tree structure of wavelet packet tables makes the search algorithms efficient. This efficiency adds to the attractiveness of wavelet packet libraries in applications.

Now we will outline a general algorithm for finding the best basis. Suppose that the data $(\mathbf{w}_{J,0})$ are of length 2^J and that the depth of the packet table is $\leq J$. Let $\mathcal{C}$ be an additive cost and let $\mathcal{C}(\mathbf{w}_{j,n})$ be the cost of the crystal $\mathbf{w}_{j,n}$. Let $\mathbf{w}_{j-1,2n}$ and $\mathbf{w}_{j-1,2n+1}$ be the children crystals of $\mathbf{w}_{j,n}$.

The algorithm leading to the best basis can be described as follows.

- **STEP 0.** Order all crystals $\mathbf{w}_{j,n}$ from the wavelet packet table into a library $\mathcal{L}$. The set $\mathcal{B}$ of crystals comprising the best basis is empty.

- **STEP 1.** Consider the first (in the order of STEP 0) crystal $\mathbf{w}_{j,n}$ in $\mathcal{L}$ as a **candidate** for $\mathcal{B}$. If there are no crystals available in $\mathcal{L}$ then **STOP**.

- **STEP 2.** If $j = 0$ or $\mathcal{C}(\mathbf{w}_{j,n}) \leq \mathcal{C}(\mathbf{w}_{j-1,2n}) + \mathcal{C}(\mathbf{w}_{j-1,2n+1})$, then go to **STEP 3**. Else discard $\mathbf{w}_{j,n}$ from $\mathcal{L}$ and return to **STEP 1**.

- **STEP 3.** Move the crystal $\mathbf{w}_{j,n}$ to $\mathcal{B}$. Exclude all crystals in $\mathcal{L}$ for which $\mathbf{w}_{j,n}$ is an "ancestor." Return to **STEP 1**.

In the search for the best basis, comparisons are always made between two adjacent generations in the library tree. Therefore, the complexity of the search is proportional to the number of nodes in the tree, which is $O(n)$ for inputs of length n.

Example 5.4.2 To illustrate the best basis selection, we consider $n = 128$ observations ($J = 7$) from the sea-level data set (see Percival and Mofjeld [337]).

A wavelet packet table of depth 4 was calculated using the DAUB2 wavelet. The best basis was selected using the entropy cost function. Fig. 5.10 is produced by S+WAVELETS' EDA-plot. The MRA subspaces are indexed according to Mallat's convention (see page 52). The top left panel gives the cost table. The top right panel gives the corresponding best basis. The bottom left panel gives the box plots for the coefficients from each level, the DWT, and the best basis. Finally, the bottom right panel gives the energy plot. This plot compares how the energy is "disbalanced" in the original data, the DWT, the single best level (which corresponds to a basis), and the best basis.

Notice that the best basis does not include crystals from levels $j = 6$ and $j = 5$.

144 SOME GENERALIZATIONS

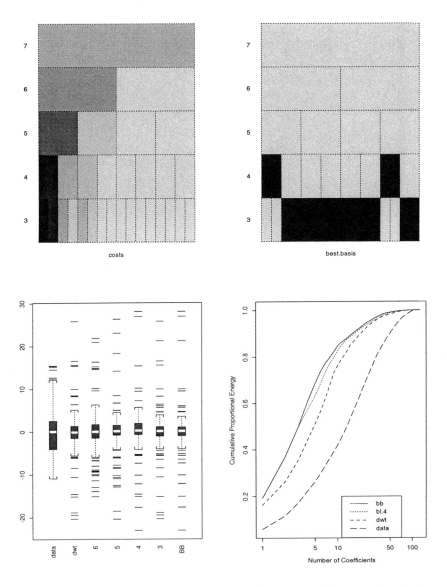

Fig. 5.10 S+WAVELETS' EDA-plot for wavelet-packet-processed the sea-level data sequence of length $n = 128$.

5.5 ε-DECIMATED AND STATIONARY WAVELET TRANSFORMATIONS

Critical sampling defines wavelet expansions of the form $\sum_{j,k\in\mathbb{Z}} d_{jk} 2^{j/2}\psi(2^j x - k)$. Observe that when $j = 0$, the shifts are of size 1. In the next finer level, $j = 1$, the shifts are of size $\frac{1}{2}$, and so on. Though such level-dependence of sampling rates gives good time-scale resolution, as previously discussed, it is responsible for non-stationarity of wavelet transformations. In other words, the wavelet transformation of a shifted data set is not a shift of the wavelet transformation of the original data set.

To preserve stationarity one can retain the finest sampling rate at all scales, which can be done by an overcomplete transformation

$$\sum_{j,k\in\mathbb{Z}} d_{jk}^* \psi(2^j(x-k)). \tag{5.20}$$

Operationally, this is equivalent to not performing decimation steps in the process of wavelet decomposition. Because of this, the expansion in (5.20) is sometimes called *non-decimated*.

Stationary wavelet transformation was introduced by Dutilleux [138] by utilizing the *à trous* algorithm (see the footnote on page 55). Fast algorithmic implementation was discussed in Shensa [375]. Coifman and Donoho [88], Mallat and Hwang [278], Nason and Silverman [311], and Pesquet, Krim, and Carfantan [339] are related references on the theory and application of stationary transformations.

5.5.1 ε-Decimated Wavelet Transformation

First, we outline some notation and then introduce the ε-decimated wavelet transformation. In this section, we will use the compact notation introduced by Nason and Silverman in [311].

Let $\mathcal{S}^k : \ell(\mathbb{Z}) \mapsto \ell(\mathbb{Z})$ be the shift operator defined coordinate-wise as

$$(\mathcal{S}^k \underline{a})_n = a_{n+k}$$

and let

$$\mathcal{D}_0 = [\downarrow 2] \text{ and } \mathcal{D}_1 = [\downarrow 2]\,\mathcal{S}$$

be a pair of decimation operators, see (4.5). Operators $\mathcal{D}_0$ and $\mathcal{D}_1$ decimate by retaining values at even and odd indices, respectively. The operator $\mathcal{D}_0$ was used earlier; a single step in a wavelet decomposition was defined as an action of filters **H** and **G** followed by decimation,

146   SOME GENERALIZATIONS

$$\underline{c}^{(j-1)} = \mathcal{D}_0 \mathbf{H} \underline{c}^{(j)} \text{ and } \underline{d}^{(j-1)} = \mathcal{D}_0 \mathbf{G} \underline{c}^{(j)}.$$

The reconstruction step was

$$\underline{c}^{(j)} = \mathcal{R}(\underline{c}^{(j-1)}, \underline{d}^{(j-1)}) \; [= \mathcal{R}_0(\underline{c}^{(j-1)}, \underline{d}^{(j-1)})].$$

Decimation by $\mathcal{D}_0$ was simply a convention. An orthogonal decomposition can be obtained by applying $\mathcal{D}_1$ in (5.21),

$$\underline{c}_1^{(j-1)} = \mathcal{D}_1 \mathbf{H} \underline{c}^{(j)} \text{ and } \underline{d}_1^{(j-1)} = \mathcal{D}_1 \mathbf{G} \underline{c}^{(j)},$$

with the reconstruction step

$$\underline{c}^{(j)} = \mathcal{R}_1(\underline{c}_1^{(j-1)}, \underline{d}_1^{(j-1)}).$$

Vectors $\underline{c}_1^{(j-1)}$ and $\underline{d}_1^{(j-1)}$ are different from $\underline{c}^{(j-1)}$ and $\underline{d}^{(j-1)}$, but the underlying transformation is still orthogonal.

By permuting $\mathcal{D}_0$ and $\mathcal{D}_1$ in the sequence of consecutive decimations in a wavelet decomposition, one obtains the so-called ϵ-decimated wavelet transformation which is also an orthogonal transformation. The name ϵ-decimated comes from the binary representation $\epsilon_0 \epsilon_1 \ldots \epsilon_{J-1}$ of a number ϵ that defines the decomposition.

The following result connects the standard and ϵ-decimated discrete wavelet transformations.

Theorem 5.5.1 *[311] Let ϵ be an integer whose binary representation is*

$$(\epsilon)_2 = \epsilon_0 \epsilon_1 \ldots \epsilon_{J-1}.$$

Then the coefficients in $\underline{c}^{(j)}$ and $\underline{d}^{(j)}$ in the ϵ-decimated transformation are all shifted versions of the coefficients in the ordinary DWT applied to the shifted sequence $\mathcal{S}^\epsilon \underline{c}^{(J)}$.

Proof: Fix any j and let s_1 and s_2 be integers with binary representations $\epsilon_0 \epsilon_1 \ldots \epsilon_{j-1}$ and $\epsilon_j \epsilon_{j+1} \ldots \epsilon_{J-1}$, respectively.

In the standard DWT,

$$\underline{d}^{(j)} = \mathcal{D}_0 \mathbf{G} (\mathcal{D}_0 \mathbf{H})^{J-j-1} \underline{c}^{(J)}.$$

In the ϵ-decimated case,

$$\begin{aligned}
\underline{d}^{(j)} &= \mathcal{D}_{\epsilon_j} \mathbf{G}\, \mathcal{D}_{\epsilon_{j+1}} \mathbf{H} \ldots \mathcal{D}_{\epsilon_{J-1}} \mathbf{H}\, \underline{c}^{(J)} \\
&= \mathcal{D}_0 \mathcal{S}^{\epsilon_j} \mathbf{G}\, \mathcal{D}_0 \mathcal{S}^{\epsilon_{j+1}} \mathbf{H} \ldots \mathcal{D}_0 \mathcal{S}^{\epsilon_{J-1}} \mathbf{H}\, \underline{c}^{(J)} \quad [\text{because } \mathcal{D}_1 = \mathcal{D}_0 \mathcal{S}^1] \\
&= \mathcal{D}_0 \mathbf{G}\, (\mathcal{D}_0 \mathbf{H})^{J-j-1}\, \mathcal{S}^{s_2} \underline{c}^{(J)} \quad [\text{because } \mathcal{S}\mathcal{D}_0 = \mathcal{D}_0 \mathcal{S}^2].
\end{aligned}$$

Now apply the operator $\mathcal{S}^{s_1}$. We obtain

$$\begin{aligned}
\mathcal{S}^{s_1} \underline{d}^{(j)} &= \mathcal{S}^{s_1} \mathcal{D}_0 \mathbf{G}\, (\mathcal{D}_0 \mathbf{H})^{J-j-1}\, \mathcal{S}^{s_2} \underline{c}^{(J)} \\
&= \mathcal{D}_0 \mathbf{G}\, (\mathcal{D}_0 \mathbf{H})^{J-j-1}\, \mathcal{S}^{\epsilon} \underline{c}^{(J)},
\end{aligned}$$

since $\epsilon = 2^{J-j} s_1 + s_2$. Thus, $\underline{d}^{(j)}$ shifted by an amount s_1 is the jth detail sequence of the standard DWT applied to the original data shifted by an amount ϵ. The corresponding result for $\underline{c}^{(j)}$ can be derived in a similar manner.

Therefore, it follows that the ϵ-decimated transformation can be obtained as a standard DWT of the shifted argument. The choice of ϵ corresponds to the choice of the "origin" with respect to which basis functions are defined.

5.5.2 Stationary (Non-Decimated) Wavelet Transformation

Next we describe the *stationary* or *non-decimated* wavelet transformation.

For quadrature mirror filters $\underline{h}$ and $\underline{g}$, we define up-sampled filters $\underline{h}^{[r]}$ and $\underline{g}^{[r]}$ in a recursive way:

$$\begin{aligned}
\underline{h}^{[0]} &= \underline{h}, \quad \underline{g}^{[0]} = \underline{g} \\
\underline{h}^{[r]} &= [\uparrow 2]\, \underline{h}^{[r-1]}, \quad \underline{g}^{[r]} = [\uparrow 2]\, \underline{g}^{[r-1]}.
\end{aligned}$$

For example, the dilated filter $\underline{h}^{[r]}$ is obtained by inserting zeroes between the taps in $\underline{h}^{[r-1]}$.

Let $\mathbf{H}^{[r]}$ and $\mathbf{G}^{[r]}$ be convolution operators with filters $\underline{h}^{[r]}$ and $\underline{h}^{[r]}$, respectively. A stationary wavelet transformation is defined as a sequential application of operators (convolutions) $\mathbf{H}^{[j]}$ and $\mathbf{G}^{[j]}$.

Definition 5.5.1 *Let* $\underline{a}^{(J)} = \underline{c}^{(J)}$ *and*

$$\begin{aligned}
\underline{a}^{(j-1)} &= \mathbf{H}^{[J-j]} \underline{a}^{(j)}, \\
\underline{b}^{(j-1)} &= \mathbf{G}^{[J-j]} \underline{a}^{(j)}.
\end{aligned}$$

The stationary wavelet transformation of $\underline{c}^{(J)}$ *is* $\underline{b}^{(J-1)}, \underline{b}^{(J-2)}, \ldots, \underline{b}^{(J-j)}, \underline{a}^{(J-j)}$, *for some* $j \in \{1, 2, \ldots, J\}$, *which is the depth of the transformation.*

148   SOME GENERALIZATIONS

If the length of an input vector $\underline{c}^{(J)}$ is 2^J, then for an arbitrary $0 \leq m < J$, $\underline{a}^{(m)}$ and $\underline{b}^{(m)}$ are of the same length.

The operator $(\mathbf{H}^{[j]}, \mathbf{G}^{[j]})$ is not orthogonal but $(\mathcal{D}_0 \mathbf{H}^{[j]}, \mathcal{D}_0 \mathbf{G}^{[j]})$ and $(\mathcal{D}_1 \mathbf{H}^{[j]}, \mathcal{D}_1 \mathbf{G}^{[j]})$ are each orthogonal. The first pair of transformations produces values at even indices in $\underline{a}^{(J-j-1)}$ and $\underline{b}^{(J-j-1)}$ and the second produces the values at odd indices. Let $\mathcal{R}_0^{[j]}$ and $\mathcal{R}_1^{[j]}$ be their inverse transformations. Then,

$$\underline{a}^{(j)} = \overline{\mathcal{R}}^{[J-j]}(\underline{a}^{(j-1)}, \underline{b}^{(j-1)}),$$

for $\overline{\mathcal{R}}^{[j]} = (\mathcal{R}_0^{[j]} + \mathcal{R}_1^{[j]})/2$.

Because $\mathcal{D}_1 = \mathcal{D}_0 \mathcal{S}$ and $\mathcal{S}$ commutes with $\mathbf{H}^{[j]}$ and $\mathbf{G}^{[j]}$, it follows that $\mathcal{R}_1^{[j]} = \mathcal{S}^{-1} \mathcal{R}_0^{[j]}$. Since

$$\mathcal{D}_0^r \mathbf{H}^{[r]} = \mathbf{H} \mathcal{D}_0^r \text{ and } \mathcal{D}_0^r \mathbf{G}^{[r]} = \mathbf{G} \mathcal{D}_0^r,$$

it follows that

$$\mathcal{D}_0^r \mathcal{R}_0^{[r]} = \mathcal{R}_0 \mathcal{D}_0^r.$$

It is easy to see that the stationary wavelet transformation contains the coefficients of ϵ-decimated transformations for every choice of ϵ.

Let s_1 and s_2 be as in the proof of Theorem 5.5.1. Let $\underline{d}_\epsilon^{(j)}$ be the jth level of detail obtained in the ϵ-decimated transformation of $\underline{c}^{(J)}$. Then, for any j,

$$\begin{aligned}
\mathcal{S}^{-s_1} \mathcal{D}_0^{J-j} \mathcal{S}^\epsilon \underline{b}^{(j)} &= \mathcal{D}_0^{J-j} \mathcal{S}^{s_2} \underline{b}^{(j)} \\
&= \mathcal{D}_0^{J-j} \mathbf{G}^{[J-j-1]} \mathbf{H}^{[J-j-2]} \ldots \mathbf{H}^{[0]} \mathcal{S}^{s_2} \underline{c}^{(J)} \\
&= \mathcal{D}_0 \mathbf{G} \, \mathcal{D}_0^{J-j-1} \mathbf{H}^{[J-j-1]} \ldots \mathbf{H} \mathcal{S}^{s_2} \underline{c}^{(J)} \\
&= \mathcal{D}_0 \mathbf{G} \, \mathcal{D}_0 \mathbf{H} \, \mathcal{D}_0^{J-j-2} \mathbf{H}^{[J-j-2]} \ldots \mathbf{H} \mathcal{S}^{s_2} \underline{c}^{(J)} \\
&\cdots \\
&= \mathcal{D}_0 \mathbf{G} \, (\mathcal{D}_0 \mathbf{H})^{J-j-1} \mathcal{S}^{s_2} \underline{c}^{(J)} = \underline{d}_\epsilon^{(j)}.
\end{aligned}$$

Example 5.5.1 Let $\phi_j(x) = \phi_{j,0}(x)$ and $\psi_j(x) = \psi_{j,0}(x)$. If the data sequence $\underline{c}^{(J)}$ is associated with the function $f(x) = \sum_k c_k^{(J)} \phi_J(x - 2^{-J} k)$ then the kth coordinate of $\underline{b}^{(j)}$ is equal to

$$b_{jk} = \int \psi_j(x - 2^{-J} k) f(x) \, dx.$$

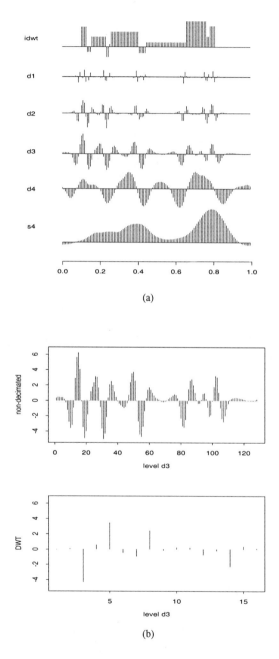

Fig. 5.11 (a) Non-decimated wavelet transformation of a sampled blocks signal, $n = 128$. (b) Level $d3$ for the non-decimated and the standard wavelet transformations. The SYMM4 wavelet was used in the decompositions.

Thus, the coefficient b_{jk} gives information at the scale 2^{J-j} and the location k. There is no restriction on the steps of increase of k by the scale. Therefore, the stationary transformation "fills the gaps" between the coefficients in any particular ϵ-decimated transformation. See further discussion and some applications in [311].

Example 5.5.2 Fig. 5.11 depicts a nondecimated wavelet transformation of the `blocks`. The `blocks`, like the `doppler`, is a test-signal from the battery of four signals proposed by Donoho and Johnstone. Formulae given of the test signals are given in [126]. See also Donoho et al. [132]. The signal is sampled at $n = 128$ equally spaced points. The non-decimated decomposition was performed by the SYMM4 wavelet. Notice that all levels have 128 coefficients, which is the size of the original data. The decomposition with four detail spaces, plus the "smooth" space, denoted by $d1 - d4$ and $s4$, respectively, will have $5 \cdot 128 = 640$ coefficients.

5.6 PERIODIC WAVELET TRANSFORMATIONS

Wavelet bases are usually defined on $\mathbb{L}_2(\mathbb{R})$. However, in some applications it is natural to use an orthogonal system that spans $\mathbb{L}_2([a, b])$, where $[a, b] \subset \mathbb{R}$. Examples include the estimation of functions with compact support. Without loss of generality, it is assumed that $[a, b] \equiv [0, 1]$. The Haar basis is a natural orthonormal basis on $[0, 1]$; it is only necessary to restrict the shift parameter: $0 \leq k < 2^j$, where j is determined by the index in V_j.

Restricting the shift parameter k for other wavelet bases does not work, because the support of some basis functions will be necessarily shared by $[0, 1]$ and $\mathbb{R}\setminus[0, 1]$.

To alleviate this problem, several solutions have been proposed. All proposed constructions preserve the original basis functions when their supports lie between the boundary points. Close to the boundaries the basis functions are modified to preserve orthogonality and provide good behavior at the boundaries. Unfortunately, an intuitive connection between the data and the wavelet coefficients for such modified wavelet bases can be lost and unwanted artifacts can be introduced. The most satisfactory solution to this problem was suggested by Cohen, Daubechies, and Vial [86] where boundaries are treated with a *pre-conditioning* step. See also Cohen et al. [85]. The pre-conditioning step involves applying a linear transform onto the data near the boundaries in such a way that polynomial data have the coefficients of a polynomial in V_0 space. That ensures that smooth functions will have small coefficients near the boundaries.

Next, we provide some results on *periodized wavelets*, modifications of standard wavelet bases constrained to $\mathbb{L}_2([0, 1])$. More details can be found in Jaffard and Laurencot [213] and Restrepo, Leaf, and Schlossnagle [352]. Unless f is already a periodic function, periodized wavelets will introduce edge effects at the endpoints $x = 0$ and $x = 1$.

Periodized wavelets are defined by the following transformation of the scaling and wavelet functions:

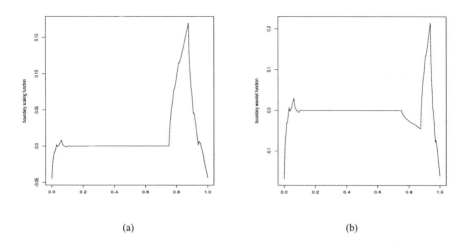

Fig. 5.12 Periodized DAUB2 wavelets: (a) scaling function, and (b) wavelet function.

$$\phi_{jk}^{per}(x) = \sum_m \phi_{jk}(x-m), \qquad (5.21)$$

and

$$\psi_{jk}^{per}(x) = \sum_m \psi_{jk}(x-m). \qquad (5.22)$$

Fig. 5.12 depicts a periodized scaling function [panel (a)] and wavelet function [panel (b)] for the DAUB2 family.

The functions from (5.21-5.22) generate a curtailed multiresolution ladder:

$$V_0^{per} \subset V_1^{per} \subset V_2^{per} \subset \cdots \qquad (5.23)$$

in which the spaces V_j^{per} are spanned by ϕ_{jk}^{per}. As expected,

$$W_j^{per} \oplus V_j^{per} = V_{j+1}^{per},$$

and the spaces W_j^{per} are spanned by ψ_{jk}^{per}.

The periodized kernel of V_j^{per} is

$$\tilde{\mathbb{K}}_j(x,y) = \sum_{k=0}^{2^j-1} \phi_{j,k}^{per}(x)\phi_{j,k}^{per}(y).$$

Negative values for j are not necessary in the multiresolution sequence (5.23) because of the following results.

Result 1. $\phi^{per}(x) = \phi_{00}^{per}(x) = 1.$

Indeed, $\phi^{per}(x) = \sum_n \phi(x-m) = 1$ follows from the Strang-Fix condition, see Example 3.5.1.

Result 2. If $j \leq 0$, then $\phi_{jk}^{per}(x) = 2^{-j/2}.$

The proof is obtained by induction. For $j = -1$ we have

$$\begin{aligned}\phi_{-1,k}^{per}(x) &= \sum_m \phi_{-1,k}(x-m) \\ &= \sum_m 2^{-1/2}\phi\left(\frac{x}{2} - \frac{m}{2} - k\right) \\ &= 2^{-1/2}\left[\sum_l \phi\left(\frac{x}{2} - (l+k)\right) + \sum_l \phi\left(\left(\frac{x}{2} - \frac{1}{2}\right) - (l+k)\right)\right]\end{aligned}$$

Since: $\{\frac{m}{2}, m \in \mathbb{Z}\} = \{l, l \in \mathbb{Z}\} + \{l + \frac{1}{2}, l \in \mathbb{Z}\}$

$$= 2^{-1/2}(1+1) = 2^{1/2}.$$

Result 3. $(\forall m \in \mathbb{Z})\ \phi_{j,k+m\cdot 2^j}^{per}(x) = \phi_{j,k}^{per}(x),\ j \geq 0.$

Result 4. Let f be a continuous, 1-periodic function. Let

$$\xi_0(x) = 1, \xi_1(x) = \psi_{00}^{per}(x), \ldots, \xi_n(x) = \psi_{jk}^{per}(x), \ldots, n = 2^j + k, \ldots$$

(compare to Haar's basis on [0,1]).

Then, there exists a sequence $\{d_n, n = 0, 1, \ldots\}$ so that

$$\sum_{n=0}^{N} d_n \xi_n(x) \to f(x),$$

when $N \to \infty$, uniformly on $[0, 1]$.

It is interesting that this result does not hold in the same generality for Fourier expansions. For details, dual results, and discussion see Daubechies [104], pages 305-308.

Mallat's cascade algorithm can be easily modified for the periodic case. For example, (4.3) for the periodic wavelets becomes

$$c_{j-1,l} = \sum_{k=0}^{2^j-1} \sum_m h_{k-2l-2^j m} c_{j,k}.$$

Walter and Cai [442] construct periodic wavelets "from scratch", i.e., without periodizing standard wavelet bases but directly from Fourier considerations. They provide several examples of periodic wavelets that can not be obtained by periodizing the usual wavelets.

We conclude the section with the remark that periodic wavelet transformations are equivalent to ordinary wavelet transformations of functions that are periodic on [0,1].

5.7 MULTIVARIATE WAVELET TRANSFORMATIONS

In practice, many signals are multidimensional. Examples include measurements in geophysics, medicine, astronomy, economics, and so on. Thus, it is natural to generalize wavelet transformations to accommodate such types of signals.

The simplest way to generate a d-dimensional wavelet basis is to multiply the functions from one-dimensional bases,

$$\psi_{j_1,j_2,\ldots,j_d;k_1,k_2,\ldots,k_d}(x_1,\ldots,x_d) = \psi_{j_1,k_1}(x_1) \cdots \psi_{j_d,k_d}(x_1),$$
$$j_1,\ldots,j_d,k_1,\ldots,k_d \in \mathbb{Z}.$$

The functions $\psi_{j_1,j_2,\ldots,j_d;k_1,k_2,\ldots,k_d}(x_1,\ldots,x_d)$ provide a complete orthonormal basis of $\mathbb{L}_2(\mathbb{R}^d)$ but have the disadvantage of "mixing" the levels $j_1,\ldots,j_d$. At the expense of having $2^d - 1$ different wavelet functions at each level j one can define a basis for which $j_1 = \cdots = j_d \, (=j)$. This construction is essentially due to Mallat [275].

There are more complicated constructions in which the variables are "not separable," i.e., multivariate wavelets are not defined as products of univariate wavelet functions. These more complicated schemes allow, for instance, the construction of wavelets adapted to certain tesselations of the space, see discussion in Jaffard and Meyer [214] and Kovačević and Vetterli [245]. For examples of bivariate nonseparable compactly supported orthonormal continuous wavelets see He and Lai [190].

For simplicity, we consider only tensor product constructions of multivariate wavelets. The examples apply to the most common (and important) case of $d = 2$.

We will first provide a general definition and then illustrate the concepts on applications in image processing.

By using d univariate orthogonal multiresolution analyses,

$\cdots \subset V_{-2,(i)} \subset V_{-1,(i)} \subset V_{0,(i)} \subset V_{1,(i)} \subset V_{2,(i)} \subset \cdots \subset \mathbb{L}_2(\mathbb{R}), \quad i = 1, 2, \ldots, d,$

one can define d-dimensional multiresolution analysis

$$\cdots \subset \mathbf{V}_{-2} \subset \mathbf{V}_{-1} \subset \mathbf{V}_0 \subset \mathbf{V}_1 \subset \mathbf{V}_2 \subset \cdots \subset \mathbb{L}_2(\mathbb{R}^d), \tag{5.24}$$

in which

$$\mathbf{V}_j = \bigotimes_{i=1}^{d} V_{j,(i)} \subset \mathbb{L}_2(\mathbb{R}^d).$$

The resulting d-dimensional multiresolution analysis corresponds to one d-variate scaling function

$$\phi(x_1, \ldots, x_d) = \prod_{i=1}^{d} \phi_{(i)}(x_i)$$

and $2^d - 1$ d-variate wavelet functions

$$\psi^{(l)}(x_1, \ldots, x_d) = \prod_{i=1}^{d} \xi_{(i)}(x_i), \text{ with } \xi = \phi \text{ or } \psi, \text{ but not all } \xi = \phi.$$

Any function $f \in \mathbb{L}_2(\mathbb{R}^d)$ can be represented as

$$\begin{aligned} f(x_1, \ldots, x_d) &= \sum_{\mathbf{k}} c_{j_0; \mathbf{k}} \phi_{j_0; \mathbf{k}}(x_1, \ldots, x_d) \\ &+ \sum_{j \geq j_0} \sum_{\mathbf{k}} \sum_{l=1}^{2^d - 1} d_{j; \mathbf{k}}^{(l)} \psi_{j; \mathbf{k}}^{(l)}(x_1, \ldots, x_d), \end{aligned}$$

where $\mathbf{k} = (k_1, \ldots, k_d) \in \mathbb{Z}^d$ and

$$\phi_{j_0;\mathbf{k}}(x_1,\ldots,x_d) = 2^{jd/2} \prod_{i=1}^{d} \phi_{(i)}(2^j x_i - k_i) \tag{5.25}$$

$$\psi_{j_0;\mathbf{k}}^{(l)}(x_1,\ldots,x_d) = 2^{jd/2} \prod_{i=1}^{d} \xi_{(i)}(2^j x_i - k_i) \tag{5.26}$$

with $\xi = \phi$ or ψ, but not all $\xi = \phi$.

We focus on the two-dimensional case because of its simplicity and importance. The functions

$$\begin{aligned}\phi_{j;k_1,k_2}(x,y) &= \phi_{j,k_1}(x) \cdot \phi_{j,k_2}(y) \\ &= 2^j \phi(2^j x - k_1) \cdot \phi(2^j y - k_2), \ k_1, k_2 \in \mathbb{Z}\end{aligned}$$

span $\mathbf{V}_j$. There are three $(2^2 - 1)$ detail spaces upgrading $\mathbf{V}_j$ to $\mathbf{V}_{j+1}$,

$$\begin{aligned}\mathbf{V}_{j+1} &= V_{j+1,(1)} \otimes V_{j+1,(2)} = (V_{j,(1)} \oplus W_{j,(1)}) \otimes (V_{j,(2)} \oplus W_{j,(2)}) \\ &= V_{j,(1)} \otimes V_{j,(2)} \oplus (V_{j,(1)} \otimes W_{j,(2)}) \\ &\quad \oplus (W_{j,(1)} \otimes V_{j,(2)}) \oplus (W_{j,(1)} \otimes W_{j,(2)}) \\ &= \mathbf{V}_j \oplus \mathbf{W}_j^{(h)} \oplus \mathbf{W}_j^{(v)} \oplus \mathbf{W}_j^{(d)}.\end{aligned} \tag{5.27}$$

In the two-dimensional case, the function $f(x,y)$ sampled on a regular rectangular grid in $\mathbb{R}^2$ will usually be associated with an image. Superscripts $\{h,v,d\}$ in (5.27) stand for *horizontal*, *vertical* and *diagonal*, since the detail spaces $\mathbf{W}_j^{(h)}, \mathbf{W}_j^{(v)}$, and $\mathbf{W}_j^{(d)}$ tend to emphasize coefficient-cliques describing horizontal, vertical and diagonal features of the image, see Example 5.7.2 and related discussion.

The spaces $\mathbf{W}_j^{(h)}, \mathbf{W}_j^{(v)}$, and $\mathbf{W}_j^{(d)}$ are spanned by translations (with respect to both arguments) of

$$\begin{aligned}\psi_{j;k_1,k_2}^{(h)}(x,y) &= \phi_{j,k_1}(x)\psi_{j,k_2}(y), \\ \psi_{j;k_1,k_2}^{(v)}(x,y) &= \psi_{j,k_1}(x)\phi_{j,k_2}(y), \text{ and} \\ \psi_{j;k_1,k_2}^{(d)}(x,y) &= \psi_{j,k_1}(x)\psi_{j,k_2}(y).\end{aligned}$$

The scaling relations in the two-dimensional case are

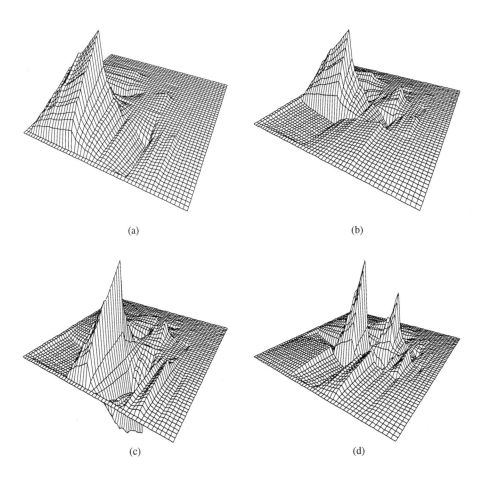

Fig. 5.13 Two-dimensional scaling function and wavelets corresponding to the DAUB2 filter. (a) $\phi_{0;0,0}(x,y)$, (b) $\psi^{(h)}_{0;0,0}(x,y)$, (c) $\psi^{(v)}_{0;0,0}(x,y)$, and (d) $\psi^{(d)}_{0;0,0}(x,y)$.

$$\phi(x,y) = 2\sum_{k,l} h_{k,l}\phi(2x-k, 2y-l),$$

$$\psi^{(h)}(x,y) = 2\sum_{k,l} g^{(h)}_{k,l}\phi(2x-k, 2y-l),$$

$$\psi^{(v)}(x,y) = 2\sum_{k,l} g^{(v)}_{k,l}\phi(2x-k, 2y-l), \text{ and}$$

$$\psi^{(d)}(x,y) = 2\sum_{k,l} g^{(d)}_{k,l}\phi(2x-k, 2y-l),$$

where

$$h_{k,l} = h_k h_l, \; g^{(h)}_{k,l} = h_k g_l, \; g^{(v)}_{k,l} = g_k h_l, \text{ and } g^{(d)}_{k,l} = g_k g_l.$$

Fig. 5.13 depicts two-dimensional wavelet functions associated with the level $j = 0$. Let $f(x,y) \in \mathbf{V}_{j_0}$. Then,

$$f(x,y) = \sum_{k,l} c_{j_0;k,l}\phi_{j_0;k,l}(x,y),$$

for

$$c_{j_0;k,l} = \int f(x,y)\,\phi_{j_0;k,l}(x,y)\,dx\,dy.$$

By using the detail spaces, any $\mathbb{L}_2(\mathbb{R}^2)$ function $f(x,y)$ can also be represented as

$$f(x,y) = \sum_{k,l} c_{j_0;k,l}\phi_{j_0;k,l}(x,y) + \sum_{i}\sum_{j \geq j_0}\sum_{k,l} d^{(i)}_{j;k,l}\psi^{(i)}_{j;k,l}(x,y),$$

since

$$\mathbb{L}_2(\mathbb{R}^2) = \mathbf{V}_{j_0} \oplus \bigoplus_{j=j_0}^{\infty}\left[(\oplus_{i \in \{h,v,d\}} W_{j,(i)}) \otimes V_{j,(i)}\right] = \mathbf{V}_{j_0} \oplus \bigoplus_{j=j_0}^{\infty}(\oplus_{i \in \{h,v,d\}}\mathbf{W}^{(i)}_j).$$

Mallat's fast cascade algorithm in the two-dimensional case becomes

[decomposition]

$$c_{j-1;k,l} = \sum_{m,n} h_{m-2k,n-2l} \cdot c_{j;m,n},$$

$$d^{(i)}_{j-1;k,l} = \sum_{m,n} g^{(i)}_{m-2k,n-2l} \cdot c_{j;m,n}, \text{ for } i \in \{h,v,d\}$$

[reconstruction]

$$c_{j;m,n} = \sum_{k,l} h_{m-2k,n-2l} \cdot c_{j-1;m,n} +$$

$$\sum_{i \in \{h,v,d\}} \sum_{k,l} g^{(i)}_{m-2k,n-2l} \cdot d^{(i)}_{j-1;m,n}.$$

To illustrate a multivariate wavelet transformation, we give several examples.

Example 5.7.1 Any gray-scale image can be digitized and represented as a matrix A whose entries a_{ij} correspond to the intensities of gray in the pixel at position (i,j). For simplicity reasons, we assume that A is a square matrix of dimension $2^J \times 2^J$, J integer. The process of discrete wavelet transformation goes as follows. The operators $\mathcal{H}$ and $\mathcal{G}$ are applied on the rows of the matrix A. Two resulting matrices are obtained: $\mathcal{H}_r A$ and $\mathcal{G}_r A$, both of dimension $2^J \times 2^{J-1}$. The subscript r suggest that the operators are applied on the rows of A. In the next step, operators $\mathcal{H}$ and $\mathcal{G}$ are applied on the columns of matrices $\mathcal{H}_r A$ and $\mathcal{G}_r A$, which results in four matrices $\mathcal{H}_c\mathcal{H}_r A, \mathcal{G}_c\mathcal{H}_r A, \mathcal{H}_c\mathcal{G}_r A$ and $\mathcal{G}_c\mathcal{G}_r A$ of dimension $2^{J-1} \times 2^{J-1}$. The matrix $\mathcal{H}_c\mathcal{H}_r A$ is "an average", while the matrices $\mathcal{G}_c\mathcal{H}_r A, \mathcal{H}_c\mathcal{G}_r A$ and $\mathcal{G}_c\mathcal{G}_r A$ are the detail matrices. Fig. 5.14 illustrates one-step-decomposition of the "Lenna" image. The process continues recursively with the matrix $\mathcal{H}_c\mathcal{H}_r A$ as an input. If the initial image is the square whose side is a power of 2, the decomposition process can be carried out until a single number is obtained.

Example 5.7.2 This example motivates the superscripts (h) - horizontal, (v) - vertical, and (d) - diagonal, for the three detail spaces in a two-dimensional multiresolution.

In Fig. 5.15, panel (a) shows a square with diagonals. Panel (b) shows its Haar wavelet decomposition with three levels. Only the horizontal, vertical, and diagonal features in the image are preserved in detail sub-matrices s1-d1, d1-s1, and d1-d1, corresponding to $\mathcal{G}_c\mathcal{H}_r, \mathcal{H}_c\mathcal{G}_r$ and $\mathcal{G}_c\mathcal{G}_r$, respectively.

Example 5.7.3 Wavelet packets can also be used as building blocks in versatile two-dimensional bases. This example shows that the best wavelet packet basis compresses with a smaller error than a comparable discrete wavelet transformation. The SYMM8

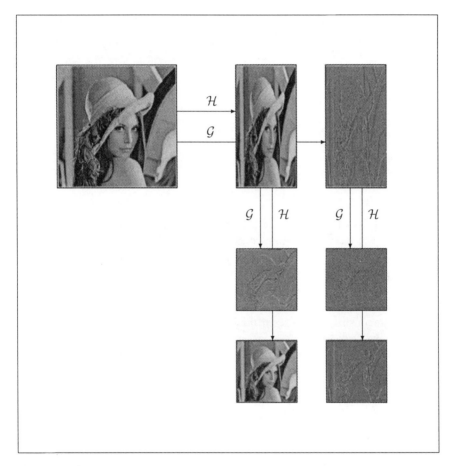

Fig. 5.14 One step in the wavelet decomposition of the Lenna image.

160 SOME GENERALIZATIONS

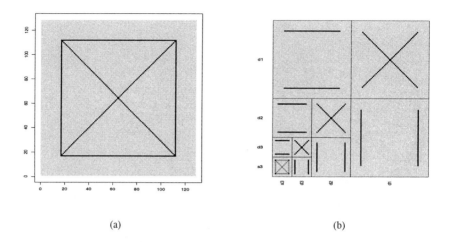

Fig. 5.15 (a) The box image. (b) The three-step decomposition of the box image. Notice favoring of the horizontal, vertical and diagonal directions in detail spaces.

wavelet was used and the wavelet packet table was of depth 5. Fig. 5.16(a) gives the original image of Reverend Thomas Bayes (1702-1761) with superimposed tilling determined by the best basis. The vertical stripes visible in the original image are artifacts of scanning from the paper media.

Panel (b) in Fig. 5.16 shows relative MSE (energy losses) as a function of compression ratio. The vertical line is positioned at a 20:1 compression ratio. Panels (c) and (d) give reconstructed images from the 5% best coefficients of the best wavelet packet basis and discrete wavelet transformation, respectively. Panels (e) and (f) give details from panels (c) and (d).

The best wavelet packet has slightly better MSE performance [panel (b)]. It also better preserves the vertical artifacts in the original image.

5.8 DISCUSSION

On ideas based on de-correlating properties of wavelets, Sweldens [392] has introduced the *lifting scheme* — a different construction of orthogonal and biorthogonal wavelets, wavelet packets, and wavelets on general surfaces. His construction does not require Fourier transformations. The lifting scheme has minimal time and space calculational complexity. It is intuitive and leads to the construction of *second generation* wavelets, which are basis functions that are not necessarily translates and dilates of a single function. The lifting scheme method enables multiscale constructions that retain all desirable aspects of the "first generation" wavelets, while being adaptable to irregular sampling and general meshes, subdivisions, custom bound-

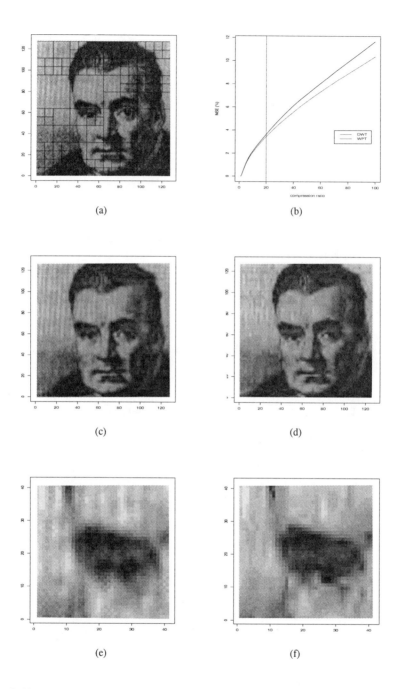

Fig. 5.16 (a) Figure and tilling induced by the best two-dimensional wavelet packet. (b) The percent of the mean square error compared with the compression ratio. (c) and (d) Reconstruction from 5% coefficients in the best basis and standard wavelet transformation, respectively. (e) and (f) Details from the panels (c) and (d).

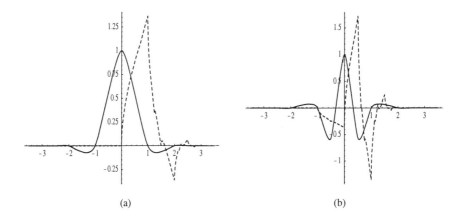

Fig. 5.17 Autocorrelation functions for ϕ [η, panel (a)] and ψ [ν, panel (b)] from the DAUB2 basis.

aries, and other domain constraints. Constructing wavelets using lifting consists of three simple phases: the first step splits the data into two subsets, *even* and *odd,* the second step calculates the wavelet coefficients (high pass) as the failure to predict the odd set based on the even, and finally, the third step updates the even set using the wavelet coefficients to compute the scaling function coefficients (low pass). The predict-phase ensures polynomial cancellation in the high-pass and the update-phase ensures preservation of moments in the low-pass.

Stationary wavelet representations, as discussed in Section 5.5.2, may have several drawbacks such as lack of symmetry, or complexity of representations of simple signals. Saito and Beylkin [363] and Beylkin and Saito [33] introduced stationary decompositions of functions for which the building atoms are autocorrelations of the standard scaling and wavelet functions. The family of such functions, which they call a *shell*, is overly redundant, but shift-invariant.

The autocorrelation functions of Saito and Beylkin are defined as

$$\eta(x) = \int \phi(y)\phi(y-x)\,dy, \text{ and}$$
$$\nu(x) = \int \psi(y)\psi(y-x)\,dy, \quad (5.28)$$

where ϕ and ψ are compactly supported scaling and wavelet functions that generate an orthogonal MRA. Fig. 5.17 depicts autocorrelation functions η and ν for the DAUB2 basis.

Saito and Beylkin [363] discuss fast algorithms for the decomposition of functions in the autocorrelation shell consisting of

$$\eta_{j_0,k}(x) = 2^{j_0/2}\eta(2^{j_0}(x-k)),$$
$$\nu_{jk}(x) = 2^{j/2}\nu(2^j(x-k)), \; j \geq j_0, k \in \mathbb{Z}.$$

The scaling-like equations hold for η and ν and provide a basis for an efficient direct reconstruction of signals from the autocorrelation shell coefficients. The authors in [363] also discuss reconstructions of signals from their zero-crossings, by utilizing autocorrelation shells.

Wavelet expansions in which the dilation factor M differs from 2 are possible. For example, if $M = 3$, the multiresolution subspaces are self-similar with factor 3, that is, $f(x) \in V_j$ iff $f(3^{-j}x) \in V_0$. The space V_1 is three times more wiggly than V_0 and two detail spaces, W_0^1 and W_0^2 say, are necessary to complement V_0 to V_1. Transfer functions m_0, m_1 and m_2 can be introduced via

$$\Phi(\omega) = m_0\left(\frac{\omega}{3}\right)\Phi\left(\frac{\omega}{3}\right)$$
$$\Psi^\ell(\omega) = m_\ell\left(\frac{\omega}{3}\right)\Phi\left(\frac{\omega}{3}\right), \; \ell = 1, 2.$$

For the details on construction see Daubechies [104], Burrus, Gopinath, and Guo [49], and Gopinath and Burrus [178]. Kovačević and Vetterli [246] explore the case when the sampling rate (in scale) is rational.

5.9 EXERCISES

5.1. Prove the biorthogonal counterparts of (3.16) and (3.17):

$$\sum_k h_k = \sum_k \tilde{h}_k = \sqrt{2},$$

and

$$\sum_k \tilde{h}_k h_{k-2l} = \delta_l.$$

5.2. How many vanishing moments have the primary and dual scaling functions in the biorthogonal coiflet family defined by

$$\underset{\sim}{h} = \frac{1}{16\sqrt{2}}(-1, 0, 9, 16, 9, 0, -1) \text{ and}$$

$$\underset{\sim}{\tilde{h}} = \frac{1}{256\sqrt{2}}(-1, 0, 18, -16, -63, 144, 348, 144, -63, -16, 18, 0, -1).$$

(Hint: Check for $\sum_k k^m h_k = 0$, $m \geq 1$ and $\sum_k k^m \tilde{h}_k = 0$, $m \geq 1$. In both filters, the index $k = 0$ corresponds to the middle tap.)

5.3. [11] Let f be an m-differentiable function and let $f^{\{j\}}(k) = 2^{j/2}\langle f, \phi_{jk}\rangle$, where ϕ is the coiflet scaling function with $L > m + 1$ vanishing moments. Prove

$$|f^{\{j\}}(k) - f(k2^{-j})| \leq C \cdot 2^{-jm},$$

where C depends only on L.

5.4. [311] If $\phi(x)$ is an orthogonal scaling function for which $\psi(x)$ has $N > 1$ vanishing moments, prove $\mathcal{M}_2 = \mathcal{M}_1^2$. $[\mathcal{M}_k = \int x^k \psi(x)\, dx]$.

5.5. Prove

$$\begin{aligned}
(i) && \mathcal{D}_1 &= \mathcal{D}_0 \mathcal{S}, \\
(ii) && \mathcal{R}_1 &= \mathcal{S}^{-1} \mathcal{R}_0, \\
(iii) && \mathcal{S}\mathcal{D}_0 &= \mathcal{D}_0 \mathcal{S}^2, \\
(iv) && \mathcal{S}\mathcal{H} &= \mathcal{H}\mathcal{S}, \\
(v) && \mathcal{S}\mathcal{G} &= \mathcal{G}\mathcal{S}.
\end{aligned}$$

5.6. The *Rademacher system* is obtained by adding up all Haar wavelets for a fixed scale, i.e.,

$$\begin{aligned}
r_0(x) &= \phi(x) \\
r_1(x) &= \psi(x) \\
r_{n+1}(x) &= \sum_{k=0}^{2^j-1} \psi(2^j x - k).
\end{aligned}$$

Prove that the Rademacher system is not complete by exhibiting a function in $\mathbb{L}_2([0, 1])$, which cannot be expressed as $\sum_n a_n r_n$.

5.7. Identify all crystals and sequences of operators $\mathcal{H}$ and $\mathcal{G}$ applied on $\mathbf{w}_{3,0}$ for the wavelet packet in Fig. 5.9(b).

5.8. Find the support of $(\mathcal{W}_{j,n,k}(x))_+$, for $j = 4, n = 1, k = 2$ in the Walsh basis.

5.9. Show that if the intervals $\mathcal{I}_{j,n}$ and $\mathcal{I}_{j',n'}$ are disjoint then

$$\int \mathcal{W}_{j,n,k}(x)\mathcal{W}_{j',n',k'}(x)\,dx = 0.$$

5.10. Let the packet table have depth $J = 3$. Show that one can form $N(3) = 25$ different orthogonal bases. Show that $N(4) = 41^2$ and $N(5) = 2306^2$.

5.11. (a) List all crystals corresponding to the shaded rectangles in Fig. 5.9(a).

(b) Express the crystals from level $j = 0$ in Fig. 5.9(b) in terms of $\mathcal{H}, \mathcal{G}$ and $\mathbf{w}_{5,0}$.

5.12. (a) Prove $\psi_{j,k}^{per}(x) = 0$, $j < 0$.

(b) Prove that the periodized system $\{\xi_n,\ n = 0, 1, 2, \ldots\}$ on page 152 is an orthonormal basis of $\mathbb{L}_2[0, 1]$.

(c) Multivariate periodized multiresolution analysis of $\mathbb{L}_2([0, 1]^2)$ can be constructed as a tensor product of univariate periodized multiresolution analyses. Discuss the construction.

5.13. Prove that from $\phi_0^{per}(x) = \sum_{n \in \mathbb{Z}} \phi(x - n) = 1$ and the scaling equations, it follows that

$$\phi_1^{per}(x) + \phi_1^{per}\left(x - \frac{1}{2}\right) = \sqrt{2}.$$

5.14. [442] $\phi_j^{per}(x)$ is 1-periodic function and has the Fourier expansion

$$\phi_0^{per}(x) = \sum_k \alpha_{jk} e^{2\pi i k x},$$

with coefficients $\alpha_{jk} = \int_0^1 \phi_j^{per}(x) e^{-2\pi i k x}\,dx$.

Prove

(a) $\sum_n |\alpha_{jn}|^2 e^{-2\pi i k n 2^{-j}} = \delta_{0,k}$,

(b) $\alpha_{1,0} = \frac{1}{\sqrt{2}}$, $\alpha_{1,2k} = 0$,

(c) $\alpha_{m+1,2k} = \frac{1}{\sqrt{2}} \cdot \alpha_{m,k}$, $k = 0, \pm 1, \pm 2, \ldots$

5.15. If $f(x)$ is 1-periodic function on $\mathbb{R}$, i.e., $f(x) = f(x + 1)$, $x \in \mathbb{R}$, show

$$\int_0^1 f(x)\psi_{jk}^{per}(x)\,dx = \int_{\mathbb{R}} f(x)\psi_{jk}(x)\,dx.$$

166 SOME GENERALIZATIONS

5.16. Find the wavelet pair $\psi_1(x)$ and $\psi_2(x)$ for the pair of scaling functions in (5.11), Example 5.2.2. The functions $\psi_1(x)$ and $\psi_2(x)$ are linear on half-intervals, orthogonal to $\phi_1(x)$ and $\phi_2(x)$ and to each other and discontinuous.

5.17. Find the three-level Haar transformation of the matrix

$$A = \begin{bmatrix} 1 & 2 & 3 & 4 & 4 & 3 & 2 & 1 \\ 2 & 3 & 4 & 5 & 5 & 4 & 3 & 2 \\ 3 & 4 & 5 & 6 & 6 & 5 & 4 & 3 \\ 4 & 5 & 6 & 7 & 7 & 6 & 5 & 4 \\ 4 & 5 & 6 & 7 & 7 & 6 & 5 & 4 \\ 3 & 4 & 5 & 6 & 6 & 5 & 4 & 3 \\ 2 & 3 & 4 & 5 & 5 & 4 & 3 & 2 \\ 1 & 2 & 3 & 4 & 4 & 3 & 2 & 1 \end{bmatrix}.$$

Organize the output as a matrix of the same size arranging "detail" and "smooth" submatrices as in Fig. 5.15.

5.18. [363] Let $\eta(x)$ and $\nu(x)$ be the autocorrelation functions as in (5.28). Let $\hat{\eta}(\omega)$ and $\hat{\nu}(\omega)$ be the corresponding Fourier transformations. Prove

(a) $\eta(k) = \nu(k) = \delta_{0k}$,

(b) $\hat{\eta}(\omega) = |m_0\left(\frac{\omega}{2}\right)|^2 \hat{\eta}\left(\frac{\omega}{2}\right)$, and

(c) $\hat{\nu}(\omega) = |m_1\left(\frac{\omega}{2}\right)|^2 \hat{\nu}\left(\frac{\omega}{2}\right)$.

Prove also,

(d) $\eta(x) = \eta(2x) + \frac{1}{2}\sum_{l=1}^{L/2} a_{2l-1}[\eta(2x - 2l + 1) + \eta(2x + 2l - 1)]$, and

(e) $\nu(x) = \eta(2x) - \frac{1}{2}\sum_{l=1}^{L/2} a_{2l-1}[\eta(2x - 2l + 1) + \eta(2x + 2l - 1)]$,

where $a_k = 2\sum_{l=0}^{L-1-k} h_l h_{l+k}$, $k = 1, \ldots, L-1$, $a_{2k} = 0$, and L is the number of non-zero taps in the corresponding wavelet filter $\underline{h}$.

(f) By inspecting (d) and (e), find the supports of η and ν.

5.19. Marr's wavelet [see Example 3.1.3] is often approximated by the so-called "dog" (difference of two Gaussians) function,

$$1.6 f(1.6\, x; \sigma) - f(x; \sigma),$$

where $f(x; \sigma) = \frac{1}{\sqrt{2\pi}\sigma} e^{-x^2/(2\sigma^2)}$. Prove that $\eta(x)$ and $\nu(x)$ in (5.28) satisfy

$$\nu(x) = 2\,\eta(2\,x) - \eta(x).$$

6
Wavelet Shrinkage

> It is error only, and not truth, that shrinks from inquiry.
> Thomas Paine, 1737-1809

In this chapter, we discuss wavelet shrinkage estimation of discretely sampled functions. Wavelet shrinkage usually refers to reconstructions obtained from the shrunk wavelet coefficients. We will provide the motivation for the shrinkage and reasons for its success when applied in the wavelet domain. Our focus will be mainly on a particular class of shrinkage rules — the *thresholding* rules. As we will argue in this chapter, thresholding rules are easy to calculate and produce surprisingly good estimators.

Although in everyday life the notion of *shrinkage* may carry a negative connotation, it is not so in the domain of statistical estimation. Many good estimators are some sort of shrinkage estimators. For example, most Bayesian, minimax, and Gamma-minimax estimators are shrinkage estimators. A branch of modern statistical theory, *Shrinkage Estimation,* was founded when Stein [385] demonstrated that under squared error loss, the sample mean is not an admissible estimator of the multivariate normal mean, when the dimension of the problem exceeds 2. The procedure that beats the sample mean is a shrinkage rule. After this shocking result, subsequent research found that the Stein-type phenomenon applies to different models and a variety of shrinkage rules were proposed. For a nice overview of Stein-type estimation,

168 WAVELET SHRINKAGE

see Berger [29], Efron and Morris [140], Maatta and Casella [271], Robert [353], and Strawderman [389], as well as references cited therein.

Shrinking and truncating the data directly or the coefficients in their Fourier series expansions is an old technique in signal and image processing. For non-local bases, such as trigonometric, shrinking the coefficients can affect the global shape of the reconstructed function and introduce unwanted artifacts. In the context of function estimation by wavelets, the shrinkage has an additional feature; it is connected with smoothing (denoising) because the measures of smoothness of a function depend on the magnitudes of its wavelet coefficients.

Mallat (École Politecnique, Paris), Healy and collaborators (Dartmouth College), Coifman (Yale University), DeVore (University of South Carolina), Lucier (Purdue University), and Donoho and Johnstone (Stanford University) simultaneously applied thresholding in the wavelet domain to problems of signal processing, approximation, and statistics. Statistical optimality of wavelet shrinkage (from both the asymptotic minimax and exact risk view points) was explored in the work of Donoho, Johnstone, and their collaborators A series of manuscripts is available via the web at:

ftp://www-stat.stanford.edu/reports/donoho.

Some recent references can be found in Donoho [123] and Donoho and Johnstone [130].[1]

6.1 SHRINKAGE METHOD

Consider the simplest paradigmatic regression model

$$y_i = f(x_i) + \sigma \epsilon_i = f_i + \sigma \epsilon_i, \ i = 1, \ldots, n, \quad (6.1)$$

where the x_i's are *equally spaced points* and the ϵ_i's are zero-mean random variables. Unless otherwise specified, it will be assumed that ϵ_i's are independent normal $\mathcal{N}(0, 1)$ random variables. The goal of nonparametric regression is to estimate the unknown function f from the observations y_i, $i = 1, \ldots, n$. In vector notation, the model (6.1) can be written as

$$\underline{y} = \underline{f} + \sigma \underline{\epsilon}, \quad (6.2)$$

where $\underline{y} = (y_1, \ldots, y_n)$, $\underline{f} = (f_1, \ldots, f_n)$ and $\underline{\epsilon} = (\epsilon_1, \ldots, \epsilon_n)$.

The discrete estimator $\hat{\underline{f}} = (\hat{f}_1, \ldots, \hat{f}_n)$ will be judged by its mean-square error (MSE),

---

[1] The original Donoho and Johnstone paper on minimax estimation by wavelet shrinkage, the first paper in the series drafted in October 1990, recently appeared in The Annals of Statistics, [130].

$$\mathrm{MSE}(\hat{f}, f) = \frac{1}{n} E\|\hat{f} - f\|_{\ell_2}^2$$
$$= \frac{1}{n} \sum_{i=1}^{n} E\left[\hat{f}(x_i) - f(x_i)\right]^2. \qquad (6.3)$$

The function f will be restricted since it is well-known that a single method cannot work well for all functions without any restrictions. It is standard to narrow attention to functions f belonging to a particular smoothness space; some additional requirements on the estimator can be imposed, as well:

[SMOOTHNESS] *The estimator $\hat{f}$ should be, with high probability, as smooth as f.*

[ADAPTATION] *The estimator $\hat{f}$ achieves almost minimax risk over one of a wide range of smoothness classes, including the classes in which linear estimators do not achieve the minimax rate.*

These vague requirements can be formalized with the help of Besov spaces, see Section 2.5. Besov spaces are popular in wavelet nonparametrics because of their generality and exceptionally expressive power.

Let W be a matrix of appropriate dimension associated with an orthogonal wavelet transformation. Observe that the "wavelet image" of (6.1) is

$$d_i = \theta_i + \sigma \epsilon_i', \ i = 1, \ldots, n, \qquad (6.4)$$

or, in the vector form,

$$\underline{d} = \underline{\theta} + \sigma \underline{\epsilon}',$$

with $\underline{d} = (d_1, \ldots, d_n) = W \cdot \underline{y}$, $\underline{\theta} = (\theta_1, \ldots, \theta_n) = W \cdot \underline{f}$, and $\underline{\epsilon}' = (\epsilon_1', \ldots, \epsilon_n') = W \cdot \underline{\epsilon}$. Due to orthogonality of W, $\epsilon_i' \sim \mathcal{N}(0, 1)$ and $\mathrm{MSE}(\hat{f}, f) = \mathrm{MSE}(\hat{\underline{\theta}}, \underline{\theta})$.

From now on, we will write $\underline{\epsilon}$ instead of $\underline{\epsilon}'$ since these two vectors have identical stochastic structure.

How do we obtain an estimator $\hat{f}$ using wavelets? Donoho and Johnstone [126] and Donoho et al. [132] propose a simple recipe based on thresholding in the wavelet

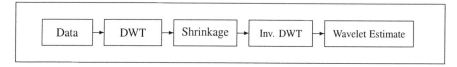

Fig. 6.1 Wavelet-shrinkage paradigm.

domain. Their wavelet estimation procedure has three main steps.

- **STEP 1.** Transform the observations $y_i, i = 1, \ldots, n$ to the wavelet domain by applying a discrete wavelet transformation. The result is a sequence of wavelet coefficients $d_i, i = 1, \ldots, n$.
- **STEP 2.** Estimate σ. Use this estimator to threshold (or shrink) the wavelet coefficients.
- **STEP 3.** Invert the thresholded (shrunk) coefficients, recovering the estimator of the function, $\widehat{f_i}$.

In the rest of this chapter, we focus mainly on STEP 2 of the wavelet regression paradigm (Fig. 6.1) – the shrinkage in the wavelet domain. In the following section, we discuss the linear regression estimators which can also be related to the shrinkage estimators. If the true function is $f(x) = \sum_j \sum_k \theta_{jk} \psi_{jk}(x)$, and the "noisy" wavelet coefficients corresponding to θ_{jk} are d_{jk}, the shrinkage rule

$$\hat{\theta}_{jk} = d_{jk} \mathbf{1}(j < j_0), \tag{6.5}$$

defines the estimator

$$\hat{f}(x) = \sum_{j<j_0} \sum_k d_{jk} \psi_{jk}(x) = \sum_k c_{j_0,k} \phi_{j_0,k}(x).$$

This shrinkage estimator is, in fact, the projection of f onto the multiresolution subspace V_{j_0}. Note that the rule in (6.5) does not depend on the magnitude of d_{jk}.

6.2 LINEAR WAVELET REGRESSION ESTIMATORS

Linear wavelet regression estimators fall into the class of *projection* estimators where the projection operators involve wavelet-based kernels. We briefly discuss some properties of wavelet kernels and give large sample results of the linear estimators. The reader can consult the work of Antoniadis [12], Antoniadis, Grégoire, and McKeague [15], Huang [205], and Walter [439] for more details.

6.2.1 Wavelet Kernels

From Theorem 2.2.3, it follows that

$$\mathbb{K}(x, y) = \sum_k \phi(x - k)\phi(y - k) \tag{6.6}$$

is a reproducing kernel of V_0. By self-similarity of multiresolution subspaces,

$$\mathbb{K}_j(x, y) = 2^j \mathbb{K}(2^j x, 2^j y) \tag{6.7}$$

is a reproducing kernel of V_j.

Thus, the projection of f on the space V_j is given by

$$\text{Proj}_{V_j} f(x) = \int 2^j \mathbb{K}(2^j x, 2^j y) f(y)\, dy.$$

The detail spaces W_j are reproducing kernel Hilbert spaces as well, and $\text{Proj}_{W_j} f(x) = \int 2^j \mathbb{Q}(2^j x, 2^j y) f(y)\, dy$, where $\mathbb{Q}(x, y) = \sum_k \psi(x - k)\psi(y - k)$.

Lemma 6.2.1 *For any j, $\int \mathbb{K}_j(x, y)\, dy = 1$.*

From Exercise 3.17, $\sum_k \phi(x - k) = 1$. Also,

$$\begin{aligned}
1 &= \Phi(0) = \int \phi(y)\, dy = \sum_k \phi(2^j x - k) \int \phi(y)\, dy \\
&= \sum_k \phi(2^j x - k) \cdot 2^j \int \phi(2^j y - k)\, dy = \int 2^j \mathbb{K}(2^j x, 2^j y)\, dy.
\end{aligned}$$

The following result, useful in exploring wavelet-kernel estimators, is proved in Kelly, Kon, and Raphael [232]. The upper bounds on $|\mathbb{K}_j|$ depend on the tail behavior of scaling function.

Theorem 6.2.1 *Let $\mathbb{K}_j(x, y)$ be the wavelet kernel of the space V_j generated by the scaling function ϕ.*

(a) If ϕ has an exponential decay, i.e., $\phi(x) \leq e^{-a|x|}$ for some positive a, then $|\mathbb{K}_j(x, y)| \leq C 2^j e^{-a 2^j |x-y|/2}$.

(b) If ϕ has algebraic decay, i.e., $\phi(x) \leq C_N/(1 + |x|)^N$ for some $N > 1$, then $|\mathbb{K}_j(x, y)| \leq C_N 2^j/(1 + 2^j|x - y|)^N \leq C_N 2^j$ for $N > 1$.

A plot of $\mathbb{K}(x, y)$ for the DAUB2 family is given in Fig. 6.2.

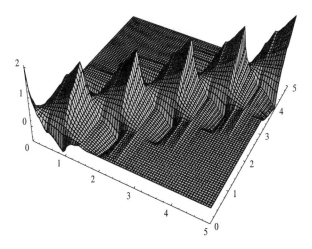

Fig. 6.2 "Sturgeon's Back": $\mathbb{K}(x, y)$ for the DAUB2 family.

6.2.2 Local Constant Fit Estimators

Consider the standard regression model

$$Y_i = m(X_i) + \epsilon_i, \quad i = 1, 2, \ldots, n.$$

Two versions of this model can be defined:

(i) *fixed design* in which the X_i's are non-random (denoted by x_i) and satisfy the condition $0 \leq x_1 \leq x_2 \leq \cdots \leq x_n \leq 1$; and

(ii) *random design* in which (X_i, Y_i) are i.i.d. bivariate random variables with conditional expectation $m(x) = E(Y_i | X_i = x)$.

For model (i), Antoniadis, Grégoire, and McKeague [15] suggest the waveletized Gasser-Müller [168, 169] kernel estimator,

$$\hat{m}(x) = \sum_{i=1}^{n} Y_i \int_{A_i} \mathbb{K}_j(x, y) \, dy, \qquad (6.8)$$

where $\mathbb{K}_j$ is defined as in (6.7). The A_i are intervals that partition $[0, 1]$ so that $x_i \in A_i$. One way of defining the intervals $A_i = [s_{i-1}, s_i)$ is by taking $s_0 = 0, s_n = 1$, and $s_i = (x_i + x_{i+1})/2, i = 1, \ldots, n-1$. Note that the sum of the weights $\int_{A_i} \mathbb{K}_j(x, y) \, dy$ is unity, by Lemma 6.2.1, and no normalizing denominator is needed.

For model (ii), Antoniadis, Grégoire, and McKeague [15] propose a wavelet version of the Nadaraya-Watson estimator (see Nadaraya [306, 307], and Watson [453]).

The Nadaraya-Watson wavelet estimator of $m(x)$ is given by

$$\tilde{m}(x) = \frac{\sum_{i=1}^{n} Y_i \mathbb{K}_j(x, X_i)}{\sum_{i=1}^{n} \mathbb{K}_j(x, X_i)}.$$

When $j \to \infty$ and $n2^{-j} \to \infty$, $\tilde{m}(x)$ is consistent for all x such that $E(Y|X = x^*)$ is bounded for x^* in a neighborhood of x.

Theorem 6.2.2 *If m is continuous at x, $j \to \infty$, and $\max |x_i - x_{i-1}| = o(2^{-j})$, then $\hat{m}(x)$ is mean-square consistent, i.e., $E[\hat{m}(x) - m(x)]^2 \to 0$, $j \to \infty$.*

Antoniadis, Grégoire, and McKeague [15] obtain several results on the asymptotic variance and asymptotic normality of $\hat{m}_d(x)$ – a piecewise-constant approximation at resolution 2^{-m} to $\hat{m}(x)$. (The estimators $\hat{m}(x)$ and $\hat{m}_d(x)$ coincide at dyadic points $[2^m x]/2^m$).

The best rate for the MSE of $\hat{m}_d(x)$ is $n^{-2\nu^*/(2\nu^*+1)}$, is attained when $m = \log_2 n/(2\nu^* + 1)$, with $\nu^* = \min(3/2, \nu, \gamma + 1/2) - \epsilon$. Constants ν and γ are the Sobolev and Lipschitz exponents of m. $\epsilon = 0$ if $\nu \neq 3/2$, and $\epsilon > 0$ if $\nu = 3/2$. The best rate is comparable to that of standard kernel estimators, see Gasser and Müller [168].

Theorem 6.2.3 *If $n2^{-m} \to \infty$ and $n2^{-2m\nu^*} \to 0$, then*

$$\sqrt{n2^{-m}}(\hat{m}_d(x) - m(x))$$

is asymptotically normal $\mathcal{AN}(0, \sigma^2 w_0^2 \kappa(x))$, where κ is some Lipschitz function for which $\max_i |s_i - s_{i-1} - \frac{\kappa(s_i)}{n}| = o(\frac{1}{n})$, for $A_i = [s_{i-1}, s_i)$, and $w_0^2 = \sum_{k \in \mathbb{Z}} \phi^2(k)$.

Remark 6.2.1 Since $\mathbb{K}_j(x, y)$ depends on x, the estimator in (6.8) adapts automatically according to local features in the data. The resolution level j in (6.8) acts as a tuning parameter, similarly to the bandwidth in the kernel-based smoothing procedures. However, since j takes integer values, the task of selecting an "optimal" j is easier than selecting the bandwidth in kernel-based methods. See also discussions in Antoniadis [12] and Härdle et al. [188].

The authors in [15] suggest cross-validation criterium in selecting j. Let $\hat{m}_{-i}$ be the leave-one-out estimator of $m(x)$, obtained with the ith observation removed from the data set. Then, the optimal j minimizes

$$\text{CV}(j) = \frac{1}{n} \sum_i [Y_i - \hat{m}_{-i}(x_i)]^2.$$

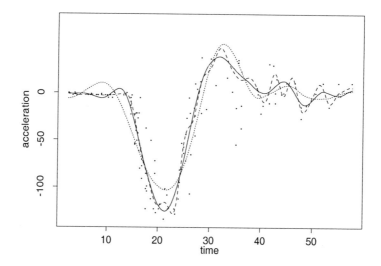

Fig. 6.3 [12] The motorcycle accident data together with estimators $\hat{m}$, for $j = 3$ (dotted line), $j = 4$ (solid line), and $j = 5$ (dashed line). The decomposing wavelet was the DAUB8.

Fig. 6.3 gives three regression estimators for the motorcycle accident data of size 133. The cross-validation criterium in (6.9) selected $j = 4$ as an optimal level.

Both Nadaraya-Watson and Gasser-Müller estimators are local constant fit estimators, (see discussion in Fan and Gijbels [146]). Fan and Gijbels note that the Nadaraya-Watson estimator can suffer from a large bias, particularly in the regions where the regression function has large derivative. The Gasser-Müller estimator corrects the bias of the Nadaraya-Watson estimator, but at the expense of increasing its variability. Both Nadaraya-Watson and Gasser-Müller estimators can have large biases when estimating a curve at a boundary region.

One possibility is to consider the local polynomial models

$$m(z) = \sum_{j=0}^{p} \beta_j (z - x)^j,$$

for z in a neighborhood of x. The estimator is defined by the vector

$$\hat{\underline{\beta}} = (\hat{\beta}_0, \ldots, \hat{\beta}_p)$$

obtained by maximization of

$$\sum_{i=1}^{n}\left[Y_i - \sum_{j=0}^{p}\beta_j(X_i - x)^j\right]^2 \mathbb{K}_j(X_i, x).$$

The estimator for the kth derivative of $m(x)$ ($k = 0, 1, \ldots, p$) is $\widehat{m^{(k)}}(x) = k!\hat{\beta}_k$.

Linear wavelet estimators can also be derived in the context of regularization methods. Some references are DeVore and Lucier [114] Antoniadis [11], Amato and Vuza [9], Chambolle et al. [67], and Fan [145].

A popular approach to regression problems (see Wahba [435], in the context of smoothing splines) is to find a function m from a smoothness space $\mathcal{F}$, which minimizes the discrete functional

$$\frac{1}{n}\sum_{i=1}^{n}(Y_i - m(x_i))^2 + \lambda \int (m^{(\nu)}(x))^2 \, dx, \qquad (6.9)$$

where $m^{(\nu)}(x)$ is νth derivative of m. The "curvature" term $\int (m^{(\nu)}(x))^2 \, dx$ is a penalty term for lack of smoothness. Antoniadis [11] and Amato and Vuza [9] find that for $m \in \mathbb{B}_{2,2}^s([0, 1])$ the minimizer of (6.9) is

$$\widehat{m}(x) = \sum_{k=0}^{2^{J_0}-1} c_{J_0,k}\phi_{J_0,k}(x) + \sum_{j=J_0}^{n}\sum_{k=0}^{2^j-1} d_{jk}^*\psi_{jk}(x),$$

where $d_{jk}^* = \frac{d_{jk}}{1+\lambda 2^{2sj}}$.

When $\lambda = O(n^{-\frac{2s}{1+2s}})$, the MSE of $\widehat{m}$ is optimal, i.e., behaves as $O(n^{-\frac{2s}{1+2s}})$. For a detailed discussion, the reader is directed to Antoniadis [12].

It was pointed out by many researchers that linear methods are not efficient when signals have considerable time-inhomogeneity such as varying degrees of smoothness. Non-linear estimators can improve the efficiency and achieve better rates. For a nice discussion on this issue the reader is directed to the introductory section of Donoho and Johnstone [130]. See also Fan et al. [147] and Hall and Patil [185].

6.3 THE SIMPLEST NON-LINEAR WAVELET SHRINKAGE: THRESHOLDING

The simplest wavelet non-linear shrinkage technique is thresholding. In the language of approximation theorists, the thresholding is sometimes called the *sampling operator*. It is common for all thresholding rules to set to 0 the coordinates of a vector $\underline{d}$, which is subjected to thresholding, if they are smaller in absolute value than a fixed

176 WAVELET SHRINKAGE

non-negative number – the threshold λ.

Depending on how the coordinates of $\underline{d}$ are processed when they are larger than λ one can define different thresholding policies. The two most common thresholding policies are *hard* and *soft*. The analytic expressions for the hard- and soft-thresholding rules are

$$\delta^h(d, \lambda) = d \, \mathbf{1}(|d| > \lambda), \quad \lambda \geq 0, d \in \mathbb{R}, \tag{6.10}$$

and

$$\begin{aligned}\delta^s(d, \lambda) &= (d - \text{sgn}(d) \cdot \lambda) \, \mathbf{1}(|d| > \lambda), \\ &= \text{sgn}(d)(|d| - \lambda)_+, \quad \lambda \geq 0, d \in \mathbb{R}, \end{aligned} \tag{6.11}$$

respectively. The rules are depicted in Fig. 6.4.

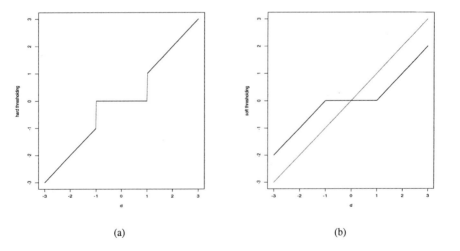

(a) (b)

Fig. 6.4 (a) Hard- and (b) soft-thresholding rules for $\lambda = 1$.

The rules $\delta^h(d, \lambda)$ and $\delta^s(d, \lambda)$ had been known in statistics before the discovery of wavelets. They were used as restricted estimators of an unknown location parameter θ in the estimation problem $d \sim \mathcal{N}(\theta, 1)$. In that context, the rule $\delta^s(d, \lambda)$ is a special case of the *limited translation estimates* proposed by Efron and Morris [140] and explored in detail by Bickel [34]. Bickel calls $\delta^h(d, \lambda)$ the *pretest* estimate and $\delta^s(d, \lambda)$ the *E-M* estimate. Given a fixed MSE at 0, Bickel demonstrated that the maximum of the MSE of $\delta^h(d, \lambda)$ uniformly exceeds that of $\delta^s(d, \lambda)$ (see Table I in [34], page 519). In digital image processing, hard- and soft- thresholding rules are time honored, as well. The soft-thresholding rule is sometimes called *clipping*, see

Jain [215].

6.3.1 Variable Selection and Thresholding

Thresholding rules are intimately connected with model selection in the traditional multivariate regression problems. Suppose that we have a linear model

$$Y = X\beta + \epsilon.$$

If the design matrix X is orthogonal, then the standard least-squares estimator $\hat{\beta} = (X'X)^{-1}X'Y$ becomes $\hat{\beta} = X'Y$, the orthogonal transformation of the observations. The stepwise elimination of the variables in the model corresponds to elimination of the components with the smallest absolute value of t-statistics in $\hat{\beta}$. This amounts to eliminating the component in $\hat{\beta}$ with the smallest absolute value. Since the design is orthogonal, the values of the remaining components in $\hat{\beta}$ do not change in the process of elimination. In the case of orthogonal wavelet transformations, the stepwise deletion procedure corresponds to the hard-thresholding rule.

Soft-thresholding rules can be interpreted as the *posterior mode* in the following Bayesian inference problem. Let $\underline{d} = \underline{\theta} + \underline{\epsilon}$, and $\underline{\epsilon} \sim MV\mathcal{N}(\underline{0}, \sigma^2 \cdot I_n)$, σ^2-known. Let the prior on the components of θ be a double exponential with scale $\frac{\sigma^2}{\lambda}$. The rule that maximizes the posterior minimizes

$$\frac{1}{2}\sum_{i=1}^{n}(d_i - \theta_i)^2 + \lambda \sum_{i=1}^{n}|\theta_i| \qquad (6.12)$$

and the extremal problem can be solved component-wise. The solution is

$$\hat{\theta}_i = \operatorname{sgn}(d_i) \cdot (|d_i| - \lambda)_+.$$

This problem is related to the *lasso* method, see Tibshirani [404].

If we view (6.12) as the penalized least-square, and the penalty $\lambda \sum_{i=1}^{n}|\theta_i|$ is replaced by $\lambda \sum_{i=1}^{n}p(\theta_i)$, where $p(\theta) = |\theta|\mathbf{1}(|\theta| \leq \lambda) + \frac{\lambda}{2}\mathbf{1}(|\theta| > \lambda)$, the solution is the hard-thresholding rule $\theta_i = |d_i|\mathbf{1}(|d_i| > \lambda)$. More comprehensive discussion on the MAP shrinkage rules is postponed until Chapter 8.

6.3.2 Oracular Risk for Thresholding Rules

The optimal thresholding rules (in sense of minimizing the risk in the described normal-location model) are obtained by choosing the threshold $\lambda = \sigma$. Of course, σ is generally not known, and the optimal risk remains unattainable.

Assume the model (6.4). Suppose we have an *oracle* that tells us which of θ_i's are close to 0. The oracle suggest only two actions: "keep" or "kill" the observation with the index i. So, the $\hat{\theta}_i$ suggested by the oracle is $\hat{\theta}_i = d_i \partial_i$, where $\partial_i = 0$ or 1.

The ideal $\underline{\partial}$ in this case is given coordinate-wise by $\partial = \mathbf{1}(|d_i| > \sigma)$. The *diagonal projection* (DP) estimator is $\underline{\hat{\theta}} = \{d_i \mathbf{1}(|d_i| > \sigma), \ i = 1, \ldots, n\}$, and its risk is expressed by

$$R(DP, \underline{\theta}) = E \|\underline{\hat{\theta}} - \underline{\theta}\|^2 = \sum_i \min(|\theta_i|^2, \sigma^2), \qquad (6.13)$$

which can be readily derived by discussing the two possible cases: $|d_i| < \sigma$ and $|d_i| \geq \sigma$. The oracular risk (6.13) is ideal and unattainable if σ is not known. However, it is useful as a benchmark for evaluating other rules.

The following result of Donoho and Johnstone [126] shows that the risk of the soft-thresholding rule with a threshold $\lambda = \sigma \sqrt{2 \log n}$ is close [up to a multiple $(1 + 2 \log n)$] to the oracular risk $R(DP, \underline{\theta})$.

Theorem 6.3.1 *Assume the model (6.4). Let $\hat{\theta}_i^* = \delta^s(d_i, \sigma\sqrt{2 \log n})$. Then,*

$$R(\underline{\hat{\theta}}^*, \underline{\theta}) \leq (2 \log n + 1) \left[\sigma^2 + R(DP, \underline{\theta}) \right]. \qquad (6.14)$$

Proof: It suffices to prove the univariate case. Assume $d \sim \mathcal{N}(\theta, 1)$ and $\hat{\theta}^* = \mathrm{sgn}(d)(|d| - \lambda)_+$. We will demonstrate that if $\delta \leq 1/2$ and $\lambda = \sqrt{2 \log \delta^{-1}}$,

$$E(\hat{\theta}^* - \theta)^2 \leq (2 \log \delta^{-1} + 1)[\delta + (\theta^2 \wedge 1)].$$

Then, the result of Theorem 6.3.1 will follow by taking an arbitrary scale σ, $\delta = 1/n$ and summing the risks ($E \|\underline{\hat{\theta}}^* - \underline{\theta}\|^2 = \sum_i E \|\hat{\theta}_i^* - \theta_i\|^2$).

$$\begin{aligned} E(\hat{\theta}^* - \theta)^2 &= 1 - 2 P(|d| < \lambda) + E(d^2 \wedge \lambda^2) & \text{[see Exercise 6.2]} \\ &\leq 1 + \lambda^2 & \text{[since } d^2 \wedge \lambda^2 \leq \lambda^2\text{]} \\ &\leq (1 + 2 \log \delta^{-1})(1 + \delta) & \text{[since } \delta > 0\text{]}. \end{aligned}$$

On the other hand, because $d^2 \wedge \lambda^2 \leq d^2$,

$$E(\hat{\theta}^* - \theta)^2 \leq 2 P(|d| \geq \lambda) + \theta^2.$$

Consider the function $g(\theta) = 2P(|d| \geq \lambda)$. It is an even function that can be majorized as

$$g(\theta) \leq g(0) + \sup_\theta \frac{g''(\theta)}{2!} \theta^2,$$

by utilizing the Taylor expansion argument. Also,

$$g(0) = 4\Phi(-\lambda) \leq \delta(2\log \delta^{-1} + 1) \quad [\text{since } \Phi(-x) \approx \phi(x)/x, \, x \text{ large}],$$

and $\sup |g''| \leq 4\sup_x |x\phi(x)| \leq 4\log \delta^{-1}$ for all $\delta \leq 1/2$.
Therefore,

$$\begin{aligned}E(\hat{\theta}^* - \theta)^2 &\leq \theta^2 + \delta(2\log \delta^{-1} + 1) + (2\log \delta^{-1})\theta^2 \\ &= (2\log \delta^{-1} + 1)(\theta^2 + \delta).\end{aligned}$$

Gao [163] extended Theorem 6.3.1 to i.i.d. random errors with exponential tails. Averkamp and Houdré [20] further extended Theorem 6.3.1 to some broad classes of non-normal errors. One of their results is given in Theorem 6.3.2.

Theorem 6.3.2 *[20] Let $d_i = \theta_i + z_i$, $i = 1,\ldots,n$, where θ_i are parameters of interest and z_i are symmetric random variables with distributions F_i and $Ez_i^2 = \sigma^2$. Then, the equation*

$$2(n+1)\int_\lambda^\infty (x-\lambda)^2 F_i(dx) = \lambda^2 + \sigma^2,$$

has a unique positive solution $\lambda_{n,i}$. Let $\lambda_n \geq \sup_i \lambda_{n,i}$, let $\Lambda_n = \frac{\lambda_n^2 + \sigma^2}{\sigma^2(1+1/n)}$, and let $\hat{\theta}_i^* = \delta^s(d_i, \sigma\sqrt{2\log n})$. Then,

$$\sup_{\theta \in \mathbb{R}^n} \frac{E\|\hat{\theta}^* - \theta\|^2}{\sigma^2 + R(DP, \theta)} \leq \Lambda_n,$$

Since the z_i are linear combinations of the original noise, the $\lambda_{n,i}$ are quite difficult to compute. The authors in [20] propose an upper bound on $\lambda_{n,i}$ which depends on the original noise only for several broad families of error distributions (in the time domain).

See also Johnstone [221], Johnstone and Silverman [222] and Kovac and Silverman [244] for an extension to correlated normal errors in the time domain. We will provide more references and some discussion on general errors in Section 6.7.

6.3.3 Why the Wavelet Shrinkage Works

In this section, we will provide several examples of wavelet thresholding. We start with a simple example which demonstrates that thresholding in the wavelet domain induces smoothing in the time domain.

Example 6.3.1 We illustrate the idea of smoothing by thresholding by an example that can be done by hand. Let $(1, 0, -3, 2, 1, 0, 1, 2)$ be "noisy data". The discrete wavelet transformation (Haar wavelet) results in $(\frac{1}{\sqrt{2}}, -\frac{5}{\sqrt{2}}, \frac{1}{\sqrt{2}}, -\frac{1}{\sqrt{2}}, 1, -1, -\sqrt{2}, \sqrt{2})$, see Fig. 6.5(a).

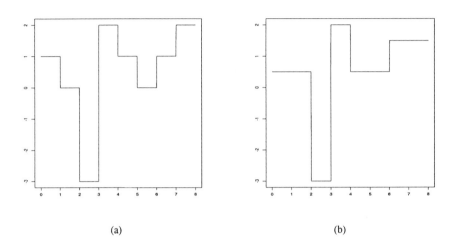

(a)   (b)

Fig. 6.5 The original [panel (a)] and the "denoised" sequence [panel (b)].

Suppose that the threshold λ is 0.9, and that the policy is "hard." The thresholded vector is $(0, -\frac{5}{\sqrt{2}}, 0, 0, 1, -1, -\sqrt{2}, \sqrt{2})$. The graph of the "denoised data," after reconstruction, is given in Fig. 6.5(b).

As discussed in Example 1.2.4, wavelet transformations tend to concentrate the energy in data. This concentration can be measured by a variety of measures of "disbalance," and it is intimately connected with the cost functionals used in the selection of best basis. Most of the cost functionals are related to concentration of energies; examples are entropy, concentration in ℓ_p, and logarithm of energy. For example, the threshold cost functional in (5.18) is directly connected with thresholding procedures. When the result of a transformation is better concentrated, then the thresholding is more efficient since the same number of coefficients retained after thresholding contains more energy.

The following two examples illustrate the point.

Example 6.3.2 Early in this century, economists became interested in measuring inequality (diversity, disbalance) of incomes or wealth in populations. In 1905, Lorentz [266] introduced a graphical representation of wealth inequalities that is now known as the *Lorentz Curve*. Let $\underline{x} = (x_1, \ldots, x_n)$ be a vector in which x_i represents the wealth of the individual i, $i = 1, \ldots, n$.

The Lorentz curve is a plot of points $(p, L_n(p))$, where $0 \le p \le 1$ and $L_n(p) =$

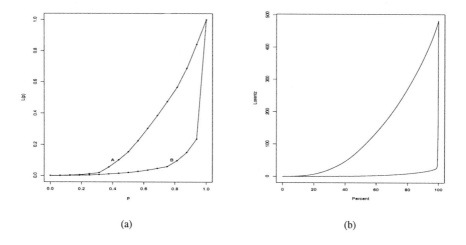

Fig. 6.6 (a) Two distributions; (b) energies of a noisy signal and its DAUB6 transformation.

$\sum_{i=1}^{\lfloor np \rfloor} x_{(i)} / \sum_{i=1}^{n} x_i$, is normed cumulative, up to index $\lfloor np \rfloor$, of increasingly ordered non-negative measurements x_i.

Lorentz's idea was that $\underset{\sim}{x}$ represents a more balanced distribution than $\underset{\sim}{y}$ if and only if the Lorentz curve A, generated by $\underset{\sim}{x}$ is above the Lorentz curve B, generated by $\underset{\sim}{y}$, see Fig. 6.6(a).

It is insightful to compare the Lorentz curves of energy distributions in the wavelet setup. To generate an exemplary data set, we sampled the function $y = \sin(2x+x^2)$ at 1024 equally spaced points in $[-2, 2]$. Next, we added an i.i.d. normal $\mathcal{N}(0, (0.15)^2)$ random noise to the sampled set, and the resulting vector was denoted by $\underset{\sim}{y}$. The wavelet transformation $\underset{\sim}{d}$ was obtained by applying the Daubechies wavelet with 6 vanishing moments to $\underset{\sim}{y}$. Fig. 6.6(b) shows the Lorentz curves of the energy of $\underset{\sim}{y}$, the vector $(y_1^2, \ldots, y_{1024}^2)$ and its DAUB6-transformed counterpart $(d_1^2, \ldots, d_{1024}^2)$.

It is notable that the energy of the transformed signal is more disbalanced than that of the original. This phenomenon is common in wavelets.

Example 6.3.3 For two n-dimensional vectors $\underset{\sim}{x}$ and $\underset{\sim}{y}$ with non-negative components, the *Schur order* is defined as

$$\underset{\sim}{x} \prec \underset{\sim}{y} \text{ iff } \begin{cases} \sum_{i=1}^{k} x_{[i]} \leq \sum_{i=1}^{k} y_{[i]}, \; k = 1, \ldots, n-1 \\ \sum_{i=1}^{n} x_{[i]} = \sum_{i=1}^{n} y_{[i]} \end{cases} \quad (6.15)$$

where $x_{[i]}$ is the ith largest component of $\underset{\sim}{x}$. When $\underset{\sim}{x} \prec \underset{\sim}{y}$, it is said that $\underset{\sim}{x}$ is *Schur majorized* by y. For instance, $(\frac{1}{n}, \ldots, \frac{1}{n}) \prec (\frac{1}{n-1}, \ldots, \frac{1}{n-1}, 0) \prec \cdots \prec (\frac{1}{2}, \frac{1}{2}, 0, \ldots, 0) \prec (1, 0, \ldots, 0)$.

The relation $\prec$ is not a relation of (total) order. For example, the vectors $\underset{\sim}{x} = (1, 2, 3, 4.5, 4.5)/15$ and $\underset{\sim}{y} = (2, 2, 2, 4, 5)/15$ are not comparable in the Schur ordering sense, that is, neither $\underset{\sim}{x} \prec \underset{\sim}{y}$ nor $\underset{\sim}{y} \prec \underset{\sim}{x}$ holds. However, the number of indices k for which the inequality in (6.15) is satisfied can always be counted. The *Schur number* $S(\underset{\sim}{x}, \underset{\sim}{y})$, for non-negative sequences $\underset{\sim}{x}$ and $\underset{\sim}{y}$, is defined as

$$S(\underset{\sim}{x}, \underset{\sim}{y}) = \sum_{k=1}^{n} \mathbf{1}\left(\sum_{i=1}^{k} x_{[i]} \leq \sum_{i=1}^{k} y_{[i]}\right). \tag{6.16}$$

If the Schur number is close to the length of the vectors then we say the vectors are "almost" Schur ordered. Note that

$$S(\underset{\sim}{x}, \underset{\sim}{y}) + S(\underset{\sim}{y}, \underset{\sim}{x}) = n + \text{number of equalities in (6.16)}.$$

For more on the Schur ordering we direct the reader to the classical monograph of Marshall and Olkin [286].

In this example, we want to explore how wavelet transformations change the energy-balance of the input and how much that change depends on the probability distribution of the input. Simulations suggest that the Schur number is sensitive to the initial distributions.

The following simulations were performed.

• Some 200 random samples of size 1024 were generated from one of five distributions: uniform $\mathcal{U}(0, 1)$, normal $\mathcal{N}(0, 1)$, exponential $\mathcal{E}(1)$, t_3, and Cauchy $\mathcal{C}a(0, 1)$.

• These pure-noise data sets were transformed to the wavelet domain utilizing the DAUB4 wavelet. The energy components in the original data and in the transformed data are compared by calculating their Schur numbers. The results, presented in Fig. 6.7, can be summarized as follows. For the uniform and exponential distributions, the wavelet transformation gave the pronounced disbalance in energies. The normal distribution is invariant with respect to orthogonal transformations, and as we expected, Schur numbers are scattered almost uniformly between 0 and 1024. The wavelet transformation "balanced" the original energy vector when the distributions were Cauchy or t.

The simulations suggest that one should be careful with wavelet data processing when it is suspected that the noise is heavy-tailed and dominates the signal.

Because of the local nature of wavelet transformations, the wavelet coefficients are seriously affected by the outliers, particularly at high-resolution levels. Wavelet estimation procedures based on shrinkage may fail to suppress the effects of coefficients associated with outliers because of their large magnitude, as happened in the California earthquakes example from Section 1.4. Therefore, wavelet based methods are sensitive to outliers, which is a potential for their use in the outlier detection problems.

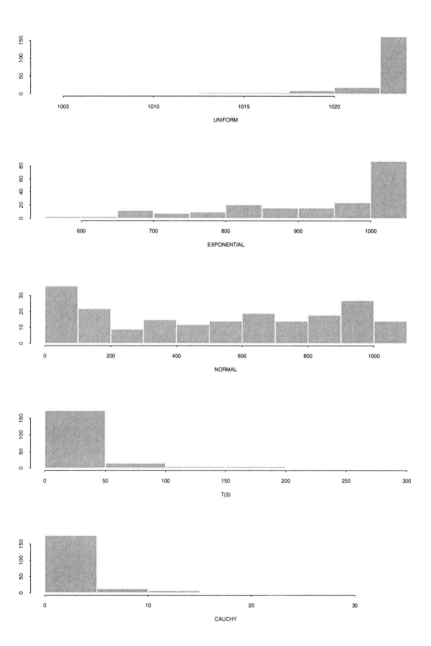

Fig. 6.7 Histograms of Schur numbers for different input distributions. The mean Schur numbers for the above 200 runs are: uniform 1022.92; exponential 918.705; normal 515.805; t_3 21.955; and Cauchy 2.845.

6.3.4 Almost Sure Convergence of Wavelet Shrinkage Estimators

The next result obtained recently shows that wavelet threshold estimators are well-behaved. Although the outcome seems intuitive, the proof is non-elementary. Analogous conclusions do not hold for the Fourier transformations, see Körner [243].

Theorem 6.3.3 *(Tao [395])* For $f \in \mathbb{L}_p(\mathbb{R}), 1 < p < \infty$. Let $d_{jk} = \langle \psi_{jk}, f(x) \rangle$ for an orthonormal wavelet basis $\{\psi_{jk}\}_{j,k \in \mathbb{Z}}$. Then,

$$\lim_{\lambda \to 0} \sum_{j,k} \delta^h(d_{jk}, \lambda) \psi_{jk}(x)$$
$$= \lim_{\lambda \to 0} \sum_{j,k} \delta^s(d_{jk}, \lambda) \psi_{jk}(x) = f(x), \text{ a.e.,}$$

where $\delta^h(d_{jk}, \lambda)$ and $\delta^h(d_{jk}, \lambda)$ are hard- and soft-thresholding rules.

This theorem can be generalized to more general shrinkage rules acting in the wavelet domain.

Let $\delta^*(x, \lambda) : \mathbb{R} \times \mathbb{R}^+ \mapsto \mathbb{R}$ be a function such that

$$|x| - |\delta^*(x, \lambda)| \leq C\lambda,$$

and

$$|\delta^*(x, \lambda)| \leq C |x|^{1+\epsilon} \lambda^{-\epsilon},$$

for fixed non-negative constants C and ϵ, and for all $x \in \mathbb{R}, \lambda > 0$.

If $f \in \mathbb{L}_p(\mathbb{R})$ for some $1 < p < \infty$, then

$$\lim_{\lambda \to 0} \sum_{j,k} \delta^*(d_{jk}, \lambda) \psi_{jk}(x) = f(x),$$

whenever x is a Lebesgue point of f and for every ψ_{jk}. See Tao and Vidakovic [396] for details.

6.4 GENERAL MINIMAX PARADIGM

The result of Donoho and Johnstone given in Theorem 6.3.1 was a decision theoretic one: a simple, soft-thresholding rule, with a threshold $\lambda = \sqrt{2 \log n} \cdot \sigma$, approached (up to a log factor) the unattainable, oracular risk (6.13).

When dealing with estimation of functions, it is well known that there is no estimator that is risk-consistent for all functions. That is, for any estimator $\hat{f}_n$, one can find a function f so that $\sup_f R_n(\hat{f}_n, f)$ does not tend to 0, when $n \to \infty$.

To develop a nontrivial theory one can restrict the class of functions under consideration. Examples of standard function classes include smoothness spaces such as Hölder, Sobolev, and Besov, and in particular *balls* of functions i.e., the classes $\mathcal{F}(M)$ in which the norms of functions from $\mathcal{F}$ are bounded above by a fixed constant M. Besov spaces are of special interest because of their richness; they allow an efficient modeling of spatial inhomogeneities.

Let $R_n(\hat{f}_n, f) = E\|\hat{f}_n - f\|$ be the risk (expected loss), where the loss norm can be quite general.

To exhibit an optimal estimator, according to the minimax paradigm, one compares the $\sup_{f \in \mathcal{F}(M)} R_n(\hat{f}_n, f)$ for a fixed $\hat{f}_n$ with the minimax risk

$$\inf_{\hat{f}} \sup_{f \in \mathcal{F}(M)} R_n(\hat{f}_n, f).$$

The estimator $\hat{f}_n^*$ is a minimax in the class $\mathcal{F}(M)$ if

$$\sup_{f \in \mathcal{F}(M)} R_n(\hat{f}_n^*, f) = \inf_{\hat{f}} \sup_{f \in \mathcal{F}(M)} R_n(\hat{f}_n, f).$$

Assume that $f \in [0, 1]$. Let $\hat{f}_n$ be a wavelet shrinkage estimator, with soft thresholding and threshold $\lambda = \sqrt{2 \log n} \cdot \sigma$. Let $\mathcal{C}(r, N)$ be the scale of all Besov spaces $\mathbb{B}_{pq}^\sigma$, such that $\frac{1}{p} < \sigma < \min(r, N)$, where r is the Hölder regularity and N is the number of vanishing moments of the decomposing wavelet ψ.

(i) The condition $\sigma > \frac{1}{p}$ is necessary to ensure that the spaces $\mathbb{B}_{pq}^\sigma$ are embedded continuously in $\mathcal{C}[0, 1]$, that is, $(\exists C)$ $\sup |f| \leq C \|f\|_{\mathbb{B}_{pq}^\sigma}$.

(ii) The condition $\sigma < \min(r, N)$ guarantees that the wavelet basis generated by ψ is an unconditional basis of $\mathbb{B}_{pq}^\sigma$.

The following result proven by Donoho [122] (see also see also Donoho et al. [132]) formalizes the [SMOOTHNESS] requirement (see page 169):

Theorem 6.4.1 *Let $\mathcal{F}$ be one of the Besov scales, i.e., $\mathcal{F} \in \mathcal{C}(r, N)$. Then there exists an increasing sequence of probabilities π_n with the property $\pi_n \to 1$, and a constant C depending on $\mathcal{F}$ and ψ but not on f and n, so that*

$$P\left(\|\hat{f}_n\|_\mathcal{F} \leq C \|f\|_\mathcal{F}, \; \forall \mathcal{F} \in \mathcal{C}(r, N) \right) \geq \pi_n. \tag{6.17}$$

Thus with a probability increasing with n, $\hat{f}_n$ is as smooth as $f \in \mathcal{F}$ [for any space $\mathcal{F}$ taken from the scale of spaces $\mathcal{C}(r, N)$]. For instance, if $f \equiv 0$, then with the probability of at least π_n, $\hat{f}_n \equiv 0$, as well.

The shrinkage methods, that minimize the risk only, can retain "annoying blips" in reconstructions of even smooth functions. Donoho [122] comments that the result in (6.17) represents the statement of visual superiority of $\hat{f}_n$ over methods based only on the risk minimization.

The second informal requirement, [ADAPTATION], from page 169, can be formalized as well.

Let the measure of loss be the $\mathbb{L}_2$-norm, $||\cdot||_2$. More general losses can be defined by $\mathbb{B}_{p'q'}^{\sigma'}$ norms, see results in Donoho et al. [132].

Theorem 6.4.2 *If the Besov ball $\mathbb{B}_{pq}^{\sigma}(M)$ arises from $C(r, N)$, so that $\frac{1}{p} < \sigma < \min\{r, N\}$, then*

$$\sup_{f \in \mathcal{F}(M)} E||\hat{f}_n^* - f||_2^2 \geq C \cdot \log(n) \cdot \inf_{\hat{f}} \sup_{f \in \mathcal{F}(M)} E||\hat{f} - f||_2^2,$$

where C depends on $\mathcal{F}$ and ψ but not on f and n. In words, the estimate $\hat{f}_n^*$ is simultaneously minimax, within a logarithmic factor, over every Besov class with range indicated.

For more general results involving the global adaptivity and for precise convergence rates (formalized by moduli of continuity) of "approaching minimaxity," we direct the reader to Donoho et al. [132]. See also Donoho and Johnstone [131].

6.4.1 Translation of Minimaxity Results to the Wavelet Domain

If the function $f \in \mathbb{B}_{pq}^{\sigma}(M)$ and the decomposing wavelet basis have regularity $r > \max(1, \sigma)$, then for some constant C the associated wavelet coefficients belong to a *Besov body*

$$\Theta_{pq}^{\sigma}(C) = \left\{ \alpha_{j_0,k}, \theta_{jk} : \left(\sum_{k=0}^{2^{j_0}-1} |\alpha_{j_0,k}|^p \right)^{1/p} + \left(\sum_{j \geq j_0} \left[2^{j(\sigma+1/2-1/p)} \left(\sum_{k=0}^{2^j-1} |\theta_{jk}|^p \right)^{1/p} \right]^q \right)^{1/q} < C \right\},$$

with the usual meaning when $q = \infty$.

For a given r-regular wavelet, the Besov seminorm is

$$|\theta|_{\tilde{b}^\sigma_{p,q}} = \left\{ \sum_{j \geq j_0} \left[2^{j(\sigma+1/2-1/p)} \left(\sum_{k=0}^{2^j-1} |\theta_{jk}|^p \right)^{1/p} \right]^q \right\}^{1/q}.$$

Since the Besov function norm is equivalent to the norm of the wavelet coefficients,

$$\|f\|_{\mathbb{B}^\sigma_{p,q}} \asymp \|\alpha_{j_0,\cdot}\|_p + |\theta|_{\tilde{b}^\sigma_{p,q}},$$

the magnitudes of the wavelet coefficients provide smoothness characterizations of Besov spaces. For details see Section 6.10 in Meyer [294] and Triebel [411].

The difficulty of estimation in the wavelet domain is measured by the minimax risk

$$R^*(\epsilon, \Theta^\sigma_{pq}(C)) = \inf_{\hat{\theta}} \sup_{\Theta^\sigma_{pq}(C)} E\|\hat{\theta} - \theta\|_2^2,$$

where ϵ is the standard deviation of the normal, zero-mean error z_i, in $d_i = \theta_i + z_i$. The minimax risk, constrained only to thresholding estimates $\hat{\theta}_\lambda = \{\delta^s(d_i, \lambda_i)\}$, is

$$R^*_T(\epsilon, \Theta^\sigma_{pq}(C)) = \inf_{\hat{\theta}_\lambda} \sup_{\Theta^\sigma_{pq}(C)} E\|\hat{\theta}_\lambda - \theta\|_2^2.$$

Donoho and Johnstone [130] show that one can find $\hat{\theta}_\lambda$ so that

$$R^*_T \leq \Lambda(p)(1 + o(1))R^*, \quad \text{as } \epsilon \to 0.$$

The constant $\Lambda(p)$ is smaller than 2.22 for all $p \geq 2$, and computational experiments indicate $\Lambda(1) \leq 1.6$. The adaptivity property of SureShrink style estimator (see page 199), is formalized by Theorem 6.4.3.

Theorem 6.4.3 *Let $\sigma > \frac{1}{p} - \frac{1}{2}$. Then,*

$$\sup_{\Theta^\sigma_{pq}(C)} E\|\hat{\theta}^* - \theta\|_2^2 \leq (1 + o(1)) \cdot R^*_T(\epsilon, \Theta^\sigma_{pq}(C)), \quad \epsilon \to 0.$$

In short, even without knowing σ, p, q, or C, one obtains results as good asymptotically as if those parameters were known. Since the minimax risk is close to the minimax threshold risk, the problem of adapting across a scale of Besov bodies is solved.

188 WAVELET SHRINKAGE

6.5 THRESHOLDING POLICIES AND THRESHOLDING RULES

In Section 6.3, we defined and explored the most popular shrinkage policies: hard and soft thresholding. It this section, we mention several other popular shrinkage policies: semisoft, non-negative garrote, and hyperbole thresholding, as well as n-degree garrote shrinkage. We also discuss the exact risk analysis of the hard- and soft-thresholding rules and some of their modifications. Shrinkage rules that are a result of a Bayesian inference in the wavelet domain, will be discussed in Chapter 8.

6.5.1 Exact Risk Analysis of Thresholding Rules

One way to evaluate and compare thresholding rules is by calculating and inspecting their bias, variance, and risk functions under the model (6.4). Through exact risk analysis, when it is possible, one can assess the rules more efficiently than by simulation.

Donoho and Johnstone [127] (in a general setup) evaluate risks of hard- and soft-thresholding rules and demonstrate that they are asymptotically Γ-minimax, when Γ is the family of all priors with means belonging to ℓ_p-balls. Bruce and Gao [43], Bruce et al. [41], Gao [166], and Marron et al. [283] conduct the exact risk analysis of hard- and soft-thresholding rules under squared error loss.

Let $X \sim \mathcal{N}(\theta, \sigma^2)$, σ^2 known, and $\delta = \delta(X, \lambda)$ be either a hard- or soft-thresholding rule with threshold λ. Without loss of generality, we may assume $\sigma^2 = 1$. The exact expressions for bias, variance and risk under squared error loss for hard- and soft-thresholding rules are, respectively:

$$\begin{aligned} M_\lambda^h(\theta) &= \theta + \theta[1 - \Phi(\lambda - \theta) - \Phi(\lambda + \theta)] + \phi(\lambda - \theta) - \phi(\lambda + \theta), \\ V_\lambda^h(\theta) &= (\theta^2 + 1)[2 - \Phi(\lambda - \theta) - \Phi(\lambda + \theta)] \\ &\quad + (\lambda + \theta)\phi(\lambda - \theta) + (\lambda - \theta)\phi(\lambda + \theta) + (M_\lambda^h(\theta))^2, \\ R_\lambda^h(\theta) &= 1 + (\theta^2 - 1)[\Phi(\lambda - \theta) - \Phi(-\lambda - \theta)] \\ &\quad + (\lambda + \theta)\phi(\lambda + \theta) + (\lambda - \theta)\phi(\lambda - \theta), \end{aligned} \quad (6.18)$$

$$\begin{aligned} M_\lambda^s(\theta) &= M_\lambda^h(\theta) - \lambda[\Phi(\lambda + \theta) - \Phi(\lambda - \theta)], \\ V_\lambda^s(\theta) &= V_\lambda^h(\theta) - \lambda[v(\lambda, \theta) + v(\lambda, -\theta)], \\ R_\lambda^s(\theta) &= 1 + \lambda^2 + (\theta^2 - \lambda^2 - 1)[\Phi(\lambda - \theta) - \Phi(-\lambda - \theta)] \\ &\quad - (\lambda - \theta)\phi(\lambda + \theta) - (\lambda + \theta)\phi(\lambda - \theta), \end{aligned} \quad (6.19)$$

where ϕ and Φ are standard normal density and cumulative distribution function and

$$v(\lambda, \theta) = [1 + \Phi(\lambda - \theta) - \Phi(\lambda + \theta)] \cdot [(2\theta - \lambda)(1 - \Phi(\lambda - \theta)) + 2\phi(\lambda - \theta)].$$

The squared bias, variance, and risk for hard- and soft-thresholding rules are plotted in Fig. 6.8. The first column of plots corresponds to hard- while the second column corresponds to soft-thresholding rules. Notice that hard thresholding produces larger variance (since the rule is discontinuous) while the soft thresholding produces larger bias (since it shrinks big coefficients uniformly towards 0 by λ. The threshold λ is 3.724 (the universal threshold for $n = 1024$, Section 6.6.2).

Marron et al. [283] have an extensive discussion on the comparison between hard and soft thresholding from a data analytic point of view.

6.5.2 Large Sample Properties of $\hat{f}$

Let
$$W = \begin{pmatrix} w_{11} & \cdots & w_{1n} \\ & \cdots & \\ w_{n1} & \cdots & w_{nn} \end{pmatrix}$$
be the matrix performing a discrete wavelet transformation, i.e., $\underline{d} = W\underline{y}$.

In terms of W, the wavelet shrinkage estimator estimator of $\underline{f}$ is
$$\hat{f} = W^{-1}(\hat{\sigma}\delta_{\lambda_k}(W\underline{y}/\hat{\sigma})).$$

Therefore, the component $\hat{f}_i$ can be written as $\sum_k w_{ki}(\hat{\sigma}\delta_{\lambda_k}(W\underline{y}/\hat{\sigma})_k)$, $i = 1,\ldots,n$. If $\hat{\sigma}^2$ is an $\mathbb{L}_2$-consistent estimator of the variance, then

$$E(\hat{f}_i) \approx \hat{\sigma} \sum_k w_{ki} M_{\lambda_k}\left(\frac{\theta_k}{\hat{\sigma}}\right), \quad i = 1,\ldots,n$$

and

$$\text{Var}(\hat{f}_i) \approx \hat{\sigma}^2 \sum_k w_{ki}^2 V_{\lambda_k}\left(\frac{\theta_k}{\hat{\sigma}}\right), \quad i = 1,\ldots,n,$$

where M_λ and V_λ are given in (6.18) and (6.19).

Under certain conditions, Brillinger [38] showed that, for each i, $\hat{f}_i$ is asymptotically Gaussian; see also Bruce and Gao [43]. Standard errors for the shrinkage estimator can be estimated by

$$s_i = \hat{\sigma}\sqrt{\sum_k w_{ki}^2 V_{\lambda_k}\left(\frac{\hat{\theta}_k}{\hat{\sigma}}\right)}.$$

For instance, an approximate 95% (pointwise) confidence interval for $\hat{f}$ is

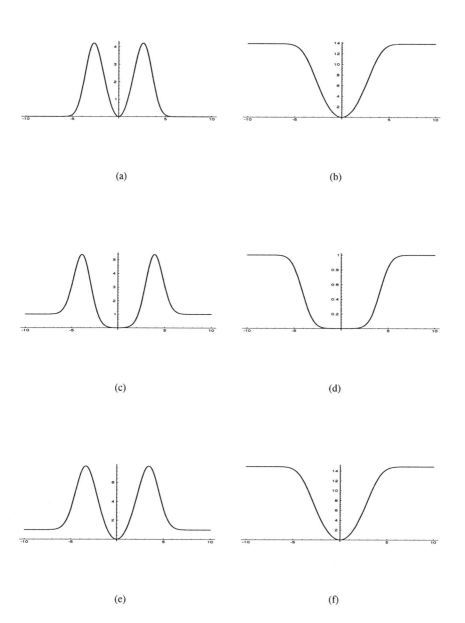

Fig. 6.8 Squared bias, (a) and (b); variance, (c) and (d); and risk, (e) and (f), for hard- and soft-thresholding rules, respectively.

$[\hat{f}_i - 2s_i, \hat{f}_i + 2s_i]$. Also, an estimator of the bias is

$$\text{bias}(\hat{f}_i) \approx \hat{\sigma} \sum_k w_{ki}^2 \left[M_{\lambda_k}\left(\frac{\hat{\theta}_k}{\hat{\sigma}}\right) - \frac{\theta_k}{\hat{\sigma}} \right].$$

6.5.3 Some Other Shrinkage Rules

In addition to hard- and soft-thresholding rules, a range of different shrinkage rules have been proposed.

Gao and Bruce [167] generalize hard- and soft-thresholding rules by introducing *semisoft* or *firm* shrinkage. The semisoft-thresholding rule depends on two non-negative parameters, λ_1 and λ_2. The rule is defined as

$$\delta^{ss}(x, \lambda_1, \lambda_2) = \text{sgn}(x)\frac{\lambda_2(|x| - \lambda_1)}{\lambda_2 - \lambda_1}\mathbf{1}(\lambda_1 < |x| \leq \lambda_2) + x\mathbf{1}(|x| > \lambda_2),$$

and plotted in Fig. 6.9(a). Note, that

$$\lim_{\lambda_2 \to \infty} \delta^{ss}(x, \lambda_1, \lambda_2) = \delta^s(x, \lambda_1),$$

and

$$\lim_{\lambda_2 \to \lambda_1} \delta^{ss}(x, \lambda_1, \lambda_2) = \delta^h(x, \lambda_1).$$

By selecting appropriate λ_1 and λ_2, Gao and Bruce [167] demonstrate that semisoft-thresholding rule has, uniformly in θ, smaller risk than the hard-thresholding rule. More precisely, for any given λ there exist λ_1 and λ_2, $\lambda_1 < \lambda < \lambda_2$ such that for all θ

$$R^{ss}_{\lambda_1, \lambda_2}(\theta) < R^h_\lambda(\theta).$$

The risk, variance, and bias functions of semisoft-thresholding rules have different shapes than those of hard- or soft-thresholding rules (see Exercises 6.9 and 6.10). Fan [145] shows that the semisoft-thresholding rule can be obtained as a solution of a variational problem, as in (6.12) for a slightly more complicated penalty function.

The non-negative garrote (*nn*-garrote) thresholding function [Fig. 6.9(b)] is given by

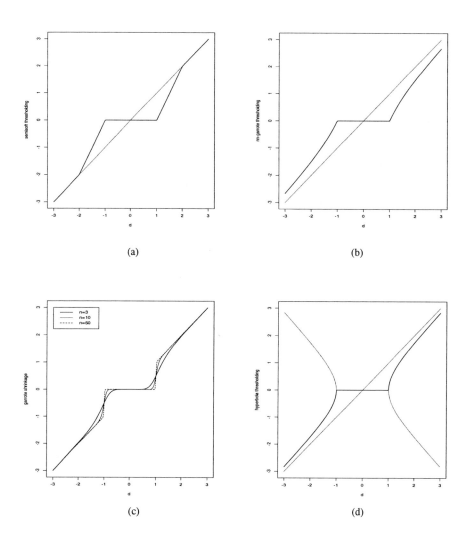

Fig. 6.9 (a) Semisoft rule with $\lambda_1 = 1$ and $\lambda_2 = 2$, (b) *nn*-garrote rule with $\lambda = 1$; (c) n-degree garrote, for $\lambda = 1$, and $n = 3, 5$, and 50; and (d) hyperbole rule, see Exercise 6.4.

$$\delta^{nng}(x,\lambda) = \left(x - \frac{\lambda^2}{x}\right)\mathbf{1}(|x| > \lambda).$$

Gao [166] demonstrated, by simulation, that nn-garrote thresholding has advantages over both hard and soft thresholding in terms of risk and sensitivity to small perturbations in the data.

Also, if $R_{oracle}(\Theta) = \sigma^2/2^J + \frac{1}{n}\sum_{j=1}^{J}\sum_k (\theta_{jk}^2 \wedge \sigma^2)$ (the usual oracular risk is increased by $\sigma^2/2^J$ because σ is estimated) is the oracle furnished risk than for an appropriately chosen $\alpha \in (0,1]$, the nn-garrote rule also achieves the ideal risk, within a factor of $\log n$, i.e.,

$$R(\hat{f}_\lambda, f) \leq C \log n \left[R_{oracle}(\Theta) + O\left(\frac{1}{n^\alpha \sqrt{\log n}}\right)\right].$$

The exact risk function for the nn-garrote rule is

$$\begin{aligned} R_\lambda^{nng}(\theta) &= E(\delta^{nng}(X,\lambda) - \theta)^2 \\ &= R_\lambda^h(\theta) + 2\theta\lambda^2 A_\lambda(\theta) + \lambda^4 B_\lambda(\theta) \\ &\quad - 2\lambda^2[1 - \Phi(\lambda - \theta) + \Phi(-\lambda - \theta)], \end{aligned}$$

where

$$\begin{aligned} A_\lambda(\theta) &= \int_\lambda^\infty \frac{\phi(x-\theta) - \phi(x+\theta)}{x}\,dx \text{ and} \\ B_\lambda(\theta) &= \int_\lambda^\infty \frac{\phi(x-\theta) + \phi(x+\theta)}{x^2}\,dx. \end{aligned} \quad (6.20)$$

The n-degree garrote shrinkage rule was considered by Breiman [37] in the context of model selection (only for $n = 1$). The rule is defined as

$$\delta^g(x, \lambda) = \frac{x^{2n+1}}{\lambda^{2n} + x^{2n}}, \quad (6.21)$$

and is depicted in Fig. 6.9(c) for selected values of n. Note that the n-degree garrote rule converges pointwise to a hard-thresholding rule when $n \to \infty$ (except for $x = \pm\lambda$ in which the rule has constant value $\pm\frac{\lambda}{2}$).

Panel (d) in Fig. 6.9 gives a graph of the hyperbole rule, see Exercise 6.4 for more details.

6.6 HOW TO SELECT A THRESHOLD

There are many proposals for threshold selection when the shrinkage policy is thresholding. A comprehensive overview is given in Nason [308]. In this section, we discuss several popular choices: percentile, universal, sure, cross-validation, false-discovery-rate, Lorentz, and block thresholds.

6.6.1 Mallat's Model and Induced Percentile Thresholding

Mallat [276] noticed that, for a variety of signals and images, the distributions of wavelet coefficients appeared similar. Typically, the distributions were symmetric about zero and had a sharp peak at zero. Mallat proposed modeling of "typical wavelet coefficients" by distributions from the exponential power family $[\mathcal{EPD}(\alpha, \beta)]$ whose densities are given by

$$f(d) = C \cdot e^{-(|d|/\alpha)^\beta}, \quad \alpha, \beta > 0. \tag{6.22}$$

The normalizing constant C in (6.22) is given by $C = \frac{\beta}{2\alpha \Gamma(1/\beta)}$. Since

$$E|D| = \frac{\alpha \Gamma(2/\beta)}{\Gamma(1/\beta)} \quad \text{and} \quad ED^2 = \frac{\alpha^2 \Gamma(3/\beta)}{\Gamma(1/\beta)},$$

the parameters α and β can be estimated by the method of moments. The estimators are

$$\hat{\beta} = G^{-1}\left(\frac{m_1^2}{m_2}\right) \quad \text{and} \quad \hat{\alpha} = \frac{m_1 \Gamma(1/\hat{\beta})}{\Gamma(2/\hat{\beta})},$$

where $m_1 = \frac{1}{n} \sum |d_i|$ and $m_2 = \frac{1}{n} \sum d_i^2$ are empirical counterparts of $E|D|$ and ED^2, and $G(x) = \frac{\Gamma^2(2/x)}{\Gamma(1/x)\Gamma(3/x)}$. The model (6.22) can be used for designing a percentile-based thresholds. The induced percentile thresholding can be either hard or soft with $\lambda = q_{1-\alpha/2}$, where $q_{1-\alpha/2}$ is the $(1 - \alpha/2) \cdot 100$-percentile of the model distribution. When no model is assumed, a percentile based on the empirical distribution, $\lambda = \hat{q}_{1-\alpha/2}$, can be used.

For an empirical analysis of percentile wavelet thresholding in the context of image analysis, see Buccigrossi and Simoncelli [46], Simoncelli and Adelson [380], and Zeppenfeldt, Börger, and Koppes [472].

Next, we give an example.

Example 6.6.1 The function $y = \sin(x) + x\cos(2x)$ is evaluated at 4096 equally spaced points in $[-3, 3]$. Three noisy versions of this function are generated by adding random normal $\mathcal{N}(0, 0.5^2)$, uniform $\mathcal{U}(0, 1)$, and t_5 noise, respectively.

After applying the DAUB4 wavelet transformation to the noisy signals, the parameters in (6.22) are estimated utilizing detail coefficients only. The estimators for

Table 6.1 Empirical moments and estimators of parameters for three simulated noisy data sets.

signal + noise	m_1	m_2	$\hat{\alpha}$	$\hat{\beta}$	$\hat{C}$
signal + $\mathcal{N}(0, 0.5^2)$	0.46693	1.41488	0.00380	0.30458	15.2
signal + $\mathcal{U}(0, 1)$	0.30288	1.51483	0.00002	0.19976	205.9
signal + t_5	1.05253	2.766	0.52098	0.69535	0.76

$E|D|, ED^2, \alpha, \beta$, and C, are given in Table 6.1.

For the signal with normal noise, the empirical model gives $\hat{q}_{0.975} = 1.97$. In other words, by thresholding with $\lambda = 1.97$ we retain approximately 5% of the detail coefficients in the decomposition.

6.6.2 Universal Threshold

We have already seen in Section 6.3 that hard- and soft-thresholding rules with threshold $\lambda = \sqrt{2 \log n}\, \sigma$ can asymptotically approach the "oracular" risk. Donoho and Johnstone [126] call λ the universal threshold.

In addition to its good asymptotic minimax properties, the universal threshold removes noise with high probability contributing to the visual quality of reconstructed signals. Such noise-cleaning arguments are based on probabilistic results concerning the maximum of n identically distributed Gaussian random variables. We start with an example in which the noise is uniformly bounded.

Example 6.6.2 To motivate the universal threshold, let us consider a slightly modified setup (6.4). Let $\underline{d} = \underline{\theta} + \sigma\underline{\epsilon}$, where $\underline{d} = (d_1, \ldots, d_n), \underline{\theta} = (\theta_1, \ldots, \theta_n)$, and $\underline{\epsilon} = (\epsilon_1, \ldots, \epsilon_n)$. Suppose that the random variables ϵ_i are uniformly bounded, $|\epsilon_i| \leq 1$, so that σ determines the size of the noise.

Let $\hat{\theta}_i = d_i \mathbf{1}(|d_i| \geq \lambda)$ and let $\lambda = \sqrt{2}\sigma$.
Then,

$$R(\hat{\theta}, \theta) \leq C \cdot R(DP, \theta), \qquad (6.23)$$

where $C \leq 6$.

Indeed, when $|d_i| > \lambda$,

$$|\theta_i| = |d_i - \sigma\epsilon_i| \geq |d_i| - |\sigma\epsilon_i| \geq (\sqrt{2} - 1)\sigma.$$

Then, $\sigma \leq (\sqrt{2} + 1)|\theta_i|$.

When $|d_i| \leq \lambda$,
$$|\theta_i| \leq |d_i| + \sigma|\epsilon_i| \leq (\sqrt{2}+1)\sigma.$$

Therefore, $|\hat{\theta}_i - \theta_i|$ is bounded by both $(\sqrt{2}+1)\sigma$ and $(\sqrt{2}+1)|\theta_i|$. By taking squares and summing, we obtain (6.23). The risk is up to a multiplicative constant close to the oracular risk.

If the noise is normal, then we do not know its size. Nevertheless we can assume that the noise is "bounded" asymptotically, $|\epsilon_i| \leq \sqrt{2\log n}$.

Several theoretical results on the maximum of n Gaussian random variables can be found in the literature. We state the result of Pickands [341] because of its generality.

Theorem 6.6.1 *Let $X_1, X_2, \ldots, X_n, \ldots$ be a stationary Gaussian process such that $EX_i = 0, EX_i^2 = 1$, and $EX_i X_{i+k} = \gamma(k)$. If $\lim_{k \to \infty} \gamma(k) = 0$, then $X_{(n)}/\sqrt{2\log n} \to 1$, a.s. when $n \to \infty$.*

Results of Theorem 6.6.1 are applicable to dependent normal noise. Some discussion for cases when the noise is dependent and/or non-normal is given in Section 6.7.

If the random variables $X_1, \ldots, X_n, \ldots$ are independent standard normals, then for large n, $P\left(|X_{(n)}| > \sqrt{c\log n}\right) \sim \frac{\sqrt{2}}{n^{c/2-1}\sqrt{c\pi \log n}}$, so the universal threshold

$$\lambda = \sqrt{2\log n}\, \hat{\sigma}, \qquad (6.24)$$

ensures with high probability that no noise is present in the data after thresholding.

There are several standard choices for the estimator $\hat{\sigma}$. Almost all methods involve the wavelet coefficients at fine scales. Often, only the finest scale is used to estimate the variance of noise. The signal-to-noise ratio (SNR)[2] is usually small at high resolutions and, if the signal is not too irregular, the finest scale should contain mainly noise. Moreover, the finest scale contains 50% of all coefficients.

Some standard estimators of σ are

$$s = \sqrt{\frac{1}{n/2-1}\sum_{i=1}^{n/2}\left[d_i^{(J-1)} - \bar{d}^{(J-1)}\right]^2}, \qquad (6.25)$$

or a more robust MAD (median absolute deviation from the median)

---

[2]The SNR is defined as a ratio of standard deviations of signal and noise. In the signal processing community, SNR is defined as $10\log_{10}\sigma^2/\sigma_n^2$, where σ^2 is the variance of the input signal and σ_n^2 is the variance of the estimated noise. The SNR defined via logarithms is measured in *decibels, dB*.

$$\hat{\sigma} = 1/0.6745 \cdot MAD[d^{(J-1)}]$$
$$= 1.4826 \cdot \text{MEDIAN}[|d^{(J-1)} - \text{MEDIAN}(d^{(J-1)})|],$$

where $d^{(J-1)}$ is the vector of finest detail coefficients associated to the multiresolution subspace W_{J-1} (if the original data "reside" in V_J). The components and the mean of $d^{(J-1)}$ are denoted by $d_i^{(J-1)}$ and $\bar{d}^{(J-1)}$, respectively.

In some situations, especially when (i) the data sets are large, and (ii) σ is overestimated, the universal thresholding tends to give under-fitted models. See discussion in Section 6.6.5.

Example 6.6.3 In this example we consider universal thresholding of the doppler signal. The setup is identical to that in Donoho and Johnstone [126]. The signal [$n = 2048$ ($J = 11$)] is scaled so that the SNR = 7 for a standard normal noise. The decomposing wavelet is SYMM8 and the boundary handling is periodic. The discrete wavelet decomposition is of depth 7, i.e., the coarsest detail space is W_4 (corresponding to $d7$ in S+WAVELETS.)

The universal threshold is $\lambda = 3.93$ and only coefficients from $W_{10} - W_5$ (corresponding to $d1 - d6$) are thresholded using the hard-thresholding rule in (6.10).

Fig. 6.10 illustrates the example. The noisy signal is given in panel (a), the threshold and wavelet coefficients from levels $d3$ and $d5$ are depicted in panels (b) and (c), and the reconstructed signal is in panel (d). Notice that only 8 coefficients "survive" thresholding in each of $d3$ and $d5$.

The average "energy" of the original signal is $\frac{1}{n}\|y\|_{\ell_2}^2 = 50.3479$, the sample standard deviation of the noise is 1.0118 [estimated using (6.25)], and the MSE of the estimator $\hat{y}$ is $\frac{1}{n}\|\hat{y} - y\|_{\ell_2}^2 = 0.11104$.

Small MSE is certainly desirable; however, it is often the case that reconstructions with the minimum MSE possess undesirable artifacts. Such artifacts æstetically distort the reconstruction, while contributing little to the MSE.

The combination of the universal threshold and soft thresholding policy is an early proposal by Donoho and Johnstone [126] known as *VisuShrink*. The feature of VisuShrink is that it *guarantees* a noise free reconstruction, but in doing so it often underfits the data by setting the threshold conservatively high. Such behavior of VisuShrink was observed by many researchers, notably by Fan et al. [148] and Nason [308].

The *optimal minimax threshold* has been proposed by Donoho and Johnstone [126] as an improvement to the universal threshold. For small-to-moderate sample sizes, the optimal minimax threshold, λ_m, can be substantially smaller than the universal threshold. If the rule is $\delta^s(\bullet, \lambda_m)$, the bound (6.14) in Theorem 6.3.1 can be sharpened, i.e., the factor $(2 \log n + 1)$ can be replaced by much smaller Λ^*.

For the standard normal noise, Table 6.2 (adapted from [124]) compares the universal and optimal minimax thresholds.

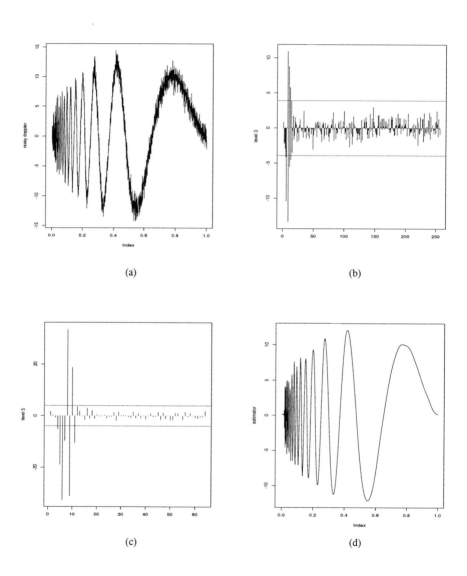

Fig. 6.10 Hard thresholding with the universal threshold. (a) Noisy signal, SNR = 7. (b) Wavelet coefficients from the level $d3$ (W_8) and universal threshold. (b) Wavelet coefficients from the level $d5$ (W_6). (d) Reconstructed signal.

Table 6.2 The optimal minimax threshold and related bounds

n	λ_m	$\sqrt{2\log n}$	Λ^*	$2\log n + 1$
64	1.474	2.884	3.124	8.318
256	1.669	3.330	4.442	11.090
1024	2.232	3.723	5.976	13.863
4096	2.594	4.079	7.728	16.635
16384	2.952	4.405	9.715	19.408

6.6.3 A Threshold Based on Stein's Unbiased Estimator of Risk

Donoho and Johnstone [128] developed a technique of selecting a threshold by minimizing Stein's unbiased estimator of risk (Stein, [386]). The original paper of Donoho and Johnstone was drafted in 1990. This threshold is implemented in an adaptive denoising procedure, *SureShrink*. The adaptation in SureShrink is achieved by specifying thresholds level-wise.

The following result gives the theoretical background for the threshold selection.

Theorem 6.6.2 *Let* $d_i \overset{i.i.d.}{\sim} \mathcal{N}(\theta_i, 1)$, $i = 1, \ldots, k$. *Let* $\hat{\underline{\theta}}$ *be an estimator of* $\underline{\theta} = (\theta_1, \ldots, \theta_k)$. *If the function* $\mathbf{g} = \{g_i\}_{i=1}^{k}$, *in the representation* $\hat{\underline{\theta}}(\underline{d}) = \underline{d} + \mathbf{g}(\underline{d})$, *is weakly differentiable, then*

$$E^\theta \|\hat{\underline{\theta}} - \underline{\theta}\|^2 = k + E^\theta \left\{ \|\mathbf{g}(\underline{d})\|^2 + 2\nabla \mathbf{g}(\underline{d}) \right\}, \quad (6.26)$$

where $\nabla \mathbf{g} = \sum_{i=1}^{k} \{\frac{\partial}{\partial d_i} g_i\}$.

It is interesting that the estimator $\hat{\underline{\theta}}$ in (6.26) can be nearly arbitrary; for instance, it can be biased, non-linear, and so on. The application of (6.26) to $\delta^s(\underline{d}, \lambda)$ gives

$$\mathrm{SURE}(\underline{d}, \lambda) = k - 2\sum_{i=1}^{k} \mathbf{1}(|d_i| \leq \lambda) + \sum_{i=1}^{k}(|d_i| \wedge \lambda)^2, \quad (6.27)$$

as an unbiased estimator of risk, i.e.,

$$E\|\delta^s(\underline{d}, \lambda) - \underline{\theta}\|^2 = E\ \mathrm{SURE}(\underline{d}, \lambda).$$

When k is large, the law of large numbers (LLN) argument states that SURE is close to its expectation, motivating the following threshold selection:

$$\lambda^{sure} = arg \min_{0 \leq \lambda \leq \lambda^U} \mathrm{SURE}(\underline{d}, \lambda). \quad (6.28)$$

The hard-thresholding rule is discontinuous and does not have a bounded weak derivative.

The threshold λ^{sure} and the soft-thresholding rule are the core of the level-dependent procedure Donoho and Johnstone call the *SureShrink*. If the wavelet representation at a particular level is not too sparse, the SURE threshold is used. Otherwise, the universal threshold is selected. The level j is considered sparse if

$$s_j^2 \leq \frac{1}{\sqrt{n_j}} \log_2 n_j^{3/2},$$

where n_j is the number of coefficients in the jth level, and $s_j^2 = \frac{1}{n_j} \sum_{k=1}^{n_j}(d_{jk}^2 - 1)$. For details see Donoho and Johnstone [128].

It is possible to derive thresholds of SURE-type for $\delta^{ss}, \delta^{nng}$, and so on, but the simplicity of the representation (6.27) is lost. Donoho and Johnstone demonstrate that the minimum in (6.28) is achieved at one of the $|d_i|$. Since there are only k such values, the algorithm to calculate λ^{sure} is fast. The computational complexity of their algorithm is $O(k \log k)$.

SureShrink is related to Akaike's AIC criterion; see a discussion in Ogden [321], page 146.

6.6.4 Cross-Validation

Cross-validation is a classical statistical procedure used in different statistical settings. For example, in density estimation or in spline smoothing, cross-validation provides an automatic procedure for choosing the bandwidth or a smoothing parameter. Some general references are Burman [48], Hall and Koch [182], and Wahba [435]. Several researchers applied cross-validation techniques in the wavelet domain.

Nason [309] applied cross-validation to the problem of threshold selection. His method utilizes the standard paradigm: minimize the prediction error generated by comparing a prediction, based on part of the data, to the remainder of the data.

We give a brief overview of Nason's twofold cross-validation procedure. More details can be found in [308, 309]. Let $y_1, y_2, \ldots, y_n$ (n is an integral power of 2) be the observations. Let $\hat{f}_\lambda^E$ be the wavelet threshold estimate based on the re-indexed subsample $y_2, y_4, \ldots, y_{n/2}$. Let $\bar{f}_{\lambda,j}^E$ be an interpolated estimator, defined as

$$\bar{f}_{t,j}^E = \begin{cases} (\hat{f}_{\lambda,j+1}^E + \hat{f}_{\lambda,j}^E)/2, & j = 1, 2, \ldots, \frac{n}{2} - 1 \\ (\hat{f}_{\lambda,1}^E + \hat{f}_{\lambda,n/2}^E)/2, & j = \frac{n}{2} \end{cases}. \quad (6.29)$$

Let $\hat{f}_\lambda^O$ be the counterpart to $\hat{f}_\lambda^E$, computed using the odd-indexed points and let the interpolants $\bar{f}_{\lambda,j}^O$ be computed as in (6.29).

A cross-validatory estimate of the mean-square error is

$$\hat{M}(\lambda) = \sum_{j=1}^{n/2} [(\bar{f}_{\lambda,j}^E - y_{2j+1})^2 + (\bar{f}_{\lambda,j}^O - y_{2j})^2].$$

The periodicity in data is assumed and $y_{n+1} = y_1$. The value $\lambda^\star$ that minimizes $\hat{M}(\lambda)$ is based on $\frac{n}{2}$ points (both estimates of f, $\hat{f}_\lambda^E$ and $\hat{f}_\lambda^O$ are based on $\frac{n}{2}$ points), and a correction for sample size is needed. The adjustment is made by a heuristic comparison with the universal threshold, see Exercise 6.3.

With this correction, Nason's cross-validatory threshold is

$$\lambda_{opt} = \left(1 - \frac{\log 2}{\log n}\right)^{-1/2} \lambda^\star. \tag{6.30}$$

Nason [309] showed that one can almost always find a unique minimizer of $\hat{M}(\lambda)$ and compared the performance of the cross-validatory threshold to the Donoho-Johnstone universal and SureShrink methods. He also reported that in the case of heavy-tailed noise the described cross-validation method did not perform well.

Wang [448] addresses the problem in which the noise is correlated and exhibits long-range dependence. By using the wavelet-vaguelette decomposition (Donoho [119], Kolaczyk [238], also Section 11.2) data can be nearly decorrelated. However, due to level-wise heterogeneity of variances in the wavelet domain an adjustment to the cross-validation procedure of Nason-type is necessary.

Weyrich and Warhola [456] (see also Amato and Vuza [9]) use generalized cross-validation criterion, previously proposed by Wahba [434] in the context of spline approximations. Let y be the data, W be the matrix of wavelet transformation, and $\tilde{D}_\lambda = \mathrm{diag}(1, 1, \ldots, 1, \tilde{\partial}_{\lambda, J-k}, \ldots, \tilde{\partial}_{\lambda, J-2}, \tilde{\partial}_{\lambda, J-1})$. The vectors $\tilde{\partial}_{\lambda,j}$ in $\tilde{D}_\lambda$ have entries $\mathbf{1}(|d_i| > \lambda)$ and their pattern matches the pattern in $W\underline{y} = (c^{(J-k)}, d^{(J-k)}, d^{(J-k+1)}, \ldots, d^{(J-1)})$. The sub-vector of 1's in $\tilde{D}_\lambda$ corresponds to smooth coefficients $c^{(J-k)}$, which are not shrunk. The generalized cross-validation function is

$$\mathrm{GCV}(\lambda) = \frac{\frac{1}{n}\|\underline{y} - \underline{y}_\lambda\|^2}{\left[\frac{1}{n}\mathrm{tr}(I - A)\right]^2}, \tag{6.31}$$

where $A = W^{-1}\tilde{D}_\lambda W$ is so-called *influence* matrix, $\underline{y}_\lambda = A\underline{y}$, and $\mathrm{tr}(M)$ is the trace of a matrix M. The cross-validation value of λ^* is defined as

$$\arg\min_{\lambda \in \mathbb{R}^+} E[\mathrm{GCV}(\lambda)].$$

An approximation, justified by the LLN, can be found by direct minimization of

(6.31). The simulation results provided by the authors are encouraging. Asymptotic optimality of Weyrich and Warhola type procedures was established by Jansen, Malfait, and Bultheel [219].

6.6.5 Thresholding as a Testing Problem

The process of thresholding of wavelet coefficients can be viewed as a testing problem. For each wavelet coefficient $d_i = \theta_i + \sigma\epsilon_i$, the hypothesis $H_0 : \theta_i = 0$ is tested against the alternative $H_1 : \theta_i \neq 0$. If the hypothesis H_0 is rejected, the coefficient d_i is retained in the model. Otherwise, it is discarded.

The universal threshold can be viewed as a critical value of a similar test in which the level is

$$\begin{aligned}
\alpha &= P(|d_i| > \sqrt{2\log n}\,\sigma\,|H_0) \\
&= 2\Phi(-\sqrt{2\log n}) \\
&\approx (n\sqrt{\pi \log n})^{-1} \quad [\text{ because } \Phi(-x) \approx \phi(x)/x \text{ when } x \text{ is large }].
\end{aligned} \quad (6.32)$$

The power of the test against the alternative $H_1 : \theta_i = \theta(\neq 0)$ is $O\left(\frac{1}{n\sqrt{\pi \log n}}\right)$ as well; see Exercise 6.11.

In testing n statistical hypotheses simultaneously, the Bonferroni procedure guarantees that the overall level of the omnibus test is α by setting the levels for individual hypotheses to be $\frac{\alpha}{n}$. Bonferroni methods are not used much in practice, since for large n the level $\frac{\alpha}{n}$ becomes unduly small resulting in loss of "strictness".

The universal thresholding is equivalent to a Bonferroni-type test. It controls the probability of even one erroneous inclusion of a coefficient. The approximate level of $(n\sqrt{\pi \log n})^{-1}$ tends to zero as n increases. Subsequent severe decreases in power are compensated by accepting almost all H_0, leading to severe underfitting, see related comments on page 197.

One way to control such dissipation of power is proposed by Abramovich and Benjamini [1, 2]. The proposal is based on the *false discovery rate* (FDR) method of Benjamini and Hochberg [25].

Let R be the number of wavelet coefficients that "survived" thresholding. If S of them are correctly kept, then $V = R - S$ are erroneously kept. The error in such a procedure can be expressed in terms of the random variable $Q = V/R$. The false discovery rate of coefficients is the expectation of Q; that is, the expected portion of coefficients erroneously kept. Following Benjamini and Hochberg [25], Abramovich and Benjamini [1] propose maximizing the number of coefficients kept, subject to condition $EQ \leq \alpha$. Their procedure has three steps:

- **STEP 1.** For each d_{jk} find its two-sided p-value, p_{jk} in testing $H_0 : \theta_{jk} = 0$,

$$p_{jk} = 2\left[1 - \Phi\left(|d_{jk}|/\sigma\right)\right].$$

- **STEP 2.** Order the p_{jk} according to their size, $p_{(1)} \leq p_{(2)} \leq \cdots \leq p_{(n)}$. Find $k = \max\{i|\ p_{(i)} < (i/n) \cdot \alpha\}$. For this k calculate

$$\lambda_k = \sigma \Phi^{-1}\left(1 - p_{(k)}/2\right).$$

- **STEP 3.** Threshold wavelet coefficients at level λ_k.

If ϵ is Gaussian, the above procedure ensures the FDR to be below α. If the input is pure noise (all $\theta_i = 0$), then controlling the FDR implies the control of the probability of including erroneously even one coefficient (Bonferroni's approach). Indeed, Q is then $Ber(p)$, $p = P(V \geq 1)$. In orthogonal regression, using FDR can give substantially smaller thresholds. When the model is sparse, FDR-like selection yields estimators with strong large sample adaptivity properties. A balancing of FDR and unbiased risk methods provides asymptotic minimax MSE adaptivity of broad generality.

Ogden and Parzen [324, 325] propose the determination of thresholds via recursive likelihood ratio tests. They find the asymptotic distribution for the statistic T representing the cumulative energy of the observed wavelet coefficients under consideration. If T is significant, the coefficient with largest energy is adopted into the model, and the test is repeated on the remaining cumulative energy. Of course, a new statistic T and the critical level have to be calculated at each iteration. The procedure is repeated until T is found not significant, and the observed coefficients constituting T are replaced by 0 in the model. Ogden and Parzen show that the overall power of their procedure remains high even if n is moderately large. For two-dimensional implementations see Ogden and Hilton [322].

6.6.6 Lorentz Curve Thresholding

Based on the Lorentz curve for the distribution of the energy of wavelet coefficients, it is possible to propose a thresholding criteria. The thresholding method is distribution-free, i.e., no assumptions on the distribution of noise are imposed. We first give a definition of the Lorentz curve for continuous random variables.

Let X be a nonnegative random variable such that $EX = \mu < \infty$. It is assumed that

(**A1**) the c.d.f. F is a strictly monotone function on the support of X, and

(**A2**) F is continuous at μ.

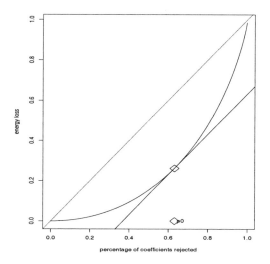

Fig. 6.11 Finding $\hat{p}_0$.

The Lorentz curve corresponding to X is defined as

$$L(p) = \frac{1}{\mu} \int_0^{\xi_p} x\, dF(x), \qquad (6.33)$$

where ξ_p is the (unique) population pth quantile. Distributions with finite mean uniquely define their Lorentz curve, and vice versa. The scale-free character of $L(p)$ allows development of goodness-of-fit tests for families of distributions depending on an unknown scale parameter (see Gail and Gastwirth [162]).

Example 6.6.4 For a uniform distribution, $L(p) = p^2$. For an exponential distribution, $L(p) = p + (1-p)\log(1-p)$. For a Pareto distribution, $[f(x) = ak^a x^{-(a+1)}, a,k > 0,\ x \geq k]$, $L(p) = 1 - (1-u)^{(a-1)/a}$, $a > 1$.

Let p_0 be the argument at which the distance between the Lorentz curve and the line $L(p) = p$ is maximum [Fig. 6.11].

The following lemma holds.

Lemma 6.6.1 *Under assumption* (**A1**),

$$p_0 = F(\mu).$$

Proof: The assumption (**A1**) is needed to ensure uniqueness of the quantile ξ_p. A change of variables in the integral (6.33) gives $L(p) = \frac{1}{\mu}\int_0^p \xi_t\, dt$. By solving the equation $(p - L(p))' = 0$, one gets $\mu = \xi_p$, i.e., $p_0 = F(\mu)$. An empirical counterpart of $p_0 = F(\mu)$ is

$$\hat{p}_0 = F_n(\bar{X}) = \frac{1}{n} \sum \mathbf{1}(X_i \leq \bar{X}).$$

The following theorem gives an almost sure convergence result for $\hat{p}_0$.

Theorem 6.6.3 *Under assumptions* **(A1)** *and* **(A2)**,

$$\hat{p}_0 \to p_0, \quad \text{a.s.}$$

Proof:
$$|F_n(\bar{X}) - F(\mu)| \leq |F_n(\bar{X}) - F_n(\mu)| + |F_n(\mu) - F(\mu)|.$$

Assumption **(A2)** implies that $|F_n(\mu) - F(\mu)| \to 0$, and the Glivenko-Cantelli theorem gives $|F_n(\bar{X}) - F_n(\mu)| \to 0$, a.s., proving the theorem.

Lorentz Curve Thresholding Rule: Threshold (replace by 0) the $\hat{p}_0 \cdot 100\%$ coefficients with smallest energy.

The point $\hat{p}_0$ is the proportion at which the gain (in parsimony) by thresholding an additional element is smaller than the loss in the energy. Both losses are measured on a scale 0–1, and are equally weighted (Fig. 6.11).

Remark 6.6.1 For different weights of losses (loss in energy is k times more serious than the gain in parsimony) one can generalize the above rule as follows: For $k > 0$ one maximizes $(kp - L(p))$ instead of $(p - L(p))$. The solution is $p_0 = F(k\mu)$ and the sample counterpart of p_0 is $\hat{p}_0 = F_n(k\bar{X}) = \frac{1}{n} \sum \mathbf{1}(X_i \leq k\bar{X})$. The almost-sure convergence result, as in Theorem 6.6.3, holds.

The described thresholding procedure is equivalent to the hard thresholding for a particular threshold level λ.

In Lorentz thresholding, only those coefficients d_i for which $d_i^2 \leq \bar{d}^2$ are thresholded. With $\bar{d}^2$ we denote $\frac{d_1^2 + d_2^2 + \cdots + d_n^2}{n} = \frac{\|\underline{d}\|^2}{n}$. It is straightforward to show that the hard Lorentz thresholding function is

$$\delta(d_i) = d_i \cdot \mathbf{1}\left(|d_i| \leq \|\underline{d}\|/\sqrt{n}\right).$$

The orthogonality of the wavelet transformations implies that $\lambda = \frac{\|y\|}{\sqrt{n}}$. Consequently, the distribution of λ will be known when the distribution of the input data y is known. For instance, if the y_i's are i.i.d. random variables with a finite second moment μ_2, then $\lambda \to \sqrt{\mu_2}$, a.s.

The "soft" thresholding of the energy (when the largest rejected energy value is subtracted from the non-thresholded elements) corresponds to the hyperbole thresholding, see Exercise 6.4.

Lorentz thresholding proved useful in problems of separating attached and detached eddy motions in turbulence signals, see Section 11.5. A related threshold selection procedure is the "top n" ("top $p\%$"). The threshold is determined so that the n coefficients ($p\%$ of the coefficients) with largest energies survive the thresholding.

6.6.7 Block Thresholding Estimators

Most of the standard wavelet methods achieve adaptivity through term-by-term thresholding of the empirical wavelet coefficients. Other coefficients have no influence on the treatment of a particular coefficient. Block thresholding rules shrink wavelet coefficients in groups rather than individually. Simultaneous decisions are made to retain or to discard all the coefficients within a block. For example, all coefficients in a block are included or excluded simultaneously depending on a function of their magnitudes. Performance of the block thresholding method can be superior to that of its term-by-term counterparts, since the block thresholding addresses the dependence of the neighboring coefficients in the wavelet domain. Block thresholding methods are shown to reduce the bias and to react more rapidly to sudden frequency changes in the signal. However, it was demonstrated that some block thresholded estimators are more sensitive with respect to the selection of threshold. The assignment of coefficients to blocks is also ad hoc and may call for some type of spinning procedure (see page 208) in order to reduce the effect of blocking.

Block thresholding was the subject of research of several authors. Cai [53] studied block thresholding via the approach of ideal adaptation with an oracle. Based on an oracle inequality, Cai proposed the following procedure:

- **STEP 1.** Transform the data into the wavelet domain via the discrete wavelet transformation: $\underline{d} = W \cdot \underline{y}$.
- **STEP 2.** At each resolution level j, group the empirical wavelet coefficients into disjoint blocks (jb) of length $L = \lfloor \log n \rfloor$.
- **STEP 3.** Within each block (jb), estimate the coefficients simultaneously via a James-Stein-type shrinkage rule

$$\hat{\theta}_{jk} = (1 - \lambda_* L \sigma^2 / S_j^2)_+ \cdot d_{jk}, \quad \text{for all } j, k \in (jb).$$

where

$$S_j^2 = \sum_{j,k \in (jb)} d_{jk}^2.$$

- **STEP 4.** Obtain the estimate of the function via the inverse discrete wavelet transformation of the processed coefficients. This estimate is called a *BlockJS* estimate.

The thresholding constant λ_* is set to $\lambda_* = 4.505$, which is an approximate solution of the equation $\lambda - \log \lambda = 3$. The value λ_* is derived from an oracle inequality introduced in Cai [53].

The estimator then enjoys good numerical performance and asymptotic optimality.

More specifically, Cai [53] showed the following asymptotic result.

Theorem 6.6.4 *Let ψ be an r-regular wavelet and let $\mathbb{B}_{pq}^{\sigma}(M), M \in \mathbb{R}$ be a Besov ball. Then, the BlockJS estimator satisfies*

$$\sup_{f \in B_{p,q}^{\sigma}} E\|\hat{f}_n - f\|^2 \leq \begin{cases} Cn^{-\frac{2\sigma}{1+2\sigma}}(1+o(1)), \\ \qquad p \in [2, \infty], \sigma \in (0, r) \\ Cn^{-\frac{2\sigma}{1+2\sigma}}(\log n)^{(2/p-1)/(1+2p)}(1+o(1)), \\ \qquad p \in [1, 2), \sigma \in (1/p, r) \end{cases}$$

for all $0 < M < \infty$, $q \in [1, \infty]$, and $p \in [2, \infty]$.

Cai and Silverman [55] propose a different thresholding method that also incorporates information on neighboring coefficients into the decision making. The coefficients are considered in overlapping blocks; the treatment of coefficients in the middle of each block depends on the data in the whole block. They investigate asymptotic and numerical performances of two particular versions of the estimator. Cai and Silverman [55] show that, asymptotically, one version of the estimator achieves the exact optimal rates of convergence over a range of Besov classes for global estimation, and attains the adaptive minimax rate for estimating functions at a point. In numerical comparisons with various methods, both versions of the estimator perform excellently.

Cai [52] and Hall et al. [186] explore other types of block thresholding estimators. Pensky [331] and Efroimovich [139] consider global thresholding methods in which the blocks are whole levels (quazilinear procedures).

6.7 OTHER METHODS AND REFERENCES

The denoising procedures in the literature usually focus on a preselected basis. Donoho [121], Donoho and Johnstone [125] and Saito [360] address the problem of ideal denoising in an orthonormal basis selected from a library of bases.

The local Karhunen-Loève and local discriminant bases suitable for feature extraction and classification problems have been developed by Coifman and Saito [92, 93], Saito [359, 362], and Saito and Coifman [364, 365]. See also Saito and Coifman [366] and Saito [361] for applications in geology and geophysical acoustic.

Saito [360], Antoniadis, Gijbels, and Grégoire [14], and Krim and Pesquet [247] approach the problem of threshold selection from the *minimum description length* (MDL) standpoint. For example, according to Saito's approximate MDL principle, the k "top" coefficients enter in the model m, if the pair (k, m) minimizes

$$\text{AMDL}(k, m) = \frac{3k}{2} \log_2 n + \frac{n}{2} \log_2 \|\underline{d}_m - \underline{d}_m^*\|^2. \qquad (6.34)$$

In (6.34), n is sample size, $\underline{d}_m$ is the vector of wavelet coefficients in the basis $\mathcal{B}_m$

from a fixed library of bases, and $\underline{d}^*_m$ is the hard-thresholded vector $\underline{d}_m$ with exactly k non-zero components. See [14] for a discussion and results on the consistency of the approximate MDL principle.

Bruce et al. [41] introduce a robust *smoother-cleaner* wavelet transformation to address the problem of heavy-tailed noise. Before applying the usual wavelet filtering in each step of the wavelet decomposition, "smooth parts" are pre-processed to eliminate the "robust residuals." Donoho and Yu [135] provide some results on median based thresholding. See also Schick and Krim [370].

Translation invariant wavelet shrinkage (*cycle-spin method*) was introduced by Coifman and Donoho [88], see also Bruce, Gao and Stuetzle [44]. Given the model (6.2), the translation invariant shrinkage estimator is

$$\hat{f}_{TI} = \frac{1}{2^J} \sum_{k=0}^{2^J-1} (WS_k)' \delta(WS_k y), \qquad (6.35)$$

where W is a matrix of the wavelet transformation,

$$S_k = \begin{pmatrix} O_{k \times (n-k)} & I_{k \times k} \\ I_{(n-k) \times (n-k)} & O_{(n-k) \times k} \end{pmatrix}$$

is a shift matrix, and δ is the shrinkage operator. $I_{k \times k}$ is the identity matrix with k rows and $O_{r \times c}$ is an $r \times c$ matrix of zeroes.

An equivalent way to introduce translation invariant shrinkage is via nondecimated wavelet transformations, see Section 5.5. If W_{TI} is a matrix of nondecimated wavelet transformation [it maps a vector of length n into a vector of length $(J+1)n$], then (6.35) is equivalent to

$$\hat{f}_{TI} = W^{\#}_{TI} \delta(W_{TI} \cdot y) \qquad (6.36)$$

where $W^{\#}_{TI} = (W'_{TI} W_{TI})^{-1} W'_{TI}$ is a generalized inverse of W_{TI}.

Coifman and Donoho [88] discuss asymptotic properties of $\hat{f}_{TI}$, and Bruce, Gao, and Stuetzle [44] demonstrate superb performance of the cycle-spin method. See also Lang et al. [252].

Wavelet shrinkage procedures for more general noise were considered by many researchers. Johnstone and Silverman [222] consider correlated normal errors in the time domain, corresponding error distributions in the wavelet domain, and optimality of thresholding procedures in their setup. The level-dependent thresholds are chosen to minimize a data based unbiased risk criterion. Neumann and Spokoiny [314] apply the classical leave-out-one cross-validation method in non-Gaussian regression. Neumann and von Sachs [315] consider asymptotic normality of empirical wavelet coefficients and apply it to stationary and locally stationary noises. By considering tail probabilities of empirical wavelet coefficients, Patil [329] (for hazard rates), Gao [164] (for spectral densities), and Delyon and Juditsky [112] (global estimation problems), developed appropriate thresholds. By utilizing the extreme value theory, point processes, and the domain of attraction theory, Wu [465, 466] developed shrinkage procedures, that possess asymptotic minimax optimality properties for a wide class of error distributions.

Generalizations of WaveShrink-type techniques to nonequispaced (NES) designs impose additional conceptual and calculational burdens. There are several proposals on how to estimate a regression function by wavelets when the design is irregular. The simplest proposal is to carry out the analysis as if the data were equally spaced. This method is known as the *coercion to equal spacing*. Another class of wavelet-based methods applicable to nonequispaced data utilizes interpolation and averaging. Based on the available data, values of the function at equally spaced points are approximated; to such approximate values the standard wavelet methods are applied. References on approximation-type method include Deslauriers and Dubuc [113], Antoniadis, Grégoire, and McKeague [15], Antoniadis, Grégoire, and Vial [17], Jansen and Bultheel [217], Hall and Turlach [187], Brown and Cai [54], Foster [156], and Lenarduzzi [256], among others. Sardy et al. [367] consider the Haar basis and propose four approaches that extend the Haar wavelet transformation to NES data. Each approach is formulated in terms of continuous wavelet functions applied to a piecewise constant interpolation of the observed data, and each approach leads to wavelet coefficients that can be computed via a matrix transformation of the original data. Pensky and Vidakovic [334] propose a linear method that utilizes a probabilistic model imposed on the design in calculating the empirical coefficients c_{jk}. Here is a brief description of the method in [334].

Let

$$(X_1, Y_1), (X_2, Y_2), \ldots, (X_n, Y_n), \tag{6.37}$$

be a sample of size n. Let $f(x)$ be the marginal density of X and let $m(x) = E(Y|X = x)$ be the regression function to be estimated. Assume that $f(x)$ does not vanish in the interior of its support and $m(x) \in \mathbb{L}_2(\mathbb{R})$. Instead of estimating $m(x)$ directly we estimate its projection on a multiresolution subspace V_J, $m_J(x) = \text{Proj}_{V_J} m(x) = \sum_{k \in \mathbb{Z}} c_{J,k} \phi_{J,k}(x)$.

Let $\widehat{m}_n(x)$ be an estimator of $m(x)$, more precisely, of $\text{Proj}_{V_J} m(x)$,

$$\widehat{m}_n(x) = \sum_k \widehat{c}_{J,k} \phi_{J,k}(x), \tag{6.38}$$

with $\widehat{c}_{J,k}$

$$\widehat{c}_{J,k} = \frac{1}{n} \sum_{i=1}^{n} \left[\frac{\phi_{J,k}(X_i)\, Y_i \cdot \mathbf{1}(\widehat{f}_n(X_i|X_{-i}) \geq \delta_n)}{\widehat{f}_n(X_i|X_{-i})} \right]. \tag{6.39}$$

X_{-i} is the sample without the ith observation, i.e., $X_1, \ldots, X_{i-1}, X_{i+1}, \ldots, X_n$, δ_n is positive and goes to zero with an appropriate rate, and

$$\widehat{f}_n(x|X_{-i}) = \frac{1}{n-1} \sum_{j=1;\, j \neq i}^{n} \mathbb{K}_n(x, X_j),$$

where $\mathbb{K}_n(x, y)$ is a bounded kernel, symmetric in its arguments.

The motivation for (6.39) is the following representation of $c_{J,k}$:

$$\begin{aligned} c_{J,k} &= 2^{J/2} \int_{-\infty}^{\infty} \phi(2^J x - k)\, m(x)\, dx \\ &= \int_{-\infty}^{\infty} \frac{2^{J/2} \phi(2^J x - k)\, m(x)}{f(x)} f(x)\, dx \\ &= E\left[\frac{\phi_{J,k}(X)\, Y}{f(X)} \right]. \end{aligned} \tag{6.40}$$

Good MISE rates of convergence of (6.38) are possible if the design density f is reasonably smooth, see [334] for details.

Example 6.7.1 In this example we consider the test function `heavisine`, one of the four Donoho and Johnstone's standard test functions. Its functional form is

$$h(x) = 4\sin(4\pi x) - \text{sgn}(x - 0.3) - \text{sgn}(0.72 - x),\ 0 \leq x \leq 1.$$

The function was sampled at 256 points with the sampling density for X

$$f(x) = \begin{cases} 0.2, & 0.6 \leq x < 0.8 \\ 1.2, & 0 \leq x < 0.6 \text{ and } 0.8 \leq x \leq 1. \end{cases} \tag{6.41}$$

The histogram of the sampled values $X_1, \ldots, X_{256}$ is shown in Fig. 6.12(a). Fig. 6.12(b) depicts two reconstructions. The solid line corresponds to the coercion method while the dotted line is the NES estimator. Notice that the coercion method in this case has an overall erratic behavior, especially in the region [0.6, 0.8] where the sampling was scarce. Both methods missed the sharp spike present in the `heavisine` test function. The decomposing wavelet was DAUB6. The histogram procedure (built in S-Plus) was used to produce the pre-estimator $\widehat{f}_n$. The projection space was V_4, i.e., $J = 4$.

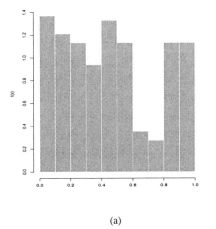 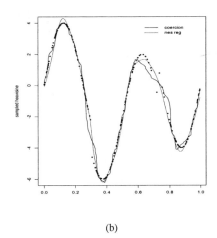

Fig. 6.12 (a) The histogram of the sampling distribution (6.41) for 256 points; (b) Graphs of the NES regression (dotted line) and coercion regression (solid line) for the `heavisine` data.

Nonlinear wavelet-based estimation of functions can be founded on grounds different than the thresholding/shrinkage paradigm. Based on a conjecture of Marr (see [282] and Chapter 8 in [295]) and the multiscale Canny edge-detection paradigm, Mallat and Zhong [280, 281] developed an algorithm for reconstructing a signal from local extrema of its (either continuous or stationary) wavelet coefficients. The authors demonstrate that the evolution of wavelet local extrema across scales well describes the local shapes and irregular structures in the signal. Such local maxima are visible as snake-like lines in the time/frequency plane.

An illustration of maximum moduli lines from which the signal can be reconstructed, is provided in Fig. 6.13. Panel (a) depicts a continuous wavelet transformation of the `doppler` signal. The signal was decomposed by the "Mexican hat" wavelet, see Example 3.1.3. The lines in panel (b) are local maxima of the absolute values (moduli) of the wavelet coefficients.

It comes as a surprise that reconstructions from "maximum moduli" lines are almost exact. Mallat and Zhong [280] state that for images, errors in maximum moduli reconstructions fall below visual sensitivity.

Carmona [60] proposes a variation on the Mallat-Zhong algorithm that approximates the signal directly in the time domain by utilizing splines and constrained approximation.

The interested reader can also consult Carmona, Hwang, and Torrésani [62, 63] for a description of singularities and for the use of ridges in maximum moduli reconstructions. From a statistical point of view, the methods based on ridges are

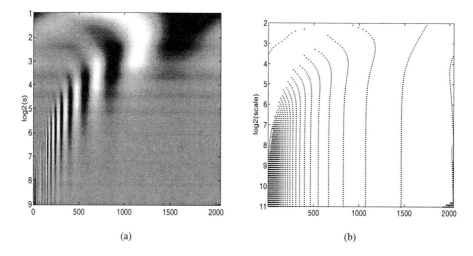

Fig. 6.13 (a) Continuous wavelet transformation of doppler signal. The size of the signal was $n = 2048$ and the decomposing wavelet was the "Mexican hat." (b) Maximum moduli lines.

interesting since they act as denoising procedures. These methods are typically insensitive to an additive white noise in the signal.

A comprehensive discussion of different algorithms performing maximum moduli reconstructions in a more general time/frequency environment and the matching software implementations can be found in Carmona, Hwang, and Torrésani [64]. See also Carmona [59] and Carmona and Hudgins [61] for some real-life applications of the maximum moduli methods.

Due to their remarkable property to "zoom in" on very short-lived phenomena in time/frequency, such as transients in signals or singularities in functions, wavelets provide an ideal tool to study localized changes.

Wang [447, 451] proposed a wavelet method to study change-points of functions in one and two dimensions. Wang's method requires no prior knowledge about the existence of change-points and can be applied to problems with an unknown number of change-points. In edge detection literature, it is almost the rule that methods that achieve optimal rates of convergence are computationally complex. On the other hand, practicable data processing algorithms are frequently suboptimal. In contrast, a wavelet-based method is optimal or nearly optimal and can be computed by fast algorithms. The change-point algorithm has also been employed in construction of the wavelet estimate of a function with jumps and cusps to reduce Gibbs errors and improve visual quality. In general the jump can be located more precisely by wavelet coefficients at higher resolution levels and at the same time detect the cusp by wavelet coefficients at lower resolution levels. See also related results in Wang [449].

Cristobal et al. [97] discuss the *sigmoidal* threshold — a variable threshold that exhibits good MSE and perceptual properties. The threshold is a function of the level j,

$$\lambda_j = \left(1 - \frac{1}{1+\exp\{n_j - j\}}\right) \cdot \sigma,$$

where n_j depends on the noise level in the scale j. The idea is to introduce a "soft" variation in the threshold values for improving the perceptual quality response. The suggested value for n_j was 2 in some applications.

Chang and Vetterli [175] and Chang, Yu, and Vetterli [176] use classification methods to develop spatial-adaptive thresholding methods for images.

Fan [144] proposes two test statistics based on the wavelet thresholding and the Neyman truncation. Fan provides extensive evidence to demonstrate that the proposed tests have higher power in detecting sharp peaks and high frequency alternations, while maintaining the same capability in detecting smooth alternative densities as the traditional tests. See also Spokoiny [383]

6.8 EXERCISES

6.1. Prove (6.13).

6.2. Prove that the ideal risk $\sum_i \min(\theta_i^2, \sigma^2)$ can be represented as $\sigma^2 N(\sigma) + C_{N(\sigma)}$, where $N(\sigma)$ is the number of θ_i such that $|\theta_i| \geq \sigma$, and C_n is the energy contained in the n largest $|\theta_i|$'s. Thus, the ideal risk is a measure of the extent to which the energy is compressed into a few wavelet coefficients.

6.3. Let $\lambda(n) = \sqrt{2 \log n}$. Show that $\lambda(n) = c_n \lambda(n/2)$ for

$$c_n = \left(1 - \frac{\log 2}{\log n}\right)^{-1/2}.$$

The constant c_n is the correction constant in (6.30).

6.4. *Project.* The hyperbole rule $\delta^{hy}(x, \lambda) = \text{sgn}(x)\sqrt{x^2 - \lambda^2} \mathbf{1}(|x| \geq \lambda)$ can be obtained by soft thresholding of energies of wavelet coefficients, d^2. It also, like the nn-garrote function, combines properties of both hard- and soft-thresholding rules. Plot the bias, variance, and risk of the hyperbole thresholding rule $\delta^{hy}(X, \lambda)$ for $X \sim \mathcal{N}(\theta, 1)$.

6.5. Since nn-garrote is weakly differentiable (in Stein's sense), prove that the $\mathbb{L}_2$-risk $R_\lambda(\theta) = E(\delta_\lambda^{nng}(X) - \theta)^2$, for $X \sim \mathcal{N}(\theta, 1)$, can be estimated unbiasedly by

$$S_\lambda(X) = 1 + (X^2 - 2)\mathbf{1}(|X| > \lambda) + \frac{\lambda^4 + 2\lambda^2}{X^2}\mathbf{1}(|X| > \lambda),$$

i.e., $ES_\lambda(X) = R_\lambda(\theta)$.

6.6. In terms of the functions $A_\lambda(\theta)$ and $B_\lambda(\theta)$ in (6.20) find the expressions for bias and variance of the *nn*-garrote thresholding rules, the functions $M_\lambda^{nng}(\theta)$ and $V_\lambda^{nng}(\theta)$, respectively. Plot the functions.

6.7. Refer to the signal with t_5-noise in Table 6.1. Assuming the model as in (6.22), compute the threshold that retains approximately 2% of the wavelet coefficients.

6.8. [54] If $X \sim \mathcal{N}(\theta, \sigma^2)$ and $\lambda = a\sigma$, prove the following bounds on the risks,

$$\begin{aligned} R(\delta^h, \theta) &\le (a^2 + 1)\sigma^2 \wedge (2\theta^2 + e^{-a^2/2}\sigma^2) \\ R(\delta^s, \theta) &\le (2a^2 + 2)\sigma^2 \wedge (2\theta^2 + 2ae^{-a^2/2}\sigma^2). \end{aligned}$$

6.9. Find the bias of the semisoft-thresholding rule.

6.10. [167] Let $X \sim \mathcal{N}(\theta, 1)$. The risk of the semisoft-thresholding rule $\delta^{ss}(x, \lambda_1, \lambda_2)$ is

$$\begin{aligned} R^{ss}_{\lambda_1,\lambda_2}(\theta) &= E[\delta^{ss}(X, \lambda_1, \lambda_2) - \theta]^2 \\ &= \theta^2[\Phi(\lambda_1 - \theta) - \Phi(-\lambda_1 - \theta)] - \Phi(\lambda_2 - \theta) + \Phi(-\lambda_2 - \theta) \\ &\quad + \Psi[\lambda_1 - \theta, \lambda_2 - \theta, r_2, r_1(\theta - \lambda_2)] \\ &\quad + \Psi[\lambda_1 + \theta, \lambda_2 + \theta, r_2, -r_1(\theta + \lambda_2)] \\ &\quad + (\lambda_2 - \theta)\phi(\lambda_2 - \theta) + (\lambda_2 + \theta)\phi(\lambda_2 + \theta) + 1, \end{aligned}$$

where $r_1 = \frac{\lambda_1}{\lambda_2 - \lambda_1}$, $r_2 = \frac{\lambda_2}{\lambda_2 - \lambda_1}$, and

$$\Psi(a, b, c, d) = (c^2 + d^2)[\Phi(b) - \Phi(a)] + c\phi(a)(ac + 2d) - c\phi(b)(bc + 2d).$$

Plot the graph of $R^{ss}_{\lambda_1,\lambda_2}(\theta)$ for $\lambda_1 = 2.86$, $\lambda_2 = 3.72$, and $|\theta| \le 7$. Superimpose the plot of $R^h_{3.33}(\theta)$. Comment.

6.11. Let $d_i = \theta_i + \sigma\epsilon_i$, $\epsilon_i \sim \mathcal{N}(0, 1)$, $i = 1, \ldots, n$. Find the power of a test for $H_0: \theta_i = 0$ versus $H_1: \theta_i = \theta$ if the critical region is $\{d_i : |d_i| > \sqrt{2\log n}\,\sigma\}$.

6.12. If the robustness of the rule δ_λ with respect to local shifts is measured by $\alpha_\lambda = \sup_{x \ne y} |\frac{\delta_\lambda(x) - \delta_\lambda(y)}{y - x}|$ find the shift sensitivity indices α^h_λ, α^s_λ, and $\alpha^{ss}_{\lambda_1,\lambda_2}$ for hard-, soft-, and semisoft-thresholding rules, respectively.

6.13. Prove $\mathbb{K}_j(x,y) = \sum_{i<j} \mathbb{Q}_i(x,y)$.

6.14. [205] A kernel $\mathbb{K}(x,y)$, as in (6.6), is said to be of order m iff it satisfies

$$\int_{-\infty}^{\infty} \mathbb{K}(x,y) y^k \, dy = \begin{cases} 1, & k = 0 \\ x^k, & k = 1, \ldots, m-1, \\ \alpha(x) \neq x^m, & k = m. \end{cases}$$

Prove that if a wavelet function $\psi(x)$ has m vanishing moments then the associated kernel $\mathbb{K}(x,y)$ is of order m, and vice versa.

6.15. In the heteroscedastic model $y_i = f(x_i) + \sigma_i \epsilon_i$, $i = 1, \ldots, n$, when the σ_i's are known, one may consider dividing both sides by σ_i, $y_i/\sigma_i = f(x_i)/\sigma_i + \epsilon_i$, $i = 1, \ldots, n$, and estimating f_i. Explain potential problems with this "idea."

6.16. [145] Show that a mixture of hard- and soft-thresholding rules

$$\hat{\theta} = (|d| - \lambda)_+ \mathbf{1}(|d| \leq 1.5\lambda) + |d| \mathbf{1}(|d| > 1.5\lambda)$$

is a minimizing solution to the variational problem

$$\frac{1}{2} \sum_{i=1}^{n} (d_i - \theta_i)^2 + \lambda \sum_{i=1}^{n} \min(|\theta_i|, \lambda).$$

7
Density Estimation

The density estimation problem has a long history and many solutions. A large body of existing literature on nonparametric statistics is devoted to the theory and practice of density estimation. General overviews can be found in Izenman [210], Scott [372], and Silverman [378]. This chapter addresses wavelet-based density estimation.

The local nature of wavelet functions promises superiority over projection estimators that use classical orthonormal bases (Fourier, Hermite, etc.). The wavelet estimators are simple, well-localized in space/frequency, and share a variety of optimality properties. The estimation procedures fall into the class of so-called projection estimators, introduced by Čencov [66], or their non-linear modifications.

It is interesting that some of the earliest contributions of wavelets in statistics were in density estimation. Doukhan [136] and Doukhan and Léon [137] first introduced linear wavelet density estimators and explored their mean-square errors.

Antoniadis and Carmona [13], Kerkyacharian and Picard [233, 234] and Walter [437, 438, 439] considered the linear wavelet estimators in Sobolev and Besov spaces, while Donoho et al. [132, 133], Delyon and Juditsky [112], among others, explored non-linear estimators and their minimax optimality in Besov spaces.

For a critical discussion of the advantages and disadvantages of wavelets in density estimation see Walter and Ghorai [443].

7.1 ORTHOGONAL SERIES DENSITY ESTIMATORS

Probability density estimation is, in some respects, related to a regression problem, yet the regression and density estimation are statistically different tasks. Consider, for example, noisy observations in both problems. In the regression setup, the noise makes the observed function less smooth, whereas in the density estimation setup, the noise smooths the observed density (as a convolution).

Given the realizations $X_1, X_2, \ldots, X_n$ of a random variable X with an unknown density f, it is often of interest to estimate f. An automatic approximation to f is the empirical "density" $f_e(x) = \frac{1}{n}\sum_{i=1}^{n} \delta(x - X_i)$. It is easy to see that f_e is an unbiased estimator of f since $Ef_e(x) = \delta \star f = f$. However, for an absolutely continuous underlying distribution, the estimator $f_e(x)$ is a poor choice. It is not smooth, moreover, it is even not a function. For these reasons, different estimators have been proposed. A number of proposals consists of approximating f_e in the class of smooth functions.

Orthogonal series density estimators, or *projection* estimators, introduced by Čencov [66], are smoothers of f_e. Čencov's idea was strikingly simple. The unknown square integrable density can be represented as a convergent orthogonal series expansion

$$f(x) = \sum_{j \in \mathcal{J}} a_j \psi_j(x), \tag{7.1}$$

where $\{\psi_j, j \in \mathcal{J}\}$ is a complete orthonormal system of functions in $\mathbb{L}_2$, and $\mathcal{J}$ is an appropriate set of indices.

From (7.1), the coefficients a_j can be expressed as

$$\begin{aligned} a_j &= \langle f, \psi_j \rangle \\ &= \int \psi_j(x) f(x)\, dx \\ &= E\psi_j(X) \qquad [\text{because } f \text{ is a density}]. \end{aligned} \tag{7.2}$$

Given the sample $\underline{X} = (X_1, X_2, \ldots, X_n)$ from the unknown distribution f, an empirical counterpart of a_j is

$$\hat{a}_j = \frac{1}{n} \sum_{i=1}^{n} \psi_j(X_i), \tag{7.3}$$

and $f(x)$ can be estimated by

$$\hat{f}(x) = \sum_{j \in \mathcal{J}} \hat{a}_j \psi_j(x). \tag{7.4}$$

The name "projection estimator" comes from the fact that the "derivative" f_e of the empirical distribution function $F_e(x) = \frac{1}{n}\sum_{i=1}^{n} H(x - X_i)$, where H is the

Heaviside unit-step function, is projected on the space $\mathcal{H}_\mathcal{J} = \overline{span}\{\psi_j, j \in \mathcal{J}\}$. The estimator (7.4) is also called *linear* since it is a linear function of the associated empirical measure F_e.

Analytically,

$$\hat{f}(x) = \sum_{j \in \mathcal{J}} \left\langle \psi_j, \frac{dF_e}{dx} \right\rangle \psi_j(x).$$

The estimator (7.4) is a special case of a kernel density estimator with kernel $\mathbb{K}_\mathcal{J}(x, y) = \sum_{j \in \mathcal{J}} \psi_j(x) \psi_j(y)$. In terms of $\mathbb{K}_\mathcal{J}$ (7.4) can be expressed as

$$\hat{f}(x) = \frac{1}{n} \sum_{i=1}^{n} \mathbb{K}_\mathcal{J}(x, X_i). \tag{7.5}$$

7.2 WAVELET DENSITY ESTIMATION

For an infinite set of indices $\mathcal{J}$, the naïve estimator (7.4) may not be well defined. It has an infinite variance and it is not consistent in the ISE sense. Moreover, in the wavelet case, $\limsup_{j \to \infty} |\hat{a}_j| = \infty$, a.s. The standard practice is to select finitely many empirical coefficients $\hat{a}_j$ and shrink them appropriately (Kronmal and Tarter [249]; Wahba [433]). Work by Delyon and Juditsky [112], Donoho et al. [133], Kerkyacharian and Picard [233, 234], and Walter [438, 439], among others, shows that wavelet density estimators can achieve optimal MISE rates for many rich smoothness families of densities. A theoretical overview of wavelet density estimation can be found in Härdle et al. [188].

7.2.1 δ-Sequence Density Estimators

Many density estimators fall into the class of δ-sequence density estimators. A comprehensive list of examples is given by Walter and Blum [441]. Wavelet density estimators are also examples of δ-sequence estimators. This classification is beneficial since the δ-sequence estimators are theoretically well explored; see Walter [438, 439].

Next, we define a δ-sequence and δ-sequence estimators and give several examples. We show that wavelet kernels are δ-sequences and provide several important theoretical results.

Definition 7.2.1 *Let D be an open interval in $\mathbb{R}$. A sequence $\delta_m(x, y)$ of bounded functions on $D \times D$ is a δ-sequence on D if for any $x \in D$ and each $\mathbb{C}^\infty$ function φ with support in D,*

$$\lim_{m\to\infty} \int \delta_m(x,y)\varphi(y)\,dy = \varphi(x). \tag{7.6}$$

The δ-sequence estimator of a density f, based on a sample $X_1, \ldots, X_n$ is,

$$\widehat{f}_{n,m}(x) = \frac{1}{n}\sum_{i=1}^{n} \delta_m(X_i, x). \tag{7.7}$$

Example 7.2.1 (a) The standard histogram is a δ-sequence density estimator. The δ-sequence is

$$\delta_m(x,y) = \frac{1}{m}\sum_{i=1}^{m} \mathbf{1}\left(\frac{i-1}{m} < x < \frac{i}{m}\right)\mathbf{1}\left(\frac{i-1}{m} < y < \frac{i}{m}\right).$$

(b) Let $Y_1, Y_2, \ldots, Y_m$ be a sequence of 0-mean random variables with a finite variance and bounded density. Let $g_m(y)$ be the density of the sample mean, $\bar{Y} = \frac{1}{m}\sum_{i=1}^{m} Y_i$. Then, the function $\delta_m(x,y) = g_m(x-y)$ is a δ-sequence. An example of such a δ-sequence is

$$\delta_m(x,y) = \sqrt{\frac{m}{2\pi}}e^{-\frac{m}{2}(x-y)^2}.$$

The δ-sequences in the previous examples are positive. Positive δ-sequences ensure uniform convergence in (7.6), and, in addition, the resulting density estimators (7.7) are non-negative. However, the convergence (in terms of MSE, IMSE) is suboptimal. The convergence rate for the histogram in Example 7.2.1(a) is bounded by $O(n^{-2/3})$. In general, positive δ-sequence estimators cannot achieve rates better than $O(n^{-4/5})$, irrespective of the smoothness of the underlying density.

Example 7.2.2 (a) The Dirichlet kernel

$$D_m(x,y) = \frac{\sin\left[(m+\frac{1}{2})(x-y)\right]}{2\pi \sin\left[\frac{1}{2}(x-y)\right]} \mathbf{1}(-\pi \leq x - y \leq \pi)$$

is a δ-sequence.

(b) Wavelet kernels are δ-sequences. Let $\mathbb{K}(x,y)$ be the reproducing kernel of the V_0 multiresolution subspace. Then

$$\delta_m(x,y) = \mathbb{K}_m(x,y) = 2^m \mathbb{K}(2^m x, 2^m y). \tag{7.8}$$

The Dirichlet kernel in Example 7.2.2(a) is not a positive δ-sequence. This enables faster convergence rates, but at the expense that the resulting estimators can be negative. Moreover, one can find continuous densities such that the corresponding δ-sequence estimator is MSE-inconsistent; that is, $E(\widehat{f}_n(x) - f(x)) \not\to 0$.

Wavelet-based δ-sequences are quasi-positive, see Walter [439] page 117 for the definition. Except for the Haar wavelet, wavelet-based δ-sequences cannot be positive (see Theorem 10.1.1). The estimators with quasi-positive δ-sequences can take negative values, but the convergence rates are higher and convergence in (7.6) is still uniform. MSE-inconsistency does not happen for wavelet estimators since for any quasi-positive δ-sequence, bias converges to 0.

As we have seen in Example 7.2.2 wavelet kernels produce δ-sequences. Following Walter [439] we give more precise conditions on the regularity of wavelets that are necessary for establishing the convergence rates of wavelet based density estimators. We first define Zak's transformation of a function:

Definition 7.2.2 *Zak's transformation of the function ϕ is given by*

$$Z\phi(x,\omega) = \sum_k e^{-i\omega k} \phi(x-k). \tag{7.9}$$

For more on Zak's transformation and its use in signal processing and wavelets, see [471, 220, 191].

Let ϕ be a scaling function and let $\Phi(\omega)$ be its Fourier transformation. Suppose that ϕ satisfies the Z_λ conditions,

[$Z_\lambda 1$] $\Phi(\omega) = 1 + O(|\omega|^\lambda)$, $\omega \to 0$,
[$Z_\lambda 2$] $Z\phi(x,\omega) = e^{-i\omega x}[1 + O(|\omega|^\lambda)]$, uniformly when $\omega \to 0$,

which are related to the derivatives of $\Phi(\omega)$ at $\omega = 0$,

Theorem 7.2.1 *Let ϕ be an r-regular scaling function satisfying the Z_λ conditions for some $\lambda > 0$. Let $\mathbb{K}_{j_0}(x,y)$ be the reproducing kernel of V_{j_0}. Then,*

$$\|\mathbb{K}_{j_0}(\cdot,y) - \delta(\cdot - y)\|_{-s} = O(2^{-\lambda j_0}), \quad \text{uniformly in } y,$$

where $\|\ \|_{-s}$ is the Sobolev $\mathbb{W}^{-s}$ norm, δ is Dirac's function, and $s > \lambda + \frac{1}{2}$.

For a proof and comprehensive discussion consult Walter [439], pages 124-128.

Corollary 7.2.1 *For $f \in \mathbb{W}^s(\mathbb{R})$, $s > \lambda + \frac{1}{2}$, and an r-regular scaling function ϕ,*

$$\|f_{j_0}(x) - f(x)\|_\infty = O(2^{-\lambda j_0}),$$

where $f_{j_0}(x) = \langle f, \mathbb{K}_{j_0}(\cdot, x)\rangle$ is the projection of f to V_{j_0}.

Indeed, since $|\langle f, g\rangle|^2 \leq \|f\|_s \cdot \|g\|_{-s}$, when $f \in \mathbb{W}^s$ and $g \in \mathbb{W}^{-s}$ (Cauchy-Schwarz inequality for the Sobolev norm),

$$\begin{aligned}|f_{j_0}(x) - f(x)| &\leq |\langle f, \delta(\cdot - y) - \mathbb{K}_{j_0}(\cdot, y)\rangle| \\ &\leq C \cdot \|\delta(\cdot - y) - \mathbb{K}_{j_0}(\cdot, y)\|_{-s} \\ &= O(2^{-\lambda j_0}).\end{aligned}$$

The following theorem gives the MSE and IMSE convergence rates of a linear wavelet density estimator, $\widehat{f}_{j_0}(x) = \frac{1}{n}\sum_{i=1}^n \mathbb{K}_{j_0}(x, X_i)$.

Theorem 7.2.2 *Let ϕ be an r-regular scaling function satisfying the Z_λ conditions. Then,*

(i) $E|\widehat{f}_{j_0}(x) - f(x)|^2 \to 0$, uniformly on compact sets as $j_0 \to \infty$ and $j_0 = O(\log n)$.

(ii) If f belongs to the Sobolev smoothness space $\mathbb{W}^s$, $s > \lambda + \frac{1}{2}$, and $j_0 \approx \frac{\log 2}{2\lambda + 1} \cdot \log n$, then

$$E|\widehat{f}_{j_0}(x) - f(x)|^2 = O(n^{-\frac{2\lambda}{2\lambda+1}}), \tag{7.10}$$

and

$$\int E|\widehat{f}_{j_0}(x) - f(x)|^2 \, dx = O(n^{-\frac{2\lambda}{2\lambda+1}}). \tag{7.11}$$

For the proofs, the reader is directed to Walter [439], page 200. It should be noted that for Meyer-type wavelets, the Z_λ conditions are satisfied for an arbitrary λ, which implies that the rates of convergence in (7.10) and (7.11) can be arbitrarily close to $O(n^{-1})$ for C^∞ densities. For theoretical results on density estimation and deconvolution problems involving Meyer-type wavelets, see Walter [440] and Walter and Shen [445].

Example 7.2.3 Histogram as a wavelet estimator. Let $\mathbb{K}_{j_0}(x, y)$ be the reproducing kernel of the multiresolution subspace V_{j_0} that is generated by the Haar wavelet. Easy calculation gives that, for the Haar wavelet,

$$\mathbb{K}(x, y) = \phi(x - \lfloor y \rfloor). \tag{7.12}$$

Because $\mathbb{K}_{j_0}(X_i, x) = \sum_i \phi(2^{j_0} X_i - \lfloor 2^{j_0} x \rfloor)$ is the count of X_i in the interval $[k2^{-j_0}, (k+1)2^{-j_0}]$ containing x, $\frac{1}{n}\sum_{i=1}^n \mathbb{K}_{j_0}(X_i, x)$ coincides with the standard definition of a histogram with nodes at $2^{-j_0} k$.

The estimator $\hat{f}_{j_0}(x) = \frac{1}{n}\sum_i \mathbb{K}_{j_0}(X_i, x) = \sum_{k \in \mathbb{Z}} c_{j_0,k} \cdot \phi_{j_0,k}(x)$ with $c_{j_0,k} = \frac{1}{n}\sum_{i=1}^n \phi_{j_0,k}(X_i)$ is not an unbiased estimator of f. The bias is

$$E\hat{f}_{j_0}(x) - f(x) = \int \mathbb{K}_{j_0}(y, x) f(y)\, dy - f(x) \neq 0,$$

since, in general, $f(y)$ does not belong to V_{j_0}. However, since $f_{j_0} = \langle f, \mathbb{K}_{j_0} \rangle \to f$ uniformly (when f is continuous and compactly supported), $\hat{f}_{j_0}$ is asymptotically unbiased.

The variance of $\hat{f}_{j_0}(x)$ is

$$E[\hat{f}_{j_0}(x) - E(\hat{f}_{j_0}(x))]^2 \leq \frac{1}{n}\int \mathbb{K}_{j_0}(y, x) f(y)\, dy \leq \frac{2^{2j_0}}{n}.$$

Theorem 7.2.3 states that $\hat{f}_{j_0}(x)$ is a.s. consistent.

Theorem 7.2.3 *Let f and ϕ be as in Theorem 7.2.2 and $r \geq 1$. Then, $\hat{f}_{j_0}(x)$ is a.s. consistent as $n \to \infty$ and $j_0 = j_0(n) = O(\log n)$. The optimal rate is*

$$O\left((n^{-\frac{\lambda}{\lambda+1}} \log \log n)^{1/2}\right) \text{ a.s.,}$$

for $j_0 \approx \frac{\log 2}{2\lambda + 1} \cdot \log n$.

7.2.2 Bias and Variance of Linear Wavelet Density Estimators

Huang [205] studied asymptotic bias and variance of linear wavelet density estimators.

Define

$$b_m(x) = x^m - \int_{-\infty}^{\infty} \mathbb{K}(x, y) y^m\, dy. \tag{7.13}$$

The functions $b_m(x)$ in (7.13) are important in expressing the asymptotic bias of linear estimators and in finding their efficiencies with respect to the standard kernel density estimators. See also Exercise 7.4.

Theorem 7.2.4 gives the bias and integrated bias for the linear estimator, $\hat{f}_j = \frac{1}{n}\sum_{i=1}^n \mathbb{K}_j(x, X_i)$.

Theorem 7.2.4 *[205] Assume that the density f belongs to the Hölder space $\mathbb{C}^{m+\alpha}$, $0 \leq \alpha < 1$, and that the wavelet-kernel $\mathbb{K}$ satisfies the following localization property: $\int_{-\infty}^{\infty} \mathbb{K}(x, y)(y-x)^{m+\alpha}\, dy < C$, for some $C > 0$. Let $j \to \infty$ and $n2^{-j} \to \infty$,* as $n \to \infty$. Then, for x fixed,

$$E\widehat{f}_j(x) - f(x) = -\frac{1}{m!}f^{(m)}(x)\, b_m(2^j x)\, 2^{-mj} + O(2^{-(m+\alpha)j}).$$

Moreover, if $f^{(m)} \in \mathbb{L}_2$, then the integrated squared bias is

$$\|E\widehat{f}_j(x) - f(x)\|^2 = \frac{B_{2m}}{(2m)!}\|f^{(m)}\|^2\, 2^{-2jm} + O(2^{-2(m+\alpha)j}),$$

where $B_{2m} = \frac{(2m)!}{(m!)^2}\int_0^1 b_m^2(x)\,dx$.

The asymptotic variance of $\widehat{f}_j$ is given in Theorem 7.2.5. Antoniadis, Grégoire, and McKeague [15] studied a related problem in the regression context, but for dyadic points.

Theorem 7.2.5 *[205] Let $f \in \mathbb{C}^1$, and f and f' be uniformly bounded. Then, for x fixed,*

$$\operatorname{Var}\widehat{f}_j(x) = \frac{2^j}{n}f(x)\,V(2^j x) + O(n^{-1}),$$

where $V(x) = \int_{-\infty}^{\infty}\mathbb{K}^2(x,y)\,dy = \mathbb{K}(x,x)$. Moreover, the integrated variance is $\frac{2^j}{n}\int_0^1 V(x)\,dx + O(\frac{1}{n})$.

7.2.3 Linear Wavelet Density Estimators in a More General Setting

The following result was proved in Kerkyacharian and Picard [233] for $\pi \geq 2$ and in Donoho et al. [133] for $\pi \geq 1$.

Let $\mathbb{B}_{\pi,q}^\sigma(M) = \{f : \int f = 1, f \geq 0, \|f\|_{\mathbb{B}_{\pi,q}^\sigma} \leq M\}$ be the class of densities with bounded Besov norm.

Theorem 7.2.6 *Let $\pi \geq 1$ and $\sigma < r$, where r is the regularity of MRA. Let $f \in \mathbb{B}_{\pi,q}^\sigma(M)$ and $j_0 = j_0(n) = \lfloor \log_2 n^{\frac{1}{1+2\sigma}}\rfloor$. Then,*

$$E_f\|\widehat{f}_{j_0} - f\|_\pi^\pi \leq C\, n^{-\frac{\pi\sigma}{1+2\sigma}}.$$

The restricted minimax risk R_n (estimators $\widehat{f}$ are confined to a class $\mathcal{C}_L$ of all linear wavelet estimators) is

$$R_n \asymp C \cdot n^{-\frac{\sigma'\pi}{1+2\sigma'}}$$

where $R_n = \inf_{\hat{f} \in \mathcal{C}_L} \sup_{f \in \mathbb{B}_{\pi,q}^\sigma(M)} E\|\hat{f}_{j_0} - f\|_\pi^\pi$, $1 \leq p, q \leq \infty$, $\pi \geq p$, $\sigma > \frac{1}{\pi}$, and $\sigma' = \sigma - \frac{1}{p} + \frac{1}{\pi}$.

The estimators discussed in this section are linear; they are essentially low-pass filters. If the density has heterogeneous regularity properties, the linear estimators can be under-performing. A precise asymptotic expression for the linear estimators is given by Masry [288] in a more general setup (observations $X_1, \ldots, X_n$ form a stationary process). See also Antoniadis and Carmona [13].

Dechevsky and Penev [110, 111] deal with a linear estimator of a cumulative distribution function and a density, based on compactly supported non-negative continuous wavelets. The estimator of a cdf is a cdf and that of a density is a density itself. Although the approach of Dechevsky and Penev gives up the orthogonality, it is efficient and computationally inexpensive. An additional advantage is the compact support of wavelets, which puts their estimator in the class of δ-sequence estimators. The authors also consider more general error measures. Their refinement concerns, in particular, an assessment of the role of the tail weight against pointwise smoothness. An exhaustive asymptotic-minimax theory includes a treatment of densities with heavy tails.

The wavelet shrinkage paradigm works well in density estimation as well. In the following section, we discuss a class of non-linear wavelet density estimators where the non-linearity is introduced by thresholding. More general shrinkage rules can result from the Bayesian inference. This topic will be postponed until Chapter 8.

7.3 NON-LINEAR WAVELET DENSITY ESTIMATORS

As in the regression problem, one may be interested in thresholding empirical wavelet coefficients to allow wavelets to innately perform adaptive fit. When the density belongs to $\mathbb{B}_{p,q}^\sigma$ space and errors are measured with respect to $\mathbb{L}_\pi$-norm ($1 \leq \pi < \infty$, $\pi > p$), non-linearity of the estimator becomes essential since the linear estimators are sub-optimal.

Thresholding should be used with care in density estimation since projecting on the subspace V_{j_1} for j_1 large and taking a threshold estimator could confound artifacts of the "comb" f_e with the local features of the "true" underlying density.

Assume that the density f has a formal expansion

$$f(x) = \sum_k \alpha_{j_0,k} \phi_{j_0,k}(x) + \sum_{j=j_0}^\infty \sum_k \theta_{jk} \psi_{jk}(x). \tag{7.14}$$

The sample counterparts of $\alpha_{j_0,k}$ and θ_{jk} are

$$c_{j_0,k} = \frac{1}{n}\sum_{i=1}^{n}\phi_{j_0,k}(X_i), \text{ and}$$

$$d_{jk} = \frac{1}{n}\sum_{i=1}^{n}\psi_{jk}(X_i),$$

respectively; see (7.2-7.3).

Following Donoho et al. [132, 133], a simple non-linear wavelet estimator can be defined via thresholding

$$\widehat{f_n}(x) = \sum_{k} c_{j_0,k}\phi_{j_0,k}(x) + \sum_{j=j_0}^{j_1}\sum_{k}\hat{\theta}_{jk}\psi_{jk}(x), \qquad (7.15)$$

where $\hat{\theta}_{jk} = \delta^h(d_{jk}, \lambda)$. The levels $j_0 = j_0(n)$ and $j_1 = j_1(n)$ are conveniently selected and $\lambda = K \cdot C(j)\, n^{-1/2}$. See the formulation of Theorem 7.3.1 and (7.16) for some possible choices of $C(j)$.

The following theorem shows that the estimator (7.15) can attain the optimal rate to within a logarithmic term; and in the "sparse" case ($\epsilon < 0$), it can obtain exactly the optimal rate.

Let $\alpha = \min\{\frac{\sigma}{1+2\sigma}, \frac{\sigma-1/p+1/\pi}{1+2\sigma-2/p}\}$, $\epsilon = \sigma p - \frac{\pi-p}{2}$, and $\sigma' = \sigma - \frac{1}{p} + \frac{1}{\pi}$.

Theorem 7.3.1 *(Donoho et al. [133]) Let $\sigma - \frac{1}{p} > 0$ and $p \wedge 1 \leq \pi \leq \infty$. If there exist constants C and K_0 such that if*

$$2^{j_0(n)} \approx \{n(\log n)^{[(\pi-p)/p]}\mathbf{1}_{\{\epsilon>0\}}\}^{1-2\alpha},$$
$$2^{j_1(n)} \approx (n/\log n)^{\alpha/\sigma'},$$

and $K \geq K_0$, then

$$\sup_{f \in \mathbb{B}_{p,q}^\sigma}(E_f\|\widehat{f_n}(x) - f(x)\|_\pi^\pi)^{1/\pi} \leq \begin{cases} C(\log n)^{(1-\epsilon/\sigma p)\alpha}\,n^{-\alpha}, & \epsilon > 0 \\ C(\log n)^{(1/2-p/q\pi)+}(\frac{\log n}{n})^\alpha, & \epsilon = 0 \\ C(\frac{\log n}{n})^\alpha, & \epsilon < 0. \end{cases}$$

For practical purposes one can simply use thresholds $\lambda_j = K \cdot \sqrt{\frac{j}{n}}$ over the range $n^{1/(2r-1)} \leq 2^j \leq \frac{n}{\log n}$, where r is the regularity of the wavelet.

Delyon and Juditsky [112] suggest a different threshold:

$$\lambda = A\sqrt{((j-j_0) \wedge 0)/n}. \qquad (7.16)$$

This selection slightly reduces the bias of the estimator, but it is sensitive to the choice of p, σ, q, and π.

Remark 7.3.1 Thresholds in density estimation problems usually behave as $O(\sqrt{j}n)$. This choice corresponds to the universal threshold since $\hat{\sigma}^2 = O(1/n)$, see Exercise 7.5.

For a comprehensive discussion on advantages of non-linear estimators, in terms of optimal convergence rates, see Section 10.4 of Härdle et al. [188]. A trade-off between variance and bias in non-linear density estimation is discussed in Hall and Patil [184]. See also Hall and Patil [183] for an asymptotic formula for the integrated mean square error of non-linear estimators.

7.3.1 Global Thresholding Estimator

Pensky [331] explored non-linear density estimation by wavelets of Meyer-type. The non-linearity is introduced by *global* block thresholding in which the blocks constitute complete levels.

The estimator achieves optimal convergence rates when $f(x)$ belongs to the Sobolev space $\mathbb{W}_s$, with $s > 0$ unknown.

Pensky defined,

$$\hat{f}_n(x) = \sum_{k \in \mathbb{Z}} c_{j_0,k} \phi_{j_0,k}(x)$$
$$+ \sum_{j=j_0}^{j_0+r} \left[\left(\sum_{k \in \mathbb{Z}} d_{jk} \psi_{jk}(x) \right) \mathbf{1} \left(\sum_{k \in \mathbb{Z}} d_{jk}^2 > \delta_j(n)^2 \right) \right], \quad (7.17)$$

where $c_{j_0,k}$ and d_{jk} are empirical wavelet coefficients. Since the Meyer scaling and wavelet functions have unbounded support, the summation with respect to k is unbounded as well. The estimator

$$\hat{f}_n^*(x) = \sum_{|k| \leq M_n} c_{j_0,k} \phi_{j_0,k}(x)$$
$$+ \sum_{j=j_0}^{j_0+r} \left[\left(\sum_{|k| \leq L_n} d_{jk} \psi_{jk}(x) \right) \mathbf{1} \left(\sum_{k \in \mathbb{Z}} d_{jk}^2 > \delta_j(n)^2 \right) \right], \quad (7.18)$$

enjoys the same rate of convergence as the estimator (7.17) for suitably chosen boundaries K_n and L_n.

Let $\varrho(\omega)$ be an even, non-negative function, non-increasing for ω large, and such that $\varrho \to 0$, as $|\omega| \to \infty$. Assume that the unknown density f belongs to the function space $\mathbb{H}_2(\varrho) = \{f(x) | \int |F(\omega)|^2 \varrho^{-2}(\omega) d\omega < \infty\}$, where $F(\omega)$ is the

228   DENSITY ESTIMATION

Fourier transformation of f. Note that $\mathbb{H}_2(\varrho)$ coincides with the Sobolev space $\mathbb{W}_2^s$ when $\varrho(\omega) = (\omega^2 + 1)^{-s/2}$.

The main result in [331], given in Theorem 7.3.2, concerns the limiting behavior of the MISE for estimators (7.17, 7.18) when

$$\varrho(\omega) = (\omega^2 + 1)^{-s/2} \exp\{-B|\omega|^s\} \tag{7.19}$$

in two cases, $B = 0$ (Sobolev space) and $B > 0$ (the space of supersmooth functions).

Theorem 7.3.2 *Let $j_0 = 2\log_2 j_0^* + 2\log_2(\ln n)$, $j_0 + r = \log_2 n$ and $\delta_j(n) = \delta_0\, 2^{j/2} n^{-1/2}$, where*

$$\delta_0 \geq 4\sqrt{\pi}\, [K_2 + 2(K_1 j_0)^{-1/2} ||\psi||_\infty ||f||_{L_2}].$$

K_1 and K_2 are the absolute constants in Talagrand's [394] inequality, and j_0^* is any positive constant, independent of n. If $f \in \mathbb{H}_2(\varrho)$, with ϱ given by (7.19), and $\gamma = max(2, s^{-1})$, then

$$E \int |\hat{f}_n(x) - f(x)|^2\, dx = \begin{cases} O(n^{-2s/(2s-1)}), & \text{if } B = 0 \\ O\left(\frac{\log^\gamma n}{n}\right), & \text{if } B > 0 \end{cases} \tag{7.20}$$

Under the assumptions of Theorem 7.3.2, the estimator $\hat{f}_n^*$ in (7.18) achieves the same MISE convergence rates as $\hat{f}_n$ in (7.20) if

$$E|X| < \infty, \quad \lim_{n \to \infty} n(\ln n)^6 M_n^{-1} < \infty, \text{ and } \lim_{n \to \infty} n^4 L_n^{-1} < \infty.$$

A different, data driven, global thresholding estimator has been proposed by Kerkyacharian, Picard, and Tribouley [235].

7.4   NON-NEGATIVE DENSITY ESTIMATORS

If the projection kernel $\mathbb{K}$ is not a positive δ-sequence, then the associated density estimation procedure may yield a density estimator that takes negative values. Although such a phenomenon is usually confined to the tails and the regions where the sample is sparse, reporting a density estimator that takes negative values might be disturbing to the user. The truncation solution opens the problem of re-normalization. In this section, we will see that it is possible to ensure that a density estimate itself is a *bona fide* density, i.e., it is non-negative and integrates to 1.

We discuss two approaches. One is the time-honored idea to first estimate a transformation φ of a density. The density estimator is then obtained by taking the inverse transformation φ^{-1}. Good and Gaskins [177], Klonias [236], Tapia and Thompson [397], and Rubin and Chen [355], among others, discuss properties of such estimators for classical orthonormal bases and the square root transformation. The log transformation was discussed by Leonard [257] in the Bayesian context and by Clutton-Brock [78], among others. Penev and Dechevsky [330] and Pinheiro and Vidakovic [343] discuss estimation of the square root of a density utilizing wavelets as decomposing bases.

The second approach is due to Walter and Shen [444]. They define a density estimator through a linear combination of biorthogonal scaling functions that are densities themselves. The linear wavelet estimator is a density as a location mixture of densities from the biorthogonal basis.

We start with non-negative density estimators obtained by estimating the square root of a density.

7.4.1 Estimating the Square Root of a Density

Estimation of the square root of a density, proposed by Good and Gaskins [177] in the context of penalized maximum likelihood estimation, was subsequently explored by many researchers. Though the square root is not the only transformation with the range of the inverse in $\mathbb{R}^+$, it is the transformation of choice for several reasons.

An important reason for estimating the square root of a density is the possibility of controlling the $\mathbb{L}_2$-norm of the estimate, i.e., to ensure that the proposed estimator integrates to 1. That is done by utilizing Parseval's identity. Once the estimator of the square root is obtained, $\widehat{\sqrt{f}}$ say, one takes $\hat{f} = (\widehat{\sqrt{f}})^2$ as an estimator of the density f.

Since f is a density, $\sqrt{f}$ is certainly in $\mathbb{L}_2$. Let

$$\sqrt{f(x)} = \sum_k \alpha_{j_0 k}\phi_{j_0 k}(x) + \sum_{j \geq j_0,\, k} \theta_{jk}\psi_{jk}(x). \tag{7.21}$$

From the Parseval identity it follows that

$$\sum_k \alpha_{j_0 k}^2 + \sum_{j \geq j_0; k} \theta_{jk}^2 = \|\sqrt{f}\|_2^2 = \int \sqrt{f} \cdot \sqrt{f} = 1. \tag{7.22}$$

If the estimators of $\alpha_{j_0 k}$ and θ_{jk} (denoted by $c_{j_0 k}$ and d_{jk}, respectively), are available and normalized as

$$\sum_k c_{j_0 k}^2 + \sum_{j=j_0}^{j_1} \sum_k d_{jk}^2 = 1,$$

then,

$$\int \widehat{f} = \int \left(\widehat{\sqrt{f}}\right)^2 = 1,$$

where

$$\widehat{\sqrt{f}} = \sum_k c_{j_0 k} \phi_{j_0 k}(x) + \sum_{j=j_0}^{j_1} \sum_{k \in \mathbb{Z}} d_{jk} \psi_{jk}(x). \tag{7.23}$$

An additional reason for this particular choice of transformation is the variance stabilizing property of the square root. The models for the empirical coefficients for the square root of a density are "closer to normal" than those for the density itself. This can be beneficial if one considers non-linear thresholding estimators. We direct the reader to a nice discussion in Fan and Gijbels [146], pages 52-53.

Suppose that the sample $X_1, X_2, \ldots, X_n$ was observed from a density f supported on a compact set. without loss of generality, consider the interval $[0, 1]$. Suppose in addition that f is bounded away from 0 on its support. Let $\mathbb{L}_2([0,1]) = \overline{\text{span}}\{\phi_{j_0,k}, \psi_{jk}, j \geq j_0, 0 \leq k \leq 2^{j_0} - 1\}$, where ϕ and ψ are periodized scaling and wavelet functions, see Section 5.6.

Pinheiro and Vidakovic [343] propose empirical counterparts of wavelet coefficients in (7.21),

$$d_{jk} = \sum_{i=1}^{n} \omega_i(X_1, \ldots, X_n) \psi_{jk}(X_i), \tag{7.24}$$

where $\omega_i(X_1, \ldots, X_n)$ are normalized weights proportional to $\left[n \sqrt{\hat{f}_n(X_i)}\right]^{-1}$, and $\hat{f}_n$ is a simple pre-estimator of the density f. The ranges of the indices are $j_0 \leq j \leq j_1$; $0 \leq k \leq 2^j - 1$ and the $c_{j_0,k}$'s are defined by analogy. The index j_0 is determined by cross-validation and j_1 by inspecting the cumulative level energies in the decomposition, see Exercise 7.7.

The estimator in (7.24) is motivated by the following equalities,

$$\theta_{jk} = \langle \sqrt{f}, \psi_{jk} \rangle = \int \psi_{jk} \sqrt{f} = \int \frac{\psi_{jk}}{\sqrt{f}} f = E \frac{\psi_{jk}(X)}{\sqrt{f(X)}},$$

where X has the density f. Apparent *circulus vitiosus* in the definition (7.24) can be justified only if the weights w_i depend on a simple and computationally inexpensive pre-estimator $\hat{f}$. One possibility is a simple estimator for f given by

$$\hat{f}_n(X_i) \propto \#\{X_j \in (X_i - r, X_i + r)\}, \quad (7.25)$$

where r is a suitably chosen radius. For this choice of $\hat{f}_n$, Pinheiro [342] proves strong consistency of the estimators in (7.24) and obtains convergence rates under mild conditions on the regularity of f.

Vannucci and Vidakovic [416] discuss selection of r and j_1. By using Beylkin's result on connections between wavelet coefficients of f and some differential functionals of f (Beylkin [32]) they show that the Fisher information functional of $\hat{f}$ is

$$\int \frac{[(\hat{f})']^2}{\hat{f}} = 4 \sum |\hat{r}_{j,k}|^2,$$

where $\hat{r}_{jk} = 2^j \sum_l r_l c_{j,k-l}$, and the r_l's are tabulated in Beylkin [32]. The level j_1 is selected as the level for which $4\sum |\hat{r}_{j,k}|^2$ is close to the minimum of th the Fisher information among all densities supported on supp(f).

Example 7.4.1 In this example we give the estimators of two test densities from the Marron-Wand family (Marron and Wand [285]): the separated bimodal and the claw, see panels (a) and (b) in Fig. 7.1. Both test densities are supported on $[-4, 4]$, which gives the minimum Fisher information of $\pi^2/64 = 0.1542$. The sample size was $n = 500$ and $r = 0.2$ in both cases. In the separated bimodal case, the DAUB10 wavelet was used; the equation $4\sum |\hat{r}_{j,k}|^2 = 0.1542$ produced $j_1 = 0$. For the claw density wavelet, the DAUB4 wavelet was used with $j_1 = 2$. The MSE was shown to be robust with respect to the selection of r.

Example 7.4.2 We illustrate the wavelet density estimator based on (7.24) in an astronomy example. Here is a brief description of the data set.

> According to the Big Bang theory, matter in the universe expanded at a tremendous rate. Gravitational forces caused the formation of galaxies. Astronomers speculate that gravitational pull led to clustering of galaxies and research indicates the presence of super-clusters of galaxies surrounded by large voids (a string-and-filament pattern). Measurements have recently become available for the distances between our galaxy and others. Distance is estimated by the red

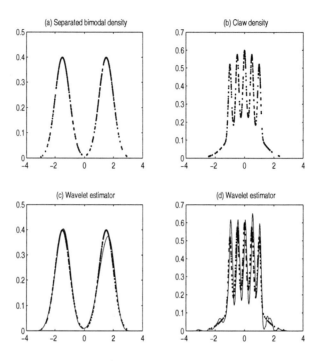

Fig. 7.1 [416] Panels (a) and (b) give plots of samples of size $n = 500$ from the separated normals and claw test densities. Panels (c) and (d) give plots of the linear wavelet estimators.

shift in the light spectrum in a fashion similar to how the Doppler effect measures the changes in speed via changes in sound. Under the expansion-universe paradigm, the furthest (from our galaxy) galaxies must be moving at greater velocities, because the distances and velocities are proportional. If, in reality, the galaxies are clumped, the velocities should have a multimodal distribution, each mode corresponding to a cluster. In the region of *Corona Borealis,* the velocities of 82 galaxies were measured. The relative measurement error is believed to be smaller than 0.5%.

Fig. 7.2(a) depicts the empirical scalogram function of the galaxy data. A scalogram is the wavelet analog of the notion of a spectrogram in Fourier analysis and will be discussed in more detail in Chapter 9. It is an easy exercise to demonstrate that the energies, depicted in Fig. 7.2(a), increase exponentially. After inspecting the shape of the scalogram, the decision to stop at level $j_1 = 4$ was made. Fig. 7.2(b) gives the estimators for maximum levels $j_1 = 3$ and $j_1 = 4$, superimposed over the histogram. In both cases, the wavelet coefficients for $\sqrt{f}$ are normalized as in (7.22). The DAUB4 basis was used.

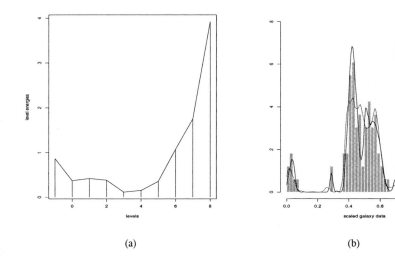

Fig. 7.2 (a) Empirical scalogram for the galaxy data. Notice an exponential growth of energies with the increase of j. (b) Wavelet density estimators for $j = 3$ [dotted line] and $j = 4$ [solid line].

Penev and Dechevsky [330] propose the pre-estimator based on order statistics.

Let $[L_{jk}, R_{jk}]$ be a domain of ψ_{jk} and let the order statistics of $X_1, \ldots, X_n$ and L_{jk}, R_{jk} be positioned as

$$X_{1:n} \leq X_{2:n} \leq \cdots \leq X_{l:n} \leq L_{jk} \leq X_{(l+1):n}$$
$$\cdots \leq X_{u:n} \leq R_{jk} \leq X_{(u+1):n} \cdots \leq X_{n:n}.$$

Penev and Dechevsky propose the following estimator for the coefficients of the square root of the density f,

$$d_{jk} = \frac{2}{\sqrt{n\pi}} \left[\psi_{jk}(\rho X_{(l+1):n} + (1-\rho)L_{jk})\sqrt{X_{(l+1):n} - L_{jk}} + \sum_{i=l+2}^{u} \psi_{jk}(\rho X_{i:n} + (1-\rho)X_{(i-1):n})\sqrt{X_{i:n} - X_{(i-1):n}} + \psi_{jk}(\rho R_{jk} + (1-\rho)X_{u:n})\sqrt{R_{jk} - X_{u:n}} \right].$$

Asymptotic properties of the estimator of $\sqrt{f}$ are independent of the end-points L_{jk} and R_{jk}, and ρ. Hence, for the asymptotic considerations one may take

$$d_{jk} = \frac{2}{\sqrt{n\pi}} \left[\sum_{i=1}^{n-1} \psi_{jk}(X_{(i+1):n}) \sqrt{X_{(i+1):n} - X_{i:n}} \right]. \tag{7.26}$$

The definition for c_{jk} is as in (7.26) with ϕ_{jk} in place of ψ_{jk}.

The authors show that c_{jk} and d_{jk} are strongly consistent estimators of α_{jk} and θ_{jk}. If f has one bounded derivative on [0,1], and the Hölder regularity of the wavelet is at least 1, then they prove the asymptotic normality of c_{jk} and d_{jk},

$$\sqrt{n}(c_{jk} - \alpha_{jk}) \Rightarrow \mathcal{N}(0, \sigma^2_{\phi_{jk}})$$
$$\sqrt{n}(d_{jk} - \theta_{jk}) \Rightarrow \mathcal{N}(0, \sigma^2_{\psi_{jk}}),$$

where

$$\sigma^2_{\psi_{jk}} = \frac{4}{\pi} - \frac{5}{4} + \frac{1}{4} \int_{L_{jk}}^{R_{jk}} \left[\frac{\psi_{jk}(x)}{\sqrt{f(x)}} \right. $$
$$\left. -(F(x))^{-1} \int_{L_{jk}}^{x} \psi_{jk}(y)\sqrt{f(y)}\,dy \right]^2 f(x)\,dx,$$

and $F(x)$ is the corresponding cdf [$F'(x) = f(x)$]. The variance $\sigma^2_{\phi_{jk}}$ is defined similarly.

The estimator proposed is,

$$\widehat{\sqrt{f}} = \sum_k c_{j_0,k}\phi_{j_0,k}(x) + \sum_{j=j_0}^{j_1} \sum_k d_{j,k}\mathbf{1}(|d_{j,k}| > \lambda_j)\,\psi_{j,k}(x),$$

where the level j_0 is determined by the cross-validation criteria, given in (7.31). The choice for j_1, $\log_2 n - \log_2(\log n)$, is standard; see Donoho et al. [132]. Finally, the suggested level-adaptive threshold, $\lambda_j \propto \sqrt{j-j_0}/\sqrt{n}$, is as in Delyon and Juditsky [112].

In addition, Penev and Dechevsky recommend the threshold proportional to $\sqrt{j-j_0} \cdot n^{-1/4}$, and demonstrate that the resulting estimator enjoys asymptotic minimax optimality properties, as discussed in Donoho at al. [133], but on somewhat more restricted spaces.

7.4.2 Density Estimation by Non-Negative Wavelets

As we will discuss in Chapter 10, Theorem 10.1.1, there is no continuous non-negative scaling function generating an orthogonal MRA. Can we estimate the unknown density via a mixture of simple, known densities whose translations form a Riesz bases of multiresolution subspaces, V_j? Do such simple densities exist? Walter and Shen [444] answer positively by an ingenious construction of a biorthogonal system whose primary scaling functions are densities.

Let ϕ be an orthogonal scaling function with compact support and let r $(0 < r \leq 1)$ be large enough to ensure that the function

$$\rho_r(x) = \sum_n r^{|n|} \phi(x - n) \tag{7.27}$$

is non-negative. Such an r is always possible to find (see Exercise 7.9). Since $\sum_k \phi(x - k) = 1$, it follows that $\rho_r(x) \to 1$, when $r \to 1$. Since the convergence is uniform on compact sets, one can always find the critical r^*, such that $(\forall x)\, \rho_r(x) \geq 0$ when $r \geq r^*$.

Let

$$\tilde{\rho}_r(x) = \frac{(1 + r^2)\phi(x) - r[\phi(x + 1) + \phi(x - 1)]}{2\pi(1 - r^2)}.$$

Then,

$$\{2^{j/2} \rho_r(2^j x - k),\ k \in \mathbb{Z}\} \text{ and } \{2^{j/2} \tilde{\rho}_r(2^j x - k),\ k \in \mathbb{Z}\}$$

are the primary and dual biorthogonal bases for V_j and $\tilde{V}_j = V_j$.

Since

$$\int \sum_n r^{|n|} \phi(x - n)\, dx = 2 \sum_{i=0}^{\infty} r^i - 1 = \frac{1 + r}{1 - r},$$

the kernel associated with V_0 can be defined as

$$\mathbb{K}_r(x, y) = \left(\frac{1 - r}{1 + r}\right)^2 \sum_{n \in \mathbb{Z}} \rho_r(x - n)\rho_r(y - n), \tag{7.28}$$

and with V_j as

$$\mathbb{K}_{r,j}(x, y) = 2^j \mathbb{K}_r(2^j x, 2^j y),\ j \in \mathbb{Z}.$$

The linear estimator based on the sample $X_1, \ldots, X_n$ is

$$\widehat{f_{r,j_0}}(x) = \frac{1}{n} \sum_{i=1}^{n} \mathbb{K}_{r,j_0}(x, X_i). \tag{7.29}$$

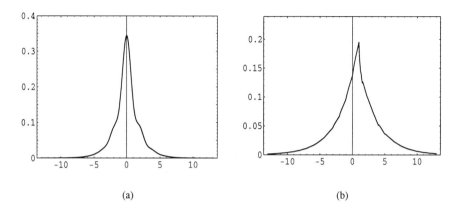

Fig. 7.3 Densities (a) $1/3\rho_{0.5}(x)$ and (b) $0.3/1.7 \cdot \rho_{0.7}(x)$.

By (7.28), $\widehat{f_{r,j_0}}(x)$ is a location mixture of "atom"-densities $\frac{1-r}{1+r}\rho_r(x)$.

Example 7.4.3 The building atoms of the linear estimator (7.29), the functions $\frac{1-r}{1+r}\rho_r(x)$, are densities. Fig. 7.3 gives graphs of the COIF12 and DAUB2 based densities with $r = 0.5$ and $r = 0.7$, respectively. The critical values of r^* for these two families are 0.2830 and 0.2579, respectively.

Walter and Shen [444] prove the following result,

Theorem 7.4.1 *Let ϕ be a scaling function belonging to the Hölder $\mathbb{C}^2$ space and let the density f belongs to the Sobolev $\mathbb{W}_2^3$ space. Then, the MSE of $\widehat{f_{r,j_0}}(x)$ satisfies*

$$\text{MSE} = O(n^{-4/5}), \tag{7.30}$$

for a proper choice of j_0, uniformly on bounded sets.

The rate in (7.30) is somewhat better than for the other non-negative estimators.

7.5 OTHER METHODS

7.5.1 Multivariate Wavelet Density Estimators

The multivariate wavelet density estimators are natural and straightforward generalizations of their univariate counterparts. They are based on the multivariate multiresolution analysis discussed in Section 5.7.

Let f be a density from $\mathbb{L}_2(\mathbb{R}^d)$. The wavelet series is

$$f(x_1, x_2, \ldots, x_n) = \sum_{\mathbf{k}} \alpha_{j_0, \mathbf{k}} \phi_{j_0; \mathbf{k}}(x_1, \ldots, x_n)$$

$$+ \sum_{j \geq j_0} \sum_{\mathbf{k}} \sum_{l=1}^{2^d - 1} \theta_{j;\mathbf{k}}^{(l)} \psi_{j;\mathbf{k}}^{(l)}(x_1, \ldots, x_n),$$

where $\mathbf{k} = (k_1, \ldots, k_d) \in \mathbb{Z}^d$ is a vector of translations, and the $\phi_{j_0;\mathbf{k}}$ and $\psi_{j;\mathbf{k}}^{(l)}$ are given by (5.25-5.26).

Given a sample $\mathbf{X}_1, \mathbf{X}_1, \ldots, \mathbf{X}_n$ from f, one estimates $\alpha_{j;\mathbf{k}}$, and $\theta_{j;\mathbf{k}}$ by

$$c_{j;\mathbf{k}} = \frac{1}{n} \sum_{i=1}^{n} \phi_{j_0;\mathbf{k}}(\mathbf{X_i}) = \frac{1}{n} \sum_{i=1}^{n} 2^{jd/2} \prod_{m=1}^{d} \phi_{(m)}(2^j X_i - k_m),$$

and

$$d_{j;\mathbf{k}}^{(l)} = \frac{1}{n} \sum_{i=1}^{n} \psi_{j;\mathbf{k}}^{(l)}(\mathbf{X_i}) = \frac{1}{n} \sum_{i=1}^{n} 2^{jd/2} \prod_{m=1}^{d} \xi_{(m)}(2^j X_i - k_m),$$

where $\xi = \phi$ or ψ, but not all $\xi = \phi$.

The linear density estimator

$$f(x_1, x_2, \ldots, x_n) = \sum_{\mathbf{k}} c_{j_0, \mathbf{k}} \phi_{j_0; \mathbf{k}}(x_1, \ldots, x_n)$$

has the form

$$\hat{f}(\mathbf{x}) = \frac{1}{n} \sum_{i=1}^{n} \mathbb{K}_j(\mathbf{X}_i, \mathbf{x}),$$

where

$$\mathbb{K}_j(\mathbf{X}, \mathbf{x}) = \prod_{m=1}^{d} \mathbb{K}_j(X_m, x_m),$$

and $\mathbb{K}_j(x, y)$ is as in (7.8). For selecting j_0, one can use a multivariate version of the CV (Tribouley [409]),

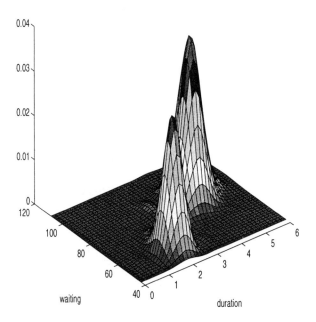

Fig. 7.4 Old Faithful Geyser bivariate density estimate. The variables are: *waiting* and *duration*.

$$\mathrm{CV}(j) = \sum_{\mathbf{k}} \left[\frac{2}{n(n-2)} \sum_{i=1}^{n} \phi_{j;\mathbf{k}}(\mathbf{X_i})^2 + \frac{n+1}{n^2(n-1)} \left(\sum_{i} \phi_{j;\mathbf{k}}(\mathbf{X}_i) \right)^2 \right].$$

The optimal level j_0 is $\arg\min_j \mathrm{CV}(j)$. For a univariate version, see (7.31). Vannucci [414] proposed a procedure for estimation of the square root of a multivariate density and gave practical recommendations for how to select the optimal level.

Example 7.5.1 [414] The Old Faithful geyser data set is frequently used as a benchmark for evaluating and comparing bivariate density estimators.

> Old Faithful is the most famous of the Yellowstone geysers. It erupts more frequently than the other big geysers, but is neither the largest nor the most regular geyser in Yellowstone. Its eruption interval is 45-90 minutes, while the eruption itself lasts 2-5 minutes and reaches a height of 100-180 feet. The data, collected in Yellowstone National Park from August 1 to August 15, 1985, have a total of 299 pairs of observations. The first variable is the *waiting* time for the eruption and the second variable is the *duration* time of the eruption.

Using the multivariate version of the estimator (7.24), for DAUB7 and the preestimator with radius $r = 0.25$, and level $j_0 = 0$, Vannucci obtained the estimator depicted in Fig. 7.4.

7.5.2 Density Estimation as a Regression Problem

Let $X_1, \ldots, X_n$ be a sample from a density supported on $[a, b]$. Let I_k, $k = 1, \ldots, N$ be intervals [bins] in a partition of $[a, b]$ each of length $\Delta = (b-a)/N$. Let x_k be the mid-point of I_k, and let $y_k = \#\{X_i \in I_k\}/(n\Delta)$. Then, because of independence of the X_i's, the bin counts $c_k = y_k n \Delta$ have the binomial $Bin(n, p_k)$ distribution with $p_k = \int_{I_k} f(x)\,dx$. When N is large (Δ is small),

$$Ey_k \approx f(x_k) \quad \text{and Var}(y_k) \approx \frac{f(x_k)}{n\Delta}.$$

Thus the density estimation problem becomes a heteroscedastic nonparametric regression problem,

$$y = f(x) + \sigma(x)\epsilon,$$

with $\sigma(x) = \frac{f(x)}{n\Delta}$, based on $\{x_k, y_k, \ k = 1, \ldots, N\}$. Fan [144] provides exact formulation and a rationale. Optimal selection of Δ was discussed by Cheng [72].

Homoscedascity is obtained by Anscombe's [10] variance stabilizing transformation to the bin counts $c_k = n\Delta y_k$,

$$y_k^* = 2\sqrt{c_k + \frac{3}{8}}.$$

Let $m^*(x)$ be a regression estimator based on pairs (x_k, y_k^*), $k = 1, \ldots, N$.
The density estimator is then

$$\hat{f}(x) = C \left[\frac{(m^*(x))^2}{4} - \frac{3}{8} \right]_+ ,$$

where C is a normalizing constant. The use of wavelets in estimating $m^*(x)$ was first discussed in Donoho [120]. Non-wavelet approaches can be found in Hjort and Jones [194] and Loader [265]. More details on the use of the Anscombe variance stabilizing transformation for Poisson regression with some data examples can be found in Donoho et al. [132].

Bock and Pliego [36] utilize the empirical distribution of properly standardized linear combinations of the observed bin-proportions in developing a percentage-type threshold. Their *V-test* density estimation method shows superiority when compared to standard WaveShrink techniques.

The binning technique was also implemented by Antoniadis, Grégoire and Nason [16] in the context of density and hazard rate estimation for right-censored data.

7.5.3 Cross-Validation Estimator

We will not estimate the density f itself, but its projection $\sum_k c_{j_0,k} \phi_{j_0,k}(x)$ on the space V_{j_0}. Then, since $\mathbb{L}_2 = \overline{\cup_j V_j}$, we require that $j_0 = j_0(n) \to \infty$, as $n \to \infty$.

Assume that the unknown density belongs to the Besov space $\mathbb{B}_{2,2}^{\sigma}$ and that the wavelet is r-regular.

For selecting j_0, we can use the cross-validation method. The following criteria was given by Tribouley [408]. It is a wavelet adaptation of a criteria considered by Rudemo [356] and Marron [284]. For

$$\mathrm{CV}(j) = \sum_k \left[\frac{2}{n(n-2)} \sum_{i=1}^n \phi_{jk}(X_i)^2 + \frac{n+1}{n^2(n-1)} \left(\sum_i \phi_{jk}(X_i) \right)^2 \right] \quad (7.31)$$

the optimal j_0 is $\arg\min_j \mathrm{CV}(j)$.

Tribouley [408] shows that when $2^j > n^{1/(2(\sigma \wedge r)+1)}$

$$\begin{aligned} \mathrm{CV}(j) &= \mathrm{MISE}(j) + T_n + o_P\left(\mathrm{MISE}(j)\right) \\ &= \mathrm{ISE}(j) + T_n + o_P\left(\mathrm{ISE}(j)\right) \end{aligned}$$

where $T_n = \int f - \frac{2}{n} \sum_i f(X-i)$ does not depend on j and o_P is a "small o in probability".

7.5.4 Multiscale Estimator

Multiscale wavelet density estimators are introduced by Wu [463, 464].

Let the sample $\underline{X} = (X_1, \ldots, X_n)$ be divided into two subsamples $\underline{Y}$ and $\underline{Z}$ of sizes n_1 and n_2, respectively, with $n_1 + n_2 = n$.

Let $\lambda_l(x, \underline{Y})$, $l = 1, \ldots, L$ be the weights based on the subsample $\underline{Y}$ such that

$$\sum_{l=1}^{L} \lambda_l(x, \underline{Y}) = 1.$$

Let $\{\phi^{\{l\}}(x) | \ 1 \leq l \leq L\}$ be the scaling functions associated with L different multiresolution analyses and let $\mathbb{K}^{\{l\}}(x, y) = \sum_k \phi^{\{l\}}(x-k)\phi^{\{l\}}(y-k)$ be the associated reproducing kernels.

For each l, define the sub-estimator

$$\hat{f}_{n_2}^{\{l\}}(x, \underline{Z}) = \frac{1}{n_2} \sum_{i=1}^{n_2} \mathbb{K}_{j_l}^{\{l\}}(x, Z_i),$$

where $j_l = 2c_l \log_2 n_2$ are indices of the projection subspaces and $c_1, \ldots, c_L$ satisfy $0 < c_1 < \cdots < c_L < \frac{1}{4}$.

The *multiscale wavelet density estimator* is defined as a linear combination of L sub-estimators with weights given by the λ_l's, i.e.,

$$\hat{f}_n(x, \underline{X}) = \sum_{l=1}^{L} \lambda_l(x, \underline{Y}) \hat{f}_{n_2}^{\{l\}}(x, \underline{Z}). \tag{7.32}$$

Wu [464] shows that when $\sum_l \text{Var}(\lambda_l) = o(n_2^{2c_L - 1})$,

$$\frac{\sqrt{n_2^{1-2c_L}} \left[\hat{f}_n(x, \underline{X}) - E(\hat{f}_n(x, \underline{X})) \right]}{\sigma(x)} \Rightarrow N(0, 1), \tag{7.33}$$

where $\sigma^2(x) = \sigma_L^2 f(x) E(\lambda_L^2(x))$ and $\sigma_L^2 = \lim_{j_L \to \infty} 2^{-j_L} \mathbb{K}_{j_L}^{\{L\}}(x, x)$.

Example 7.5.2 Consider the estimator in (7.32) with two sub-estimators and with weights designed as

$$\lambda_1(x, \underline{Y}) = \frac{1}{n_1} \exp \left\{ -\frac{|x - \bar{Y}_{n_1}|}{S_{n_1}} \right\}, \quad \lambda_2 = 1 - \lambda_1, \tag{7.34}$$

where $\bar{Y}_{n_1}$ and S_{n_1} are the sample mean and sample standard deviation of the subsample $\underline{Y}$. It can be readily shown that

$$\sum_{l=1}^{2} \text{Var}[\lambda_l(x, \underline{Y})] = o(n_2^{2c-1}), \tag{7.35}$$

for $0 < c < \frac{1}{4}$, and the condition for (7.33) is satisfied. Wu [464] reports good simulation results for a multiscale estimator with weights as in (7.34).

7.5.5 Estimation of a Derivative of a Density

Suppose $X_1, X_2, \ldots, X_n$ are iid random variables with a density f that is d-times differentiable. Prakasa Rao [350] addresses the wavelet estimation of $f^{(d)}$.

Let ϕ be a scaling function generating an r-regular multiresolution analysis and let $f^{(d)} \in \mathbb{L}_2(\mathbb{R})$. Assume that there exist $C_m \geq 0$ and $\beta_m \geq 0$ such that

$$|f^{(m)}(x)| \leq C_m |x|^{-\beta_m}, \text{ for } |x| > 1,\ 0 \leq m \leq d. \quad (7.36)$$

If $\phi_{jk}(x) = 2^{j/2}\phi(2^j x - k)$, then $\phi_{jk}^{(m)}(x) = 2^{j/2+mj}\phi^{(m)}(2^j x - k)$, $0 \leq m \leq r$.

Let f_{jd} be the orthogonal projection of $f^{(d)}$ on the multiresolution subspace V_j. Then,

$$f_{jd} = \sum_k a_{jk}\phi_{jk}(x),$$

where

$$a_{jk} = \int f^{(d)}(x)\,\phi_{jk}(x)\,dx = (-1)^d \int f(x)\,\phi_{jk}^{(d)}(x)\,dx. \quad (7.37)$$

Since, considering (7.37), the coefficient a_{jk} can be thought of as $(-1)^d E[\phi_{jk}^{(d)}(X)]$, $X \sim f$, its empirical counterpart is

$$\hat{a}_{jk} = \frac{(-1)^d}{n}\sum_{i=1}^n \phi_{jk}^{(d)}(X_i),$$

and the associated estimator of $f^{(d)}$ is

$$\hat{f}_{n,d}(x) = \sum_k \hat{a}_{jk}\phi_{jk}(x).$$

Prakasa-Rao [350] proves a result on the IMSE convergence of the estimator. A special case of this result can be stated as Theorem 7.5.1.

Theorem 7.5.1 *Under standard regularity conditions on f, $f^{(d)}$, and the multiresolution analysis, one has*

$$n^{2(s-d)/(2s-1)}E\|f^{(d)} - \hat{f}_{n,d}\|_2^2 \to \int_{-\infty}^{\infty}(\phi^{(d)})^2(x)\,dx,$$

when $n \to \infty$.

This result shows that the linear wavelet-based estimator achieves the same optimal IMSE rates as the kernel-type estimator for the dth derivative of a density, see Gasser and Müller [169].

For a wavelet-based estimation of the integral of the square of a density and its derivatives, see Prakasa Rao [350, 351] and Exercise 7.13.

7.6 EXERCISES

7.1. Let ϕ be a scaling function. Prove that $Z\phi(x,0) = 1$, where $Z\phi(x,\omega)$ is the Zak transform defined by (7.9).

7.2. Prove that $|\mathbb{K}(x,y)| \leq c\,(1+|x-y|)^{-n}$, for any $n \in \mathbb{N}$.

[Hint: Use the definition of an r-regular wavelet and the inequality $1+|a-b| \leq (1+|a|)\,(1+|b|)$.]

7.3. Prove that the reproducing kernel for the V_0 subspace generated by Haar's wavelet is, $\mathbb{K}(x,y) = \phi(x - \lfloor y \rfloor)$ [see (7.12)].

7.4. [205] Let b_m be as in (7.13). Prove

(a) b_m is 1-periodic;

(b) $x^m - \int_{-\infty}^{\infty} \mathbb{K}_j(x,y)\,y^m\,dy = 2^{-jm} b_m(2^j x)$.

(c) For the DAUBN family,

$$b_N(x) = x^N - \sum_{l=0}^{N} \binom{N}{l} a_{l,N} M_{l,N}(x),$$

where $a_{l,N} = \int_0^{2N-1} x^l \phi(x)\,dx$ and $M_{l,N} = \sum_{k=-2N+2}^{0} k^{N-l}\phi(x-k)$ [b_1 for the Haar wavelet is the first Bernoulli polynomial $B_1(x)$].

7.5. Let ψ be a compactly supported wavelet and let $d_{jk} = \frac{1}{n}\sum_{i=1}^{n} \psi_{jk}(X_i)$.

Demonstrate

(a) $\mathrm{Cov}(d_{jk}, d_{jk'}) = -\frac{1}{n}\theta_{jk}\theta_{jk'}$,

where θ_{jk}s are "true" coefficients and $|k - k'|$ is large enough.

(b) $\mathrm{Var}(d_{jk}) = O(\frac{1}{n})$,

when n is large.

7.6. Let ψ be the DAUBN wavelet function.

(i) Prove that $B_{\#N} = \min_x \max_k \{\psi_{0k}^2(x)\}$ is strictly positive.

(ii) Calculate $B_{\#2}$.

7.7. The empirical scalogram is a level-wise sum of squares of the empirical wavelet coefficients, i.e., $\mathbb{E}(j) = \sum_k (\widehat{a}_{jk})^2$, where j is a level and the $\widehat{a}_{jk}$ are given by (7.3). Prove that, if for a chosen pre-estimator $\widehat{f}_n$ in (7.3)

$$(\forall i)\ |\widehat{f}_n(X_i)| \leq C$$

holds, then for large j

$$E(j) \geq \frac{2^j \cdot B_{\#\mathbf{N}}}{nC}.$$

The level j_1 in (7.23) was determined by inspecting the cumulative energies $E(j)$. See also Fig. 7.2(a) and Exercise 7.6.

7.8. Prove that the estimator d_{jk} in (7.24) is proportional to that in (7.26) if the unknown cumulative distribution function $F_n(x)$ is pre-estimated by

$$F_n(x) = \frac{m}{n} + \frac{1}{n} \cdot \frac{x - X_{m:n}}{X_{(m+1):n} - X_{m:n}}.$$

7.9. Refer to Example 7.4.3. Prove that for any compactly supported scaling function $\phi(x)$ generating an orthogonal MRA, one can find r ($0 < r < 1$) so that $\rho_r(x)$ in (7.27) is non-negative.

7.10. Let $r^* = \inf\{r \mid \rho_r(x) \geq 0\}$ be the "critical" r. Numerically evaluate r^* for DAUB2 and COIF12 (36 tabs).

7.11. Compute $E(X)$ and $\text{Var}(X)$ if $X \sim 0.3/1.7\rho_{0.7}(x)$, as in panel (b) of Fig. 7.3. [Answer: $E(X) = 0.629642$ and $\text{Var}(X) = 15.13675$.]

7.12. Prove (7.35).

7.13. [351] The motivation for estimating the functional

$$I(f) = \int f^2(x)\,dx,$$

where f is a density, is well known. The functional $I(f)$ appears in the Pitman efficiency of the Wilcoxon test as compared to the t-test, in the asymptotic variance of the Lehmann-Hodges estimator, and in asymptotics of the IMSE for kernel-type density estimators.

Let f_j be the projection of f to V_j. Then $f_j(x) = \sum_k c_{jk}\phi_{jk}(x)$, where $c_{jk} = \langle f(x), \phi_{jk}(x) \rangle$.

Define

$$f_{j,K_n} = \sum_{k=-K_n}^{K_n} c_{jk}\phi_{jk}(x),$$

where j and K_n are sequences of positive integers, depending on n and tending to infinity when $n \to \infty$. Given a sample $X_1, X_2, \ldots, X_n$ from f, define

$$A_{jk} = \frac{1}{n(n-1)} \sum_{i=1}^{n} \sum_{l=1, l\neq i}^{n} \phi_{jk}(X_i)\phi_{jk}(X_l),$$

and

$$\hat{I}(f) = \sum_{k=-K_n}^{K_n} A_{jk}.$$

(a) Show that $\hat{I}(f)$ is an asymptotically unbiased estimator of $I(f)$, i.e., $E\hat{I}(f) \to I(f)$, $n \to \infty$.

(b) Show that

$$\operatorname{Var}(\hat{I}(f)) = -\frac{4(n-2)}{n(n-1)} \sum_{k=-K_n}^{K_n} \sum_{k'=-K_n}^{K_n} c_{jk}^2 \, c_{jk'}^2.$$

8

Bayesian Methods in Wavelets

In this chapter, we provide an overview of Bayesian inference in wavelet nonparametric problems. In most areas of application, there is a need for a shrinkage procedure to (i) adapt to data and (ii) use prior information. The Bayesian paradigm provides a natural terrain for both of these goals.

Bayesian approaches to choosing the shrinkage method are less ad hoc than some of the earlier proposals, and have been shown to be effective. It is known that, in general, Bayes rules are "shrinkers" and their shape in many cases has a desirable property for wavelet shrinkage: It can heavily shrink small arguments and only slightly shrink large arguments. If we use Bayes models for the wavelet coefficients, the resulting optimal actions can be very close to thresholding. In fact, as we saw in Section 6.3.1 hard- and soft-thresholding rules can be interpreted as Bayes actions under suitably chosen loss functions. This chapter describes several methods of Bayesian wavelet shrinkage and provides references for related research and applications.

8.1 MOTIVATIONAL EXAMPLES

Suppose, as in (6.1) and (6.4), that *noisy* measurements $y = \{y_1, \ldots, y_n\}$ are modeled as $\underset{\sim}{y} = \underset{\sim}{f} + \underset{\sim}{\epsilon}$, and that the wavelet-domain image of this model is $\underset{\sim}{d} = \underset{\sim}{\theta} + \underset{\sim}{\epsilon}$.

The goal of Bayesian wavelet shrinkage is to exhibit models on $\underset{\sim}{d}$ so that the optimal actions, resulting from appropriate Bayesian inference, mimic the actions of thresholding rules. To illustrate this idea we start with an example.

Example 8.1.1 [420] Let

$$[d|\theta, \sigma^2] \sim \mathcal{N}(\theta, \sigma^2), \quad \sigma^2 \text{ unknown},$$

be a model for a "typical" wavelet coefficient. Because of practical (computational) reasons and noninformative properties[1] the prior distribution on σ^2 is chosen to be exponential,

$$[\sigma^2] \sim \mathcal{E}(\mu) \quad (f(\sigma^2|\mu) = \mu e^{-\mu\sigma^2}).$$

The marginal model (marginal likelihood) is double exponential,

$$[d|\theta] \sim \mathcal{DE}(\theta, \frac{1}{\sqrt{2\mu}}), \tag{8.1}$$

with the density $f(d|\theta) = \frac{1}{2}\sqrt{2\mu}e^{-\sqrt{2\mu}|d-\theta|}$. This step follows from the fact that the double exponential distribution is a scale mixture of normals.

Let the prior on θ be t with location 0, scale τ, and n degrees of freedom, $[\theta] \sim t_n(0,\tau)$. Several graphs of Bayes rules, resulting from this model, are given in Fig. 8.1. The rules are odd functions and are plotted for non-negative values of the argument only. The hyperparameters μ, τ, and n can be specified by empirical Bayes arguments. The Bayes rule under the model described, for a symmetric prior on θ, i.e., $\pi(\theta) = \pi(-\theta)$, and the squared-error loss is

$$\hat{\theta}_\pi(d) = d - \frac{\Pi_1'(c) - \Pi_2'(c)}{\Pi_1(c) + \Pi_2(c)}, \tag{8.2}$$

In (8.2), Π_1 and Π_2 are the one-sided Laplace transforms of the functions $\pi(\theta + d)$ and $\pi(\theta - d)$, $\theta \in (0, \infty)$, and $c = \sqrt{2\mu}$.

It is not automatically ensured that a location model, with a prior containing a point mass at zero, will induce a shrinkage. Care is necessary in selecting all components of the model. The following example illustrates this point.

Example 8.1.2 [426] The following is a definition of DasGupta and Rubin [101].

Definition 8.1.1 *Let $f(\underline{x}|\underline{\theta})$ be a model for $\underline{X}$ and let $\underline{\theta} \sim \pi(\underline{\theta})$ be a prior on $\underline{\theta}$. Let $m(\underline{x}) = E^{\underline{\theta}}f(\underline{x}|\underline{\theta})$ be the marginal distribution of $\underline{X}$. Let $\hat{\theta}_\pi(\underline{X})$ be the Bayes rule*

---

[1] The exponential distribution minimizes the Fisher information in the class of all distributions supported on $[0,\infty)$ with a fixed first moment.

MOTIVATIONAL EXAMPLES **249**

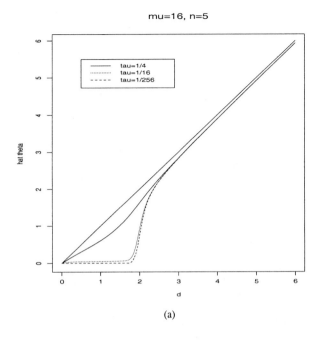

(a)

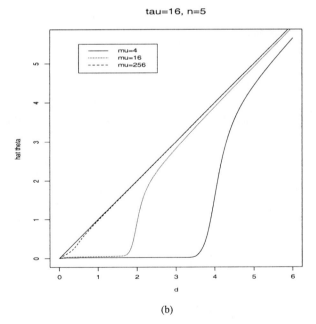

(b)

Fig. 8.1 Bayes rules for selected values of τ with $\mu = 16$ and $n = 5$ [panel (a)] and μ with $\tau = 16$ and $n = 5$ [panel (b)]. The rules are odd functions and are shown only for non-negative values of d.

with respect to the prior π. Let $\Delta = \frac{||\hat{\theta}_\pi(\underline{X})||}{||\underline{X}||}$.

The rule $\hat{\theta}_\pi$ in estimating the parameter $\underline{\theta}$ is an expander if:

$$E\Delta > 1,$$

where the expectation is taken with respect to the marginal distribution $m(\cdot)$.

Assume that an empirical wavelet coefficient is normally distributed, $d \sim \mathcal{N}(\theta, 1)$ and that the prior on θ is given by

$$\pi(\theta) = \begin{cases} 0 & \theta \neq 0, \; |\theta| \leq m \\ p & \theta = 0 \\ \frac{1-p}{2} e^{-(|\theta|-m)} & |\theta| > m \end{cases} \quad (8.3)$$

The marginal is given by

$$m_\pi(d) = p\phi(d) + \sqrt{e}\,\frac{1-p}{2}\,[e^{d+m}\Phi(-d-m-1) + e^{-d+m}\Phi(d-m-1)],$$

and the Bayes rule $\hat{\theta}_\pi(d) = d + \frac{m'(d)}{m(d)}$ is an expander for small values of d, clearly a nondesirable outcome. See also Exercise 8.2.

8.2 SMOOTH SHRINKAGE

The somewhat simplistic model in Example 8.1.1 can be replaced with more powerful and coherent Bayesian models that exploit a hierarchical structure.

In the context of wavelet regression, two approaches are discussed in more detail. The first one is Adaptive Bayesian Wavelet Shrinkage (ABWS) proposed by Chipman, Kolaczyk, and McCulloch [74]. Their approach is based on the stochastic search variable selection (SSVS) model introduced by George and McCulloch [172] with the assumption that σ^2 is known. When the wavelet coefficients are modeled independently, we can drop the indexing, and simply denote d_{jk} by d.

Chipman, Kolaczyk, and McCulloch [74] start with the model

$$[d|\theta, \sigma^2] \sim \mathcal{N}(\theta, \sigma^2). \quad (8.4)$$

The prior on θ is defined as a mixture of two normals [Fig. 8.2(a)]

$$[\theta|\gamma_j] \sim \gamma_j \mathcal{N}(0, (c_j\tau_j)^2) + (1-\gamma_j)\mathcal{N}(0, \tau_j^2), \quad (8.5)$$

where

$$[\gamma_j] \sim Ber(p_j). \tag{8.6}$$

Because the hyperparameters p_j, c_j, and τ_j depend on the level j to which the corresponding θ (or d) belongs, and can be level-wise different, the method is adaptive.

The Bayes rule under squared error loss for θ (from the level j) has an explicit form,

$$\hat{\theta}_\pi(d) = \left[P(\gamma_j = 1|d) \frac{(c_j\tau_j)^2}{\sigma^2 + (c_j\tau_j)^2} + P(\gamma_j = 0|d) \frac{\tau_j^2}{\sigma^2 + \tau_j^2} \right] d, \tag{8.7}$$

where

$$P(\gamma_j = 1|d) = \frac{p_j \pi(d|\gamma_j = 1)}{p_j \pi(d|\gamma_j = 1) + (1 - p_j)\pi(d|\gamma_j = 0)}$$

and

$$\pi(d|\gamma_j = 1) \sim \mathcal{N}(0, \sigma^2 + (c_j\tau_j)^2) \text{ and } \pi(d|\gamma_j = 0) \sim \mathcal{N}(0, \sigma^2 + \tau_j^2).$$

The shrinkage rule [(8.7), Fig. 8.2(b)] can be viewed as a smooth interpolation between two lines through the origin with slopes $\frac{\tau_j^2}{\sigma^2+\tau_j^2}$ and $\frac{(c_j\tau_j)^2}{\sigma^2+(c_j\tau_j)^2}$. The authors provide sophisticated empirical Bayes arguments for tuning the hyperparameters level-wise. They provide posterior analysis on the coefficients and function values. The simulations presented in [74] show that the ABWS method is superior to VisuShrink and SureShrink over the standard Donoho-Johnstone test functions. See also Exercise 8.3.

The approach used by Clyde, Parmigiani, and Vidakovic [82] is based on a limiting form of the conjugate SSVS prior in George and McCulloch [172]. The prior on the components of θ is a mixture of a point mass at 0, if the variable is excluded from the wavelet regression, and a normal distribution, if it is included,

$$[\theta|\gamma_j, \sigma^2] \sim \mathcal{N}(0, (1 - \gamma_j) + \gamma_j c_j \sigma^2).$$

Indicator variables, γ_j, specify which basis element, i.e., column of W, should be selected. As before, the subscript j indicates the level to which θ belongs. The set of all possible vectors γ will be referred to as the subset space. The prior distribution for σ^2 is an inverse $\tilde{\chi}^2$, i.e.,

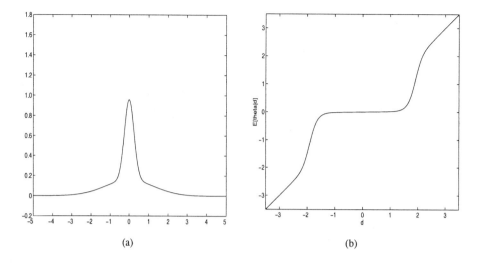

Fig. 8.2 (a) Prior on θ as a mixture of two normal distributions with different variances; (b) Shrinkage rule in [74].

$$[\lambda \nu / \sigma^2] \sim \chi_\nu^2,$$

where λ and ν are fixed hyperparameters and the γ_j's are independently distributed Bernoulli $Ber(p_j)$ random variables. The posterior mean of $\underline{\theta}|\underline{\gamma}$ is

$$E(\underline{\theta}|\underline{d},\underline{\gamma}) = \Gamma(I_n + C^{-1})^{-1}\underline{d}, \tag{8.8}$$

where Γ and C are diagonal matrices with γ_{jk} and c_{jk}, respectively, on the diagonals and 0 elsewhere. For a particular subset determined by the ones in $\underline{\gamma}$, (8.8) corresponds to the linear (affine) shrinkage.

The posterior mean is obtained by averaging over all models. Model averaging leads to the multiple shrinkage estimator of $\underline{\theta}$:

$$E(\underline{\theta}|\underline{d}) = \sum_{\underline{\gamma}} \pi(\underline{\gamma}|\underline{d}) \Gamma \left(I_n + C^{-1} \right)^{-1} \underline{d},$$

where $\pi(\underline{\gamma}|\underline{d})$ is the posterior probability of a particular subset $\underline{\gamma}$. An additional nonlinear shrinkage of the coefficients to 0 results from the uncertainty in which subsets should be selected.

Calculating the posterior probabilities of $\underline{\gamma}$ and the mixture estimates for the posterior mean of $\underline{\theta}$ above involves summing over all 2^n values of $\underline{\gamma}$. The calculational complexity of such mixing is prohibitive even for problems of moderate size, and either approximations or stochastic methods for selecting subsets γ possessing high posterior probability must be used.

In the orthogonal case, Clyde, DeSimone, and Parmigiani [79] obtain an approximation to the posterior probability of γ which is adapted to the wavelet setting in [82]. The approximation can be achieved by either conditioning on σ^2 (plug-in approach) or by assuming independence of the elements in γ.

The approximate model probabilities, for the conditional case, are functions of the data through the regression sum of squares and are given by

$$\pi(\gamma|\underline{d}) \approx \tilde{\pi}(\gamma|y) = \prod_{j,k} \rho_{jk}^{\gamma_{jk}} (1 - \rho_{jk})^{1-\gamma_{jk}} \tag{8.9}$$

$$\rho_{jk}(\underline{d}, \sigma) = \frac{a_{jk}(\underline{d}, \sigma)}{1 + a_{jk}(\underline{d}, \sigma)},$$

where

$$a_{jk}(\underline{d}, \sigma) = \frac{p_{jk}}{1 - p_{jk}} (1 + c_{jk})^{-1/2} \cdot \exp\left\{ \frac{1}{2} \frac{S_{jk}^2}{\sigma^2} \right\} \tag{8.10}$$

$$S_{jk}^2 = d_{jk}^2 / (1 + c_{jk}^{-1}). \tag{8.11}$$

The p_{jk} can be used to obtain a direct approximation to the multiple shrinkage Bayes rule. The independence assumption leads to more involved formulas. Thus, the posterior mean for θ_{jk} is approximately

$$\rho_{jk}(1 + c_{jk}^{-1})^{-1} d_{jk}. \tag{8.12}$$

Equation (8.12) can be viewed as a level dependent wavelet shrinkage rule, generating a variety of nonlinear rules. Depending on the choice of prior hyperparameters shrinkage may be monotonic, if there are no level-dependent hyperparameters, or non-monotonic; see Fig. 8.3(b). Authors report good MSE performance of approximation rules.

Clyde and George [80] propose a model in which the distributions for the error ϵ and θ are scale mixtures of normals, thus justifying Mallat's proposal (6.22) for a model of a "typical" wavelet coefficient. An empirical Bayes approach is used to estimate the prior hyperparameters, and provide analytic expressions for the shrinkage estimator based on Bayesian model averaging. The authors report an excellent denoising performance of their shrinkage method for a range of noise distributions. See also Clyde and George [81] and Johnstone and Silverman [223].

Holmes and Denison [195] propose a hyperprior on prior precision and link the induced wavelet shrinkage to some standard model selection methods. The model proposed is $\underline{y} \sim \mathcal{MVN}(W\underline{\theta}, \sigma^2 I)$ with the prior

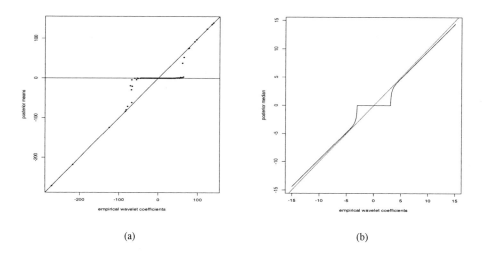

Fig. 8.3 (a) Shrinkage rule from [82] based on independence approximation (8.12). (b) Posterior median thresholding rule from [3].

$$\underline{\theta} \sim \mathcal{MVN}(\underline{0}, T^{-1}),$$

where W is the wavelet matrix and $T = \text{diag}(t_1, t_2, \ldots, t_n)$ is the precision matrix. The authors examine the relationship between T and the model fit. The mean of the posterior predictive density is $\hat{f} = W\hat{\underline{\theta}}$, which can be written as $\hat{f} = S\underline{y}$, where S is the smoothing matrix given by $S = \sigma^{-2} W(\sigma^{-2}I + T)^{-1} W'$. The degrees of freedom for a linear smoother are given by the trace of the smoothing matrix, and

$$\text{DF} = \sum_{i=1}^{n} (1 + \sigma^2 t_i)^{-1},$$

(Hastie and Tibshirani [189]), which is a natural proposal for a measure of model complexity. Smooth models with low degrees of freedom are preferred and a convenient choice for the hyperprior on t_i is

$$p(t_i | \sigma^2) \propto \exp\left\{ -c \sum_{i=1}^{n} (1 + \sigma^2 t_i)^{-1} \right\}, \qquad (8.13)$$

for some constant c, which determines how much we penalize model complexity. After analytically integrating out the wavelet coefficients θ, one obtains

Log Model Probability = Log Marginal Likelihood - (c × DF),

which has the form of many classical model choice criteria. The four criteria, depending on a selected c, are

(i) $c = 0$. The uniform prior results in the ratio of model probabilities being given by the Bayes factor (Kass and Raftery [227]).

(ii) $c = 1$. Akaike's information criteria (AIC) (Akaike [8]).

(iii) $c = 1/2 \log n$. Bayesian information criteria (BIC) (Schwartz [371]).

(iv) $c = \log n$. The risk inflation criteria (RIC) (Foster and George [155]).

See also Holmes and Mallick [197, 196] for a discussion on an efficient Bayesian variable selection method and related wavelet applications.

8.3 BAYESIAN THRESHOLDING

Bayes rules under the squared error loss and regular models are smooth and thus cannot be thresholding rules. We discuss two approaches of obtaining *bona fide* thresholding rules in a Bayesian manner. The first one is via hypothesis testing, while the second one uses weighted absolute error loss.

Donoho and Johnstone [126, 129] gave a heuristic for the selection of the universal threshold via rejection regions of suitable hypotheses tests; see the discussion in Section 6.6.5. Testing a precise hypothesis in Bayesian fashion requires a prior that has a point mass component. A method based on Bayes factors is discussed first. For details see [420].

Let

$$[d|\theta] \sim f(d|\theta).$$

After observing the coefficient d, the hypothesis $H_0 : \theta = 0$, versus $H_1 : \theta \neq 0$ is tested. If the hypothesis H_0 is rejected, θ is estimated by d. Let

$$[\theta] \sim \pi(\theta) = \pi_0 \delta_0 + \pi_1 \xi(\theta),$$

where $\pi_0 + \pi_1 = 1$, δ_0 is a point mass at 0, and $\xi(\theta)$ is a prior that describes the distribution of θ when H_0 is false.

The resulting Bayesian procedure is

$$\hat{\theta} = d\,\mathbf{1}\left[P(H_0|d) < \frac{1}{2}\right],$$

where

$$P(H_0|d) = \left(1 + \frac{\pi_1}{\pi_0}\frac{1}{B}\right)^{-1},$$

is the posterior probability of the hypothesis H_0, and $B = \frac{f(d|0)}{\int_{\theta \neq 0} f(d|\theta)\,\xi(\theta)\,d\theta}$ is the Bayes factor in favor of H_0.

For instance, let, as in Example 8.1.1,

$$[d|\theta] \sim \mathcal{DE}\left(\theta, \frac{1}{\sqrt{2\mu}}\right)$$

and

$$\pi(\theta) = \pi_0 \delta_0 + \pi_1 \xi(\theta).$$

Then, d will be thresholded if

$$\frac{\pi_0 e^{-c|d|}}{\pi_0 e^{-c|d|} + \pi_1(\Pi_1(c) + \Pi_2(c))} \geq \frac{1}{2},$$

where Π_1 and Π_2 are the one-sided Laplace transformations of $\xi(\theta - d)$ and $\xi(\theta + d)$.

Abramovich, Sapatinas and Silverman [3] use weighted absolute error loss and show that for a prior on θ

$$[\theta] \sim \pi_j \mathcal{N}(0, \tau_j^2) + (1 - \pi_j)\delta(0)$$

and for normal $\mathcal{N}(\theta, \sigma^2)$ likelihood, the posterior median is

$$\text{MEDIAN}(\theta|d) = \text{sgn}(d)\max(0, \zeta). \tag{8.14}$$

Here

$$\zeta = \frac{\tau_j^2}{\sigma^2 + \tau_j^2}|d| - \frac{\tau_j \sigma}{\sqrt{\sigma^2 + \tau_j^2}} \Phi^{-1}\left(\frac{1 + \min(\omega, 1)}{2}\right), \text{ and}$$

$$\omega = \frac{1 - \pi_j}{\pi_j} \frac{\sqrt{\tau_j^2 + \sigma^2}}{\sigma} \exp\left\{-\frac{\tau_j^2 d^2}{2\sigma^2(\tau_j^2 + \sigma^2)}\right\}.$$

The index j, as before, indicates the level containing θ (or d) facilitating adaptivity. The plot of the thresholding function (8.14) is given in Fig. 8.3(a).

Abramovich, Sapatinas, and Silverman [3] assume

$$\tau_j^2 = C_1 2^{-\alpha j} \text{ and } \pi_j = \min(1, C_2 2^{-\beta j})$$

where C_1, C_2, α, and β are non-negative hyperparameters. The hyperparameters α and β are determined from the assumption that the function to be estimated belongs to a particular Besov space, while C_1 and C_2 are determined in an empirical Bayes fashion.

The authors compare their BayesThresh rule (8.14) with several existing methods: cross-validation, false discovery rate, VisuShrink, and GlobalSure. They also reported a very good MSE performance.

8.4 MAP-PRINCIPLE

We already hinted that wavelet regularization methods in constructing linear shrinkage estimators (see page 175, also Section 6.3.1) coincide with MAP rules. The MAP rule maximizes the posterior and is typically a shrinkage rule.

Let $f(d - \theta)$ be a location model for a "typical" wavelet coefficient d and let θ be the parameter of interest with prior π. Given the observation d, the posterior distribution of θ is proportional to

$$\pi(\theta|d) \propto f(d - \theta) \cdot \pi(\theta). \tag{8.15}$$

The goal is to find the rule $\hat{\theta} = \hat{\theta}(d)$ which maximizes the posterior, i.e. the MAP rule. Let $s(\theta) = -\log \pi(\theta)$. Notice that the expression in (8.15) is maximized at the same argument at which

$$s(\theta) - \log f(d - \theta) \tag{8.16}$$

is minimized. We discuss the case when f is a normal density. If $[d|\theta] \sim \mathcal{N}(\theta, \sigma^2)$, (8.16) becomes

$$\frac{1}{2\sigma^2}|d-\theta|^2 + s(\theta). \tag{8.17}$$

If $s(\theta)$ is strictly convex and differentiable, the minimizer of (8.17) is the solution of

$$\frac{1}{\sigma^2}(\theta - d) + s'(\theta) = 0,$$

given by

$$\hat{\theta} = h^{-1}(d), \tag{8.18}$$

where

$$h(u) = u + \sigma^2 s'(u).$$

Generally, the inversion in (8.18) may not be analytically feasible. Following Hyvärinen [208], we give several examples of prior distributions on θ for which the analytical maximization is possible. The noise component is assumed normal.

Example 8.4.1 [208]
- If $\pi(\theta) = \frac{1}{\sqrt{2}}e^{-\sqrt{2}|\theta|}$, then $s'(\theta) = \sqrt{2}\,\text{sgn}(\theta)$, and $\hat{\theta}(d) = \text{sgn}(d)\max(0, |d| - \sqrt{2}\sigma^2)$.
- If

$$\pi(\theta) \propto e^{-a\theta^2/2 - b|\theta|}, \quad a, b > 0, \tag{8.19}$$

i.e., if $s'(\theta) = a\theta + b\,\text{sgn}(\theta)$, then

$$\hat{\theta}(d) = \frac{1}{1+\sigma^2 a}\text{sgn}(d)\,\max(0, |d| - b\sigma^2). \tag{8.20}$$

This shrinkage function is plotted in Fig. 8.4(a) for $a = 1$ and $b = 2$.
- Let π be a "supergaussian" probability density,

$$\pi(\theta) = C \cdot \left[\sqrt{\alpha(\alpha+1)} + \left|\frac{\theta}{b}\right|\right]^{\alpha+3}, \tag{8.21}$$

where the normalizing constant is $C = \frac{1}{2b}(\alpha+2)\left[\alpha(\alpha+1)/2\right]^{(\alpha/2+1)}$. The resulting MAP rule is

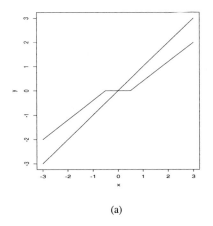 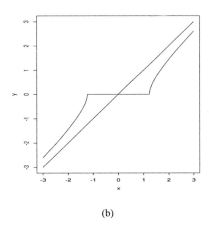

(a) (b)

Fig. 8.4 The MAP solutions for the priors in (8.19) [panel (a)] and (8.21) [panel (b)].

$$\hat{\theta}(d) = \text{sgn}(d) \max\left(0, \frac{|d| - ab}{2} + \frac{1}{2}\sqrt{(|d| + ab)^2 - 4\sigma^2(\alpha + 3)}\right), \quad (8.22)$$

where $a = \sqrt{\alpha(\alpha+1)/2}$, and $\hat{\theta}(d)$ is set to 0 if the square root in (8.22) is imaginary. The rule in (8.22) is depicted in Fig. 8.4(b).

Leporini and Pesquet [259] explore cases for which the prior is an exponential power distribution [$\mathcal{EPD}(\alpha, \beta)$] as in (6.22). If the noise also has an $\mathcal{EPD}(a, b)$ distribution with $0 < \beta < b \leq 1$, the MAP solution is a hard-thresholding rule. If $0 < \beta \leq 1 < b$ then the resulting MAP rule is

$$\hat{\theta}(d) = d - \left(\frac{\beta a^b}{b\alpha^\beta}\right)^{1/(b-1)} |d|^{(\beta-1)/(b-1)} + o(|d|^{(\beta-1)/(b-1)}). \quad (8.23)$$

The authors in [259] consider also the Cauchy noise and explore properties of the resulting rules. See also related articles by Chambolle et al. [67], Krim and Schick [248], and Simoncelli [379].

8.5 DENSITY ESTIMATION PROBLEM

The local nature of wavelet functions makes the wavelet estimator superior to projection estimators that use classical orthonormal bases (Fourier, Hermite, etc.).

Brunk [45] first proposed the shrinkage of coefficients in the projection density estimate based on linear Bayesian estimation. When an $\mathbb{L}_2$ density f has the Fourier expansion $f(x) = \sum_n a_n \xi(x)$ in some orthonormal basis $\xi_n(x), n \in \mathbb{N}$, Brunk (see also Wahba [433]) assumed normal $\mathcal{N}(0, b_n)$ and independent priors on the coefficients a_n. The suggested choice for b_n was C/m^n, where C is a constant and $m > 1$. Brunk's linear Bayes shrinkage paradigm can be extended to a wavelet setup. As an illustration we consider the normal/normal-inverse gamma model, see Chapter 9 in O'Hagan [326].

Assume the model

$$[\underline{d}|\underline{\theta}, \sigma^2] \sim \mathcal{MVN}(\underline{\theta}, \sigma^2 I) \tag{8.24}$$

for the vector $\underline{d}$ consisting of empirical coefficients $c_{j_0,k}, k \in \mathbb{Z}$, and $d_{j,k}$, $j_0 \le j \le j_1, k \in \mathbb{Z}$, in

$$\sum_k c_{j_0,k} \phi_{j_0,k}(x) + \sum_{j=j_0}^{j_1} \sum_k d_{j_0,k} \psi_{jk}(x). \tag{8.25}$$

The rationale behind the model in (8.24) is that both c_{jk} and d_{jk} are averages of n i.i.d. random variables; see the expression for empirical coefficients in (7.3). Given $\underline{\theta}$, the components of $\underline{d}$ are independent. The model in (8.24) is completed by specifying the prior distribution on the location $\underline{\theta}$ and scale σ. Let

$$[\underline{\theta}, \sigma^2] \sim \mathcal{NIG}(\alpha, \delta, 0, \Sigma), \tag{8.26}$$

where $\mathcal{NIG}(\alpha, \delta, \underline{m}, \Sigma)$ is the normal-inverse gamma distribution with density function $f(\underline{\theta}, \sigma^2 | \alpha, \delta, \underline{m}, \Sigma) = C \cdot (\sigma^2)^{\frac{\delta+p+2}{2}} \exp[-\{(\underline{\theta} - \underline{m})' \Sigma^{-1} (\underline{\theta} - \underline{m}) + \alpha\}/(2\sigma^2)]$.

Bayesian updating is straightforward because of the conjugate structure. The choice of the prior covariance matrix Σ corresponds to the choice of variances b_n in the traditional (Fourier) problems discussed in [45] and [433]. The model specified by (8.24) and (8.26) can incorporate dependence of the wavelet coefficients via appropriate specification of the prior covariance matrix Σ. If only the coefficients between levels j_0 and j_1 are considered in the estimator, their covariance structure can be described by the following covariance matrix:

$$\Sigma = \lambda_{j_0,\phi} \Sigma_{j_0,\phi} \oplus \bigoplus_{j=j_0}^{j_1} \lambda_{j,\psi} \Sigma_{j,\psi}, \tag{8.27}$$

where $\oplus$ is the direct sum operation. The Σ in (8.27), is a block diagonal matrix in which the block submatrices $\Sigma_{j_0,\phi}, \Sigma_{j,\psi}, j = j_0, \ldots, j_1$ describe correlations within their corresponding levels.

The posterior for $[\underline{\theta}, \sigma^2]$ is again the normal-inverse gamma distribution

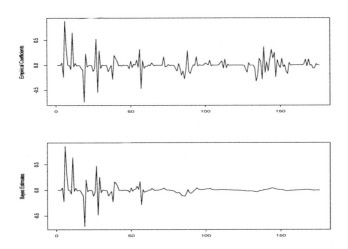

Fig. 8.5 Empirical coefficients for the galaxy data and their affine Bayes shrinkage. The coefficients are arranged as a concatenation of levels in the decomposition, starting with scaling coefficients and ending with coefficients of fine detail. Notice that the Bayes estimate preserves coarse structure and heavily shrinks detail coefficients that are responsible for overfitting.

$$[\theta, \sigma^2 | \underline{d}] \sim \mathcal{NIG}(\alpha^\star, \delta^\star, m^\star, \Sigma^\star),$$

with

$$\Sigma^\star = (I + \Sigma^{-1})^{-1}$$
$$\hat{\theta} = \Sigma^\star \underline{d} \ (= E(\theta \mid \underline{d}))$$
$$\alpha^\star = \alpha + ||\underline{d}|| + (\theta^\star)'(\Sigma^\star)^{-1}\theta^\star$$
$$\delta^\star = \delta + n.$$

The Bayes estimator of θ is $\hat{\theta} = \Sigma^\star \underline{d}$. Naturally, we estimate the density in (8.25) by replacing c_{jk} and d_{jk} from $\underline{d}$ by their affine Bayes estimators.

Example 8.5.1 In practical implementations of this method, the matrix Σ is a block diagonal matrix consisting of Laurent submatrices $\Sigma_{j_0,\phi}, \Sigma_{j_0,\psi}, \ldots, \Sigma_{j_1,\psi}$, each of appropriate dimensions, with entries $\sigma_{i,j} = \rho^{|i-j|}$, $|\rho| < 1$.

Fig. 8.5 shows shrinkage of the empirical wavelet coefficients $\underline{d}$ for the *galaxy velocities* data set (Roeder, [354]). There were 176 empirical coefficients between the levels $j_0 = 0$ and $j_1 = 6$. The following values for the λs and ρ were adopted: $\lambda_{0,\phi} = 1000, \lambda_{0,\psi} = 1000, \lambda_{1,\psi} = 100, \lambda_{2,\psi} = 10, \lambda_{3,\psi} = 1, \lambda_{4,\psi} = 0.1, \lambda_{5,\psi} =$

0.01, $\lambda_{6,\psi} = 0.001$, and $\rho = 0.8$.

Building on (8.24)-(8.27), Huerta [207] proposed the following model

$$[\underline{d}|\underline{\theta}, \sigma^2] \sim \mathcal{MVN}(\underline{\theta}, \sigma^2 I), \quad [\sigma^2] \sim \mathcal{IG}(\alpha_1, \delta_1),$$
$$[\underline{\theta}|\tau^2] \sim \mathcal{MVN}(\underline{0}, \tau^2 \Sigma), \quad [\tau^2] \sim \mathcal{IG}(\alpha_2, \delta_2).$$

The closed form for $E(\underline{\theta}|\underline{d})$ in the above model is not possible, however the full conditionals can be specified. The prior covariance matrix Σ is defined as in (8.27). The shrinkage is performed by Gibbs sampling and the simulation results obtained are promising.

Reducing of the number of hyperparameters in (8.24)-(8.27) and imposing a prior structure on some of the remaining hyperparameters in a hierarchical fashion was suggested by Vannucci and Corradi [415]. The authors applied their method, which they call BayesShrink, on the regression problem. Assuming that $\underline{\theta}$ corresponds to an autoregressive process in the time domain the authors demonstrate that the matrix Σ depends on only two hyperparameters, λ and ρ. The parameter ρ is the "autocovariance index" and λ is the precision parameter. The covariance matrix $\Sigma(\lambda, \rho) = \lambda \Sigma(\rho)$ has an interesting "finger-like" structure.

Vannucci and Corradi in [415] suggest

$$[\lambda] \sim \mathcal{IG}(p/2, q/2), \text{ and}$$
$$[\rho] \propto (C - \rho)^{r_1 - 1}(C + \rho)^{r_2 - 1}, \ |\rho| < C,$$

and simulate from the posterior distribution of $\underline{\theta}$ using the Markov chain Monte Carlo (MCMC) method.

8.6 FULL BAYESIAN MODEL

The methods in the previous sections are similar in spirit to the wavelet-shrinkage paradigm schematized in Fig. 6.1. The coefficients in the wavelet domain are shrunk in a Bayesian fashion. Full Bayesian inference in the density estimation problem would involve the likelihood incorporating the observations.

Müller and Vidakovic [304] parameterize an unknown density $f(x)$ by a wavelet series on its square root, and propose a prior model that explicitly defines geometrically decreasing prior probabilities for non-zero wavelet coefficients at higher levels of detail. Benefits of such parameterization have been discussed in Section 7.4.1.

The unknown probability density function $f(\cdot)$ is modeled by:

$$\sqrt{f(x)} = K \cdot \left[\sum_{k \in \mathbb{Z}} \xi_{j_0 k} \phi_{j_0,k}(x) + \sum_{j \geq j_0, k \in \mathbb{Z}} s_{jk} \theta_{jk} \psi_{jk}(x) \right], \quad (8.28)$$

where K is the normalization constant and $s_{jk} \in \{0, 1\}$ is an indicator variable that performs model induced thresholding. A prior on $[\theta_{jk}|s_{jk}]$ which assigns considerable prior probability mass at 0 was used.

The dependence of $f(x)$ on the vector $\theta = (\xi_{j_0,k}, s_{jk}, \theta_{jk}, j = j_0, \ldots, j_1, k \in \mathbb{Z})$ of wavelet coefficients and indicators is expressed by $f(x) = p(x|\theta)$. The sample $X = \{X_1, \ldots, X_n\}$ defines a likelihood function $p(X|\theta) = \prod_{i=1}^{n} p(X_i|\theta)$.

The model is completed by a prior probability distribution for θ. Without loss of generality, $j_0 = 0$ can be assumed. Also, any particular application will determine a maximum level of detail j_1.

$$\begin{aligned}
[\xi_{0k}] &\sim \mathcal{N}(0, \tau r_0), \\
[\theta_{jk}|s_{jk} = 1] &\sim \mathcal{N}(0, \tau r_j), \quad r_j = 2^{-j}, \\
[s_{jk}] &\sim \mathcal{B}er(\alpha^j), \quad (8.29) \\
[\alpha] &\sim \mathcal{B}eta(a, b), \\
[1/\tau] &\sim \mathcal{G}amma(a_\tau, b_\tau).
\end{aligned}$$

The wavelet coefficients θ_{jk} are non-zero with geometrically decreasing probabilities. Given that a coefficient is non-zero, it is generated from a normal distribution. When a coefficient is not included in (8.28), that is, $s_{jk} = 0$, a "pseudo-prior" $h(\theta_{jk})$ is assumed. The parameter vector θ is augmented in order to include all model parameters, i.e., $\theta = (\theta_{jk}, \xi_{jk}, s_{jk}, \alpha, \tau)$.

The scale factor r_j contributes to the adaptivity of the method. Wavelet shrinkage is controlled by both the factor r_j and geometrically decreasing prior probabilities for non-zero coefficient, α^j.

The conditional prior $p(\theta_{jk}|s_{jk} = 0) = h(\theta_{jk})$ in the model (8.29) is a pseudo-prior as discussed in Carlin and Chib [58]. The choice of $h(\cdot)$ has no bearing on the inference about $f(x)$. In fact, the model could be alternatively formulated by dropping θ_{jk} under $s_{jk} = 0$. However, this would lead to a parameter space of varying dimension. Carlin and Chib [58] argued that the pseudo-prior $h(\theta_{jk})$ should be chosen to produce values for θ_{jk} which are consistent with the data. The normal distribution $p(\theta_{jk}|s_{jk} = 0) = \mathcal{N}(\hat{\theta}_{jk}, \sigma_{jk})$, where $\hat{\theta}_{jk}$ is some rough preliminary estimate of θ_{jk}, was proposed.

The particular MCMC simulation scheme used to estimate the model (8.28, 8.29) is described. By starting with some initial values for $\theta_{jk}, j = 0, \ldots, j_1, \xi_{00}, \alpha$, and τ, the following Markov chain was implemented.

1. For each $j = 0, \ldots, j_1$ and $k = 0, \ldots, 2^j - 1$ repeat the steps **2** and **3**.

264 BAYESIAN METHODS IN WAVELETS

2. Update s_{jk}. Let θ_0 and θ_1 indicate the current parameter vector θ with s_{jk} replaced by 0 and 1, respectively. Compute $p_0 = p(y|\theta_0) \cdot (1 - \alpha^j)h(\theta_{jk})$ and $p_1 = p(y|\theta_0) \cdot \alpha^j p(\theta_{jk}|s_{jk} = 1)$. With probability $p_1/(p_0 + p_1)$ set $s_{jk} = 1$, otherwise, set $s_{jk} = 0$.

3a. Update θ_{jk}. If $s_{jk} = 1$, generate $\tilde{\theta}_{jk} \sim g(\tilde{\theta}_{jk}|\theta_{jk})$. Use, for example, $g(\tilde{\theta}_{jk}|\theta_{jk}) = \mathcal{N}(\theta_{jk}, 0.25\sigma_{jk})$, where σ_{jk} is some rough estimate of the posterior standard deviation of θ_{jk}. We will discuss alternative choices for the probing distribution $g(\cdot)$ below.

Compute
$$a(\theta_{jk}, \tilde{\theta}_{jk}) = \min\left[1, \frac{p(y|\tilde{\theta})p(\tilde{\theta}_{jk})}{p(y|\theta)p(\theta_{jk})}\right],$$

where $\tilde{\theta}$ is the parameter vector θ with θ_{jk} replaced by $\tilde{\theta}_{jk}$, and $p(\theta_{jk})$ is the p.d.f. of the normal prior distribution given in (8.29).

With probability $a(\theta_{jk}, \tilde{\theta}_{jk})$ replace θ_{jk} by $\tilde{\theta}_{jk}$; otherwise keep θ_{jk} unchanged.

3b. If $s_{jk} = 0$, generate θ_{jk} from the full conditional posterior $p(\theta_{jk}|\ldots, X) = p(\theta_{jk}|s_{jk} = 0) = h(\theta_{jk})$.

4. Update ξ_{00}. Generate $\tilde{\xi}_{00} \sim g(\tilde{\xi}_{00}|\xi_{00})$. Use, for example, $g(\tilde{\xi}_{00}|\xi_{00}) = \mathcal{N}(\xi_{00}, 0.25\rho_{00})$, where ρ_{00} is some rough estimate of the posterior standard deviation of ξ_{00}. Analogously to step **3a**, compute an acceptance probability a and replace ξ_{00} with probability a.

5. Update α. Generate $\tilde{\alpha} \sim g_\alpha(\tilde{\alpha}|\alpha)$ and compute
$$a(\alpha, \tilde{\alpha}) = \min\left[1, \frac{\prod_{jk} \tilde{\alpha}^{js_{jk}}(1-\tilde{\alpha}^j)^{s_{jk}}}{\prod_{jk} \alpha^{js_{jk}}(1-\alpha^j)^{s_{jk}}}\right].$$

With probability $a(\alpha, \tilde{\alpha})$ replace α by $\tilde{\alpha}$, otherwise keep α unchanged.

6. Update τ. Resample τ from the complete inverse Gamma conditional posterior.

7. Iterate over steps **1** through **6** until the chain is judged to have practically converged.

This algorithm implements a Metropolis chain changing one parameter at a time in the parameter vector. See, for example, Tierney [405] for a description and discussion of Metropolis chains for posterior exploration. For a practical implementation, g should be chosen such that the acceptance probabilities a are neither close to zero, nor close to one. In the implementations, $g(\tilde{\theta}_{jk}|\theta_{jk}) = \mathcal{N}(\theta_{jk}, 0.25\sigma_{jk})$ with $\sigma_{jk} = 2^{-j}$ was used. See also Müller and Vidakovic [305] for a Bayesian solution to the related problems of estimating a spectral density and NES regression.

Example 8.6.1 The wavelet-based density estimation model (8.28-8.29) is illustrated on the galaxy data set (Roeder [354]). The data is rescaled to the interval $[0, 1]$. The

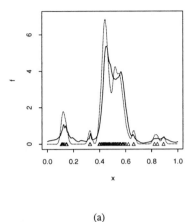 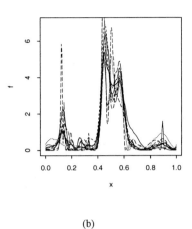

(a) (b)

Fig. 8.6 [304] (a) The estimated p.d.f. $\hat{f}(x) = \int p(x|\theta)\, dp(\theta|X)$. The dotted line plots a conventional kernel density estimate for the same data. (b) The posterior distribution of the unknown density $f(x) = p(x|\theta)$ induced by the posterior distribution $p(\theta|X)$. The lines plot $p(x|\theta_i)$ for 10 simulated draws $\theta_i \sim p(\theta|X), i = 1, \ldots, 10$.

hyperparameters were fixed as $a = 10$, $b = 10$, and $a_\tau = b_\tau = 1$. The $\mathcal{B}eta(10, 10)$ prior distribution on α is reasonably non-informative as compared to the likelihood based on $n = 82$ observations.

Initially, all s_{jk} are set to 1, and α to its prior mean $\alpha = 0.5$. The first 10 iterations as a burn-in period were discarded, then 1000 iterations of steps **1** through **6** were simulated. For each j, k step **3** was repeated three times. The maximum level of detail selected was $j_1 = 5$.

Fig. 8.6 and Fig. 8.7 depict some aspects of the analysis.

8.7 DISCUSSION AND REFERENCES

Lina and MacGibbon [263] apply a Bayesian approach to wavelet regression with complex valued Daubechies wavelets. To some extent, they exploit redundancy in the representation of real signals by the complex wavelet coefficients. Their shrinkage technique is based on the observation that the modulus and the phase of wavelet coefficients encompass very different information about the signal. A Bayesian shrinkage model is constructed for the modulus, taking into account the corresponding phase.

Simoncelli and Adelson [380] discuss the Bayes "coring" procedure in the context of image processing. The prior on the signal is Mallat's model (6.22), while the noise is assumed normal. They implement their noise reduction scheme on an oriented

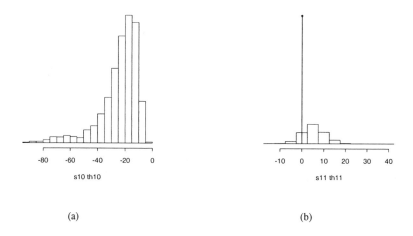

Fig. 8.7 [304] Posterior distributions $p(s_{10}\theta_{10}|X)$ [panel (a)] and $p(s_{11}\theta_{11}|X)$ [panel (b)]. While $s_{10}\theta_{10}$ is non-zero with posterior probability close to one, the posterior distribution $p(s_{11}\theta_{11}|X)$ is a mixture of a point mass at zero and a continuous part.

multiresolution representation — known as the *steerable pyramid*. The authors report that the Bayesian coring outperforms the classical Wiener filtering.

Jansen and Bultheel [218], Malfait and Roose [273], and Malfait, Jansen, and Roose [272] employed the theory of Markov random fields in wavelet denoising of images. The idea is that the thresholding of a coefficient in a two-dimensional array depends not only on its own magnitude but also on the magnitudes of its neighbors. This dependence is modeled by using masks consisting of 0 and 1 and exhibiting properties of a Markov field. Crouse, Nowak, and Baraniuk [98] also consider hidden Markov fields in a similar setup. They develop the *efficient expectation maximization* (EEM) algorithm to fit their model. Also see an overview in Nowak [318].

The shrinkage models (as in [422]) in which the hyperparameters of the prior are made time-dependent in an empirical Bayes fashion are considered in Vidakovic and Bielza [423].

Pesquet et al. [340] developed a Bayesian-based approach to the best basis problem, while preserving the classical tree search efficiency in wavelet packets and local trigonometric bases. See also Leporini [258] and Leporini, Pesquet, and Krim [260] for related results.

Kohn and Marron [237] use a model similar to one in Chipman, Kolaczyk, and McCulloch [74] in the context of the best basis selection.

An extreme dimension reduction in prediction problems, when the predictors are curves, is achieved in the wavelet domain by Brown, Fearn and Vannucci [40]. The authors utilize Bayesian model mixing in the wavelet domain in applications from the near-infrared spectroscopy. It is interesting to observe that the "most predictive" wavelet coefficients have small to moderate magnitudes. This was noticed by several researchers; in prediction problems *the bigger – the better* selection rule does not apply.

Timmermann and Nowak [406] introduce multiscale Bayesian estimators of Poisson intensities. To preserve Poisson characteristics in the wavelet domain, they utilize the non-normalized Haar transformation. The proposed mixture Beta priors on the intensity are realistic and lead to efficient shrinkage estimation. See also Nowak and Baraniuk [319] for an application in photon imaging. Independently, related results have been obtained by Kolaczyk [240, 241].

Ogden [320] develops a Bayesian wavelet method for estimating the location of a change-point. Initial treatment is for the standard change-point model (i.e., constant mean before the change and constant mean after it), but extends to the case of detecting discontinuity points in an otherwise smooth curve. The conjugate prior distribution on the change point τ is given in the wavelet domain, and it is updated by the observed empirical wavelet coefficients. See also Ogden and Lynch [323].

Ruggeri and Vidakovic [358] discuss Bayesian decision theoretic thresholding. In the set of all hard thresholding rules, they find Bayes rules under a variety of models, priors, and loss functions. They identify model-prior pairs that work well (in the sense of exhibiting a hard thresholding rule that minimizes Bayes risk) and show that in the presence of prior information on noise and signal, their procedure outperforms comparable minimax procedures. See also Ruggeri [357] and Exercise 8.1.

Lu, Huang, and Tung [268] perform linear Bayesian wavelet shrinkage in a nonparametric mixed-effect model. Their formulation is conceptually inspired by the duality between reproducing kernel Hilbert spaces and random processes as well as by connections between smoothing splines and Bayesian regressions. The unknown function f in the standard nonparametric regression formulation $[y = f(x_i) + \sigma \epsilon_i, i = 1, \ldots, n; 0 \leq x \leq 1; \sigma > 0; Cov(\epsilon_1, \ldots, \epsilon_n) = R]$ is given a prior of the form $f(x) = \sum_k \alpha_{Jk} \phi_{Jk}(x) + \delta Z(x); Z(x) \sim \sum_{j \geq J} \sum_k \theta_{jk} \psi_{jk}(x)$, where θ_{jk} are uncorrelated random variables such that $E\theta_{jk} = 0$ and $E\theta_{jk}^2 = \lambda_j$. The authors propose a linear, empirical Bayes estimator $\hat{f}$ of f that enjoys a Gauss-Markov type of optimality. Several non-linear versions of the estimator are proposed as well. Independently, and using different techniques, Huang and Cressie [202, 203] consider the same problem and derive a Bayesian estimate.

Pensky [332] proposes wavelet-based empirical Bayes estimators that are adaptive to unknown smoothness of a prior. Several conditional distributions are considered. See also Huang [204] and Exercise 8.6.

8.8 EXERCISES

8.1. [358] Consider the normal model $X \sim \mathcal{N}(\theta, \sigma^2)$, in which σ^2 is assumed known (or estimable), and the double exponential prior $\theta \sim \mathcal{DE}(0, \beta)$. Let the loss be squared error.

(a) Find the Bayes risk of the decision rule $\delta^h(x, \lambda) = x \mathbf{1}(|x| > \lambda)$.

(b) Show, that when $\sigma = \beta = 1$, the optimal λ^*, minimizing the Bayes risk, is approximately 0.94055.

8.2. Plot the graphs of $m(d)$ and $\hat{\theta}_\pi(d)$ from Example 8.1.2, for $m = 2$ and $p = 0.15$.

Approximately calculate the range of values $p \in [0, p_0]$ for which

$$\int \frac{\hat{\theta}_\pi(u)}{u} \cdot m(u)\, du > 1.$$

8.3. [74] Assume the model (8.4 - 8.6). Argue that $\text{Var}(\theta|d) \approx \tau^2$, when d is small, and $\text{Var}(\theta|d) \approx \sigma^2$, when d is large.

8.4. *Project.* Develop "The larger posterior mode shrinkage" based on the following consideration: Assume $d \sim \mathcal{N}(\theta, 1)$, $\theta \sim \mathcal{N}(0, \tau^2)$, and $\pi(\tau^2) = 1/\tau^{3/2}$. Find the posterior distribution of θ. Show that it is unimodal if $0 < d^2 < 2$ and bimodal otherwise. Show that the second mode is $\theta(d) = \left(1 - \frac{1-\sqrt{1-(2/d^2)}}{2}\right) d$.

8.5. Show that the following shrinkage rule

$$\theta(d) = \left(\frac{1}{1 - e^{-||d||^2/2}} - \frac{2}{||d||^2}\right) d$$

can be obtained as a Bayes estimator in the following hierarchical model: $\underline{d} \sim \mathcal{MVN}_p(\underline{\theta}, I_p)$, $\underline{\theta} \sim \mathcal{MVN}(0, \tau^2 I_p)$ and $\pi_2(\eta) = \eta^{2-(p/2)}$, for $\eta = 1/(1 + \tau^2)$.

8.6. [204]. Let $(X_1, \theta_1), (X_2, \theta_2), \ldots, (X_n, \theta_n)$ be n i.i.d. copies of (X, θ), where X, conditionally on θ, has the uniform $\mathcal{U}(0, \theta)$ distribution. The prior on θ, $G(\theta)$ is unknown but it is known that it is supported on $[0, m)$. Suppose a new observation $X_{n+1} = x$ is obtained.

(a) If the marginal distribution $f_G(x) = \int_x^m 1/\theta \, dG(\theta)$ were known, show that a Bayes estimate for θ_{n+1} would be

$$\theta_G(x) = x + \frac{1 - F_G(x)}{f_G(x)}.$$

(b) Let $\theta_B(x) = x \vee \left[\left(x + \frac{1 - F_n(x)}{\hat{f}_j^L(x)}\right) \wedge m\right]$ be the wavelet version of $\theta_G(x)$, where $F_n(x)$ is the sample c.d.f. and $\hat{f}_j^L(x)$ is the linear wavelet density estimator, as in (7.7, 7.8).

If $f_G(x)$ belongs to $\mathbb{W}^s$, the MRA is r-regular $r > s$, and the conditions

(i) $0 < \lim_{x \to 0+} f_G(x) < \infty$ and

(ii) $E(1/f_G^2(X)) < \infty$

hold, then the regret risk satisfies

$$0 \leq R(G, \theta_B) - R(G, \theta_G) = O\left(\frac{1}{n}\right) + O\left(\frac{1}{nh}\right) + o(h^{2s}),$$

where $h = 2^{-j}$, $j \in \mathbb{Z}$.

Which rate of $h \to 0$ leads to the optimal regret rate of $O(n^{-\frac{2s}{2s+1}})$?

9
Wavelets and Random Processes

In this chapter, we give an overview of wavelet methods used in the theory of random processes. The emphasis will be on stationary time series, although some applications of wavelets to non-stationary processes will also be mentioned.

The chief motivation for modeling a time series is the need for forecasting. To that end, an analysis of time series can be performed in the time domain as well as in the frequency domain. Wavelets provide additional insight in the analysis of time series via scale analysis. The notion of frequency in Fourier analysis can be related to the notion of scale in multiscale analysis; and often Fourier-based tools for exploring time series have their wavelet counterparts (for instance, wavelet spectra, wavelet periodogram, and scalogram). Self-similarity properties of some processes, such as fBm or ARIMA, can be well described by wavelet methods. Wavelets also generate novel random processes and functions, suitable for multiscale and tree-like modeling.

Applications of wavelets in the smoothing of periodograms (sample counterparts of spectral densities) is a task of a different nature. It is related to the nonparametric procedures covered in Chapters 7 and 8, with some distributional specificities of its own.

9.1 STATIONARY TIME SERIES

A stochastic process is a family of random variables $\{X_t, t \in T\}$ defined on a common probability space. The set $T \subseteq (-\infty, \infty)$ is thought of as a time parameter set. The process is called a *continuous parameter process* if T is continuous (an interval, for instance), and a *discrete parameter process* if T is discrete, usually a subset of integers, $T \subseteq \mathbb{Z}$.

In most of this chapter, we discuss applications of wavelets in time series that are discrete parameter random processes. We first give some basic definitions.

Let $\{X_t,\ t \in \mathbb{Z}\}$ be a random process such that the autocovariance function

$$\gamma_X(r, s) = \text{Cov}(X_r, X_s) = E(X_r - EX_r)(X_s - EX_s) \tag{9.1}$$

is finite for any pair of r and s in T. The random process (time series) $\{X_t,\ t \in T\}$ with parameter set $T = \mathbb{Z} = \{0, \pm 1, \pm 2, \dots\}$ is said to be (weakly or wide-sense) stationary if

(i) $E|X_t|^2 < \infty$,
(ii) $EX_t = m$, for all $t \in \mathbb{Z}$, and
(iii) $\gamma_X(r, s) = \gamma_X(r + t, s + t)$ for all r, s, and $t \in \mathbb{Z}$.

The stationarity condition (iii) is often given in the form,

$$\text{Cov}(X_{t+h}, X_t) = \gamma_X(h),$$

emphasizing the independence of t. When it is clear what the underlying process is, we will write $\gamma(h)$ instead of $\gamma_X(h)$.

Example 9.1.1 (i) *White noise* is a stationary sequence Z_t such that $EZ_t = 0$ and $\gamma(h) = \sigma^2 \cdot \delta_h$; in notation, $Z_t \sim \mathcal{WN}(0, \sigma^2)$.

(ii) The *moving average* MA(q) process, defined as

$$X_t = Z_t + \theta_1 Z_{t-1} + \theta_2 Z_{t-2} + \cdots + \theta_q Z_{t-q}, \quad Z_t \sim \mathcal{WN}(0, \sigma^2),$$

has autocovariance function

$$\gamma(h) = \begin{cases} \sigma^2 \sum_{j=0}^{q-|h|} \theta_j \theta_{j+|h|}, & |h| \leq q \\ 0 & |h| > q. \end{cases} \tag{9.2}$$

(iii) The *autoregressive* AR(1) process, $X_t - \phi X_{t-1} = Z_t$, $Z_t \sim \mathcal{WN}(0, \sigma^2)$, has autocovariance function

$$\gamma(h) = \sigma^2 \frac{\phi^{|h|}}{1-\phi^2}. \tag{9.3}$$

For more on time series and random processes in general we direct the reader to the monographs of Brockwell and Davis [39], Priestley [346], and West and Harrison [455], among others.

9.2 WAVELETS AND STATIONARY PROCESSES

In this section, we present results that link continuous-time stationary random processes and wavelet methods. We consider only wavelet series and discrete wavelet transformations; for continuous wavelet decompositions we direct the reader to Averkamp and Houdré [20], Cambanis and Houdré [56], and Cambanis and Masry [57].

9.2.1 Wavelet Transformations of Stationary Processes

Let $X(t)$ be a second order stationary process with autocorrelation function $\gamma(t, s)$. The discrete wavelet transformation of $X(t)$ is a discrete random field

$$\{d_{jk},\ j, k \in \mathbb{Z}\} = \left\{ \int_{\mathbb{R}} X(t)\psi_{jk}(t)\, dt,\ j, k \in \mathbb{Z} \right\}, \tag{9.4}$$

which is well defined if the path integrals in (9.4) are defined and

$$\int_{\mathbb{R}} \sqrt{\gamma(t,t)}\, |\psi_{jk}(t)|\, dt < \infty. \tag{9.5}$$

Thus, when (9.5) is satisfied

$$E d_{jk} d_{j'k'} = \iint_{\mathbb{R}^2} \gamma(t,s) \psi_{jk}(t) \psi_{j'k'}(s)\, dt\, ds.$$

The following lemma is straightforward. Its versions can be found in many papers; see for example: [349],[19], [56], and [247].

Lemma 9.2.1 *Let $X(t)$, $t \in \mathbb{R}$ be a weakly stationary process. For $l, n \in \mathbb{Z}$ and $j \geq l$, the random sequence $\{d_{j, 2^{j-l}k+n},\ k \in \mathbb{Z}\}$ is weakly stationary as well.*

Simply speaking, any level in a wavelet transformation of a stationary random process (or sequence) is a stationary sequence. The converse, may not be true. For example,

$$X(t) = \sum_k Z_k \psi(t-k),$$

where the Z_k are uniform $\mathcal{U}(-1,1)$, is a well-defined process of second order but not necessarily stationary. The corresponding levels in the wavelet decomposition of $X(t)$ are clearly stationary since they are either all zeroes or Z_k, $k \in \mathbb{Z}$ (the level corresponding to W_0). Also see Exercise 9.3.

The following result is from Averkamp and Houdré [18].

Theorem 9.2.1 *[18] Let $X(t)$ be a second order stationary process for which $\gamma(t,s)$ is bounded and continuous in $\mathbb{R}^2$. Let d_{jk} be defined for all j,k. Then, $\{d_{jk}, k \in \mathbb{Z}\}$ is weakly stationary iff $X(t)$ is weakly stationary.*

The condition of boundedness of $\gamma(t,s)$ can be dropped if the wavelet basis is compactly supported.

9.2.2 Whitening of Stationary Processes

Let $X(t)$, $t \in \mathbb{R}$ be a stationary process and let $X_m(t)$ be its projection on the multiresolution subspace V_m, i.e.,

$$X_m(t) = \sum_k c_{mk} \phi_{mk}(t), \tag{9.6}$$

with $c_{mk} = \int X(t) \phi_{mk}(t)\, dt$. If the wavelet ϕ is r-regular,

$$E(X(t) - X_m(t))^2 \to 0 \text{ when } m \to \infty. \tag{9.7}$$

Indeed, since

$$\begin{aligned}
E(X(t) - X_m(t))^2 &= EX(t)^2 - 2EX(t)X_m(t) + EX_m(t)^2 \\
&= \gamma(0) - 2\int \gamma(t-s)\mathbb{K}_m(t,s)\, ds \\
&\quad + \iint \gamma(u-s)\mathbb{K}_m(s,t)\mathbb{K}_m(t,u)\, du\, ds
\end{aligned}$$

and

$$\int \gamma(u-s)\mathbb{K}_m(t,s)\,ds \to \gamma(u-t),$$

uniformly on bounded sets, (9.7) is implied.

The informal statement "Wavelets whiten data" can be formalized if the data on input have a stationary dependence structure. Let

$$X_m(t) = \sum_{j=-\infty}^{m-1} \sum_k d_{jk}\psi_{jk}(t),$$

be an equivalent representation of (9.6) with

$$d_{jk} = \int_{\mathbb{R}} X(t)\psi_{jk}(t)\,dt. \tag{9.8}$$

It is easy to see that

$$Ed_{jk}d_{j'k'} = \frac{1}{2\pi}\int_{\mathbb{R}} \hat{\gamma}(\omega)\Psi\left(\frac{\omega}{2^j}\right)\overline{\Psi\left(\frac{\omega}{2^{j'}}\right)} \tag{9.9}$$
$$\cdot e^{-i\omega k 2^{-j}} e^{i\omega k' 2^{-j'}} 2^{-j/2} 2^{-j'/2}\,d\omega,$$

where Ψ and $\hat{\gamma}$ are the Fourier transformations of ψ and γ, respectively. By further exploring (9.9), Walter [439] proved the result given in Theorem 9.2.2.

Theorem 9.2.2 *Let the wavelet basis be of Meyer-type such that both Ψ and $\hat{\gamma}$ belong to $\mathbb{C}^p$, $p > 1$. Then, the coefficients d_{jk} and $d_{j'k'}$ defined as in (9.8)*
(i) Are uncorrelated if $|j - j'| > 1$,
(ii) Have arbitrarily small correlation if $|j - j'| = 1$, and
(iii) Have correlation $O(|k - k'|^{-p})$ if the scales j and j' coincide.

Note that if the scale j is fixed, the d_{jk} form a stationary sequence; for details see discussion on page 274; see also Masry [288]. We remark that analogous results hold for $1/f$ processes, which will be discussed in Section 9.5.

Example 9.2.1 In this example, we simulate an AR process with $\phi = -0.6$, apply Haar's wavelet transformation to it, and demonstrate the whitening property of the transformation by inspecting spectral densities associated to different levels of the decomposition.

Fig. 9.1(a) shows spectral densities for the original autoregressive time series (bell-shaped line) and for the time series at the finest level of detail in the transformation (more flat, U-shaped line). Recall that the white noise process has uniform spectral

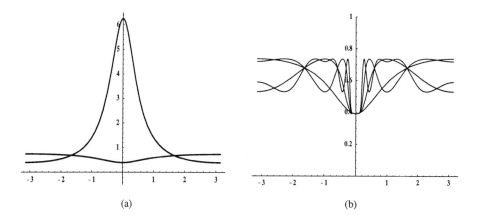

Fig. 9.1 Spectral densities for the levels in the Haar transformation of an autoregressive AR(1) process ($\phi = -0.6$). Panel (a) gives the spectral densities of the original process and the finest level of detail (more flat line). Panel (b) depicts superimposed spectral densities for "detail" processes at the first four levels.

density (Example 9.3.1). Thus, the closeness of a spectral density to the uniform distribution can be taken as a measure of closeness of the associated process to white noise. Panel (b) gives the spectral densities for the first four levels of detail.

9.2.3 Karhunen-Loève-Like Expansions

Karhunen-Loève (K-L) expansions convert continuous-time random processes to sequences of uncorrelated random variables. More formally, if $X(t), t \in [0, T]$ is a real, mean-square continuous, zero-mean process, then there exists an orthonormal system of nonrandom functions $\phi_1(t), \phi_2(t), \ldots$, and a set of uncorrelated random variables $Z_1, Z_2, \ldots$, such that

$$X(t) = \sum_i Z_i \phi_i(t),$$

where $Z_i = \int_0^T X(t)\phi_i(t)\,dt$. The functions $\phi_i(t)$ are eigenfunctions of an associated integral operator

$$(\Gamma f)(t) = \int_0^T \gamma(t, s) f(s)\, ds, \qquad (9.10)$$

where $\gamma(t, s) = EX(t)X(s)$ is the autocovariance function. The K-L transformation is "energy packing." The first k coefficients, corresponding to the k largest eigenvalues of the operator (9.10) minimize the MSE among all orthogonal (unitary in the complex case) transformations. For details and related results on K-L decompositions, see Vetterli and Kovačević [419] and Walter [439].

Example 9.2.2 The Brownian bridge process, $X(t)$, defined for $t \in [0,1]$, is a zero-mean stochastic process with autocovariance function

$$\gamma(t,s) = (t \wedge s)[1 - (t \vee s)].$$

This process can be expressed as $X(t) = B(t) - tB(1)$, $0 \le t \le 1$, where $B(t)$ is a standard Brownian (Wiener) process. The eigenvalue problem

$$(\Gamma\phi)(t) = \lambda\phi(t), \qquad (9.11)$$

where $(\Gamma\phi)(t)$ is defined in (9.10) is explicit,

$$\lambda\phi''(t) = -\phi(t), \quad \phi(0) = \phi(1) = 0,$$

with solutions $\lambda_n = \frac{1}{\pi n^2}$ and $\phi_n(t) = \sqrt{2}\sin(\pi n t)$, $n = 1, 2, \ldots$. Hence, $X(t) = \sum_n Z_n \cdot \sqrt{2}\sin(\pi n t)$ where $Z_n = \int_0^1 X(t) \cdot \sqrt{2}\sin(\pi n t)\, dt$.

Even though the K-L is an elegant theory, it has limited applicability. As a rule, it is difficult to obtain exact solutions of the eigenvalue problem (9.11). The transformation is data-dependent and approximate solutions are computationally intensive. For a discussion see Wornell [460] and Jain [215].

As we have seen in Theorem 9.2.2, wavelet transformations are approximations to K-L transformations. Even stronger results are possible. For a given (wide-sense) stationary process, one can construct a system of basis functions so that the associated coefficients are uncorrelated, a feature similar to K-L expansions. The convenient wavelet basis for such constructions are Meyer-type wavelets; see the discussion in Section 3.4.3. Unlike the standard K-L expansions, the wavelet based expansions do not require solving eigen-equations. They are applicable to time-unlimited processes as well. An additional advantage over Fourier-based approximations is that wavelet K-L-like expansions have completely uncorrelated coefficients and are more stable. Notice that, in order to construct a wavelet K-L-like expansion, one needs to know the autocovariance function in advance.

The construction described below is due to Zhang and Walter [473]. Let $\gamma(\cdot)$ be the autocovariance function of a stationary process $X(t)$ with the property $\hat{\gamma}(\omega) \ne 0$, $\omega \in \mathbb{R}$. If $\Phi(\omega)$ and $\Psi(\omega)$ are the Fourier transformations of the functions ϕ and ψ, respectively, let

$$\hat{\theta}_0(\omega) = \Phi(\omega) \cdot [\hat{\gamma}(\omega)]^{-1/2}, \qquad (9.12)$$

and

$$\hat{\theta}^0(\omega) = \Phi(\omega) \cdot [\hat{\gamma}(\omega)]^{1/2}. \tag{9.13}$$

If $Z_m = \int_{\mathbb{R}} X(t)\theta_0(t-m)\,dt$, then Z_m and Z_n are uncorrelated whenever $m \neq n$. Indeed,

$$\begin{aligned}
EZ_m Z_n &= \iint \gamma(t-s)\theta_0(t-m)\theta_0(s-n)\,dt\,ds \\
&= \frac{1}{2\pi}\int \hat{\gamma}(\omega)|\hat{\theta}_0(\omega)|^2 e^{i\omega(n-m)}\,d\omega \\
&= \frac{1}{2\pi}\int |\hat{\Phi}(\omega)|^2 e^{i\omega(n-m)}\,d\omega \\
&= \int \phi(t-n)\phi(t-m)\,dt = \delta_{m,n}.
\end{aligned}$$

Define

$$\hat{\theta}_j(\omega) = \Phi_j(\omega) \cdot [\hat{\gamma}(\omega)]^{-1/2} \quad \text{and} \quad \hat{\theta}^j(\omega) = \Phi_j(\omega) \cdot [\hat{\gamma}(\omega)]^{1/2},$$

and their shifts

$$\hat{\theta}_{jk}(\omega) = \Phi_{jk}(\omega) \cdot [\hat{\gamma}(\omega)]^{-1/2} \quad \text{and} \quad \hat{\theta}^{jk}(\omega) = \Phi_{jk}(\omega) \cdot [\hat{\gamma}(\omega)]^{1/2}.$$

Let U_0 and U^0 be the spaces spanned by $\{\theta_0(t-k),\ k \in \mathbb{Z}\}$ and $\{\theta^0(t-k),\ k \in \mathbb{Z}\}$. The spaces U_0 and U^0 are the subspaces in a biorthogonal multiresolution-like analysis. In fact, there exists a pair of linear transformations $\mathcal{R}^{-1/2}$ and $\mathcal{R}^{1/2}$, so that the $U_m = \mathcal{R}^{-1/2}(V_m)$ and $U^m = \mathcal{R}^{1/2}(V_m)$, where V_m are subspaces in a standard MRA that is generated by ϕ.

Lemma 9.2.2 *For any j, $\{\theta_{jk}(t)\} = \{\theta_j(t - 2^{-j}k)\}$ is a Riesz basis of U_m, and $\{\theta^{jk}(t)\} = \{\theta^j(t - 2^{-j}k)\}$ is a Riesz basis of U^m. The sequence of functions $\{\theta_{jk}(t), \theta^{jk}(t)\}$ is a biorthogonal system.*

By analogy, one can define wavelet-like functions

$$\hat{\xi}_j(\omega) = \Psi_j(\omega)(\hat{\gamma}(\omega))^{-1/2} \quad \text{and} \quad \hat{\xi}^j(\omega) = \Psi_j(\omega)(\hat{\gamma}(\omega))^{1/2},$$

and their translated versions

$$\hat{\xi}_{jk}(\omega) = \Psi_{jk}(\omega) \cdot [\hat{\gamma}(\omega)]^{-1/2} \quad \text{and} \quad \hat{\xi}^{jk}(\omega) = \Psi_{jk}(\omega) \cdot [\hat{\gamma}(\omega)]^{1/2},$$

which are Riesz bases of the detail-like spaces $M_j = \mathcal{R}^{-1/2}(W_j)$ and $M^j = \mathcal{R}^{1/2}(W_j)$; and $\{\xi_{jk}(t), \xi^{jk}(t)\}$ are biorthogonal sequences in both j and k. We say *wavelet-like* since the spaces U_j, U^j, M_j, and M^j are not self-similar and no common analytic form of the scaling function exists (the analytic form of the scaling function depends on j).

The space $\mathbb{L}_2(\mathbb{R})$ has a multiresolution-like decomposition,

$$\mathbb{L}_2(\mathbb{R}) = U^0 \oplus \bigoplus_{j \geq 0} M^j.$$

Theorem 9.2.3 *[473]. If the spectral density of $X(t)$ is strictly positive, then there exists a K-L-like expansion*

$$\begin{aligned} X(t) &= \sum_{k \in \mathbb{Z}} Z_{0,k} \theta^0(t-k) + \sum_{j \geq 0} \sum_{k \in \mathbb{Z}} Z'_{j,k} \xi^j(t - 2^{-j}k) \\ &= \sum_{j \in \mathbb{Z}} \sum_{k \in \mathbb{Z}} Z'_{j,k} \xi^j(t - 2^{-j}k), \end{aligned} \tag{9.14}$$

with uncorrelated expansion coefficients

$$Z_{0,k} = \langle X(t), \theta_0(t-k) \rangle \quad \text{and} \quad Z'_{j,k} = \langle X(t), \xi_j(t - 2^{-j}k) \rangle.$$

Furthermore, the sums in (9.14) converge in the mean-square sense.

Generalizations to nonstationary processes with stationary increments are possible; see discussion in Zhang and Walter [473] and Walter [439].

9.3 ESTIMATION OF SPECTRAL DENSITIES

Any statistical inference involving a time series can be conducted in the time and frequency domains. The methods are complementary and provide different insights. Spectral analysis of time series and, in particular, estimation of the spectral density are indispensable tools for exploring the frequency behavior of a time series.

It is a well-known fact that a raw periodogram is an inconsistent estimator of the spectral density. To achieve consistency – smoothing of the periodogram is needed. Various kernel smoothing procedures are traditional solutions. Wavelet shrinkage methods have been applied to the spectral density estimation in work by Gao [164, 165], Lumeau et al. [269], Moulin [302], Moulin, O'Sullivan, and Snyder

[303], and Walter [439]. The idea of using wavelets in smoothing log-periodograms was announced in Donoho [120].

In this section, we will describe Gao's method only. Gao [165] proposes a threshold that ensures that the reconstructed log-spectrum is as nearly noise-free as possible. In addition to enhancing the visual appearance of the estimator, the noise-free character leads to attractive theoretical properties over a wide range of smoothness assumptions.

An absolutely summable complex-valued function $\gamma(\cdot)$ defined on the integers is the autocovariance function of a stationary process if and only if

$$f(\omega) = \frac{1}{2\pi} \sum_{t=-\infty}^{\infty} \gamma(t) e^{-it\omega} \geq 0, \text{ for all } \omega \in [-\pi, \pi]. \tag{9.15}$$

The function $f(\omega)$ is called the spectral density associated with $\gamma(\cdot)$.

Example 9.3.1 (i) The white noise process $\mathcal{WN}(0, \sigma^2)$ has a uniform spectral density $f(\omega) = \frac{\sigma^2}{2\pi}$.

(ii) The moving average MA(1) process, defined as $X_t = Z_t + \theta Z_{t-1}$, $Z_t \sim \mathcal{WN}(0, \sigma^2)$, has the spectral density $f(\omega) = \frac{\sigma^2}{2\pi}(1 + 2\theta \cos\omega + \theta^2)$.

(iii) The autoregressive AR(1) process, $X_t - \phi X_{t-1} = Z_t$, $Z_t \sim \mathcal{WN}(0, \sigma^2)$, has the spectral density $f(\omega) = \frac{\sigma^2}{2\pi}(1 - 2\phi \cos\omega + \phi^2)^{-1}$.

A statistic used as a raw estimator of the spectral density is the periodogram.

Definition 9.3.1 *The periodogram $I(\omega)$, based on the sample $X_0, \ldots, X_{T-1}$ and defined at the Fourier frequencies $\omega_j = \frac{2\pi j}{T}$, $j = 0, 1, \ldots, [T/2]$, is*

$$I(\omega_j) = \frac{1}{2\pi T} \left| \sum_{t=0}^{T-1} X_t e^{-it\omega_j} \right|^2.$$

For any set of frequencies $\omega_1, \omega_2, \ldots, \omega_n$ such that $0 \leq \omega_1 < \cdots < \omega_n < \pi$, the $I(\omega_i)$ are asymptotically independent exponential random variables with means $2\pi f(\omega_i)$, where f is the spectral density. The periodogram is not a consistent estimator of $2\pi f(\omega)$ since its variance does not go to 0 when the sample size increases.

Smoothing the periodogram will not only help in extracting significant frequencies, but smoothed periodograms can also be consistent estimators of f. For a standard theory, see Brockwell and Davis [39] (page 351).

9.3.1 Gao's Algorithm

Let $X_0, X_1, \ldots, X_{2T-1}$, $T = 2^m$ be the observed time series. Gao's algorithm for spectral density estimation consists of four steps.

1. Calculate the log-periodogram

$$z_\ell = \log I(\omega_\ell), \; \ell = 0, 1, \ldots, T-1,$$

at the frequencies $\omega_\ell = \frac{2\pi\ell}{2T}$, where $I(\omega) = \frac{1}{2\pi} \cdot \frac{1}{2T} |\sum_{t=0}^{2T-1} X_t e^{-it\omega}|^2$.

2. Take a standard periodic wavelet transformation of z_ℓ and obtain the empirical wavelet coefficients $d_{jk}, j = 0, 1, \ldots, m-1; \; k = 0, \ldots, 2^j - 1$.

3. Apply the soft-thresholding rule $\delta^s(x, \lambda)$, with threshold λ depending on T and the level j in the following way:

   (a) If the resolution levels are fine ($j = m-1, m-2, \ldots$) then select the threshold

$$\lambda_{j,T} = \alpha_j \log T. \tag{9.16}$$

The values of α_j, for some commonly used bases, such as COIF, DAUB, SYMM, and some selected $T = 2^m$, are tabulated below.

level j	α_j	level j	α_j
$m-1$	1.29	$m-6$	0.54
$m-2$	1.09	$m-7$	0.46
$m-3$	0.92	$m-8$	0.39
$m-4$	0.77	$m-9$	0.32
$m-5$	0.65	$m-10$	0.27

   (b) If the resolution is coarse, i.e., if $j \ll m-1$, then use

$$\lambda_T = \sqrt{2 \log T \cdot \pi^2/6} = \frac{\pi}{\sqrt{3}} \sqrt{\log T}. \tag{9.17}$$

4. Take the inverse wavelet transformation and obtain the estimator g_ℓ of the log-spectrum at the Fourier frequencies ω_ℓ.

The threshold justification is based on Wahba's log-periodogram representation (Wahba, [432]),

$$z_\ell = g_\ell + \epsilon_\ell, \; \ell = 0, 1, \ldots, T-1. \tag{9.18}$$

where g_ℓ is the log-spectrum and $\epsilon_\ell \approx \log(\eta_\ell/2) + \gamma$; η_ℓ are independent random variables, distributed as χ_2^2, and $\gamma \approx 0.57721$ is the Euler-Mascheroni constant. Note that for the frequency $\omega = 0$, $I(0) \sim \chi_1^2$. When T is large the distributional deviation of $I(0)$ can be ignored. Since $E\epsilon_\ell = 0$ and $\text{Var } \epsilon_\ell = \frac{\pi^2}{6}$, the threshold (9.17) is justified by an asymptotic normality argument.

For the fine levels, the asymptotic normality does not hold since the coefficients are obtained by only a few filtering operations on the empirical wavelet coefficients. The noise at fine levels has non-Gaussian character and careful analysis is needed for its suppression. The threshold in (9.16) is a result of such an analysis.

The need for level-dependent thresholds are justified by the results in Theorem 9.3.1.

Theorem 9.3.1 *[164] Under Wahba's approximation model for log periodograms (9.18),*

$$(i) \quad P(\cup_j [\sup_k |d_{jk} - E(d_{jk})| > \lambda_{j,T}]) \to 0, \quad T \to \infty. \tag{9.19}$$

If, in addition, the supports of the wavelets are compact, for the finest level $j = m-1$,

$$(ii) \quad P(\sup_k |d_{m-1,k} - E(d_{m-1,k})| > \sqrt{2\log T}) \to 1, \quad T \to \infty. \tag{9.20}$$

The statement (9.19) guarantees that thresholds are set high so that the noise is eliminated with high probability. The statement (9.20) has the following interpretation. The large coefficients in the finest level of detail (associated with the spikes in the periodogram) surpass the threshold with high probability. Similar results hold for levels $j = m-2, m-3$, etc.

In Example 9.3.2 we apply Gao's method on Wolf's sunspot data set and compare the result to a standard smoothing procedure.

Example 9.3.2 (Wolf's sunspot numbers)

In 1848, the Swiss astronomer, Johann Rudolph Wolf, introduced a daily measurement of "sunspots." His method, which is still in use today, counts the total number of spots visible on the face of the Sun and the number of groups into which they cluster, because neither quantity alone satisfactorily measures sunspot activity. An observer computes a daily sunspot number by multiplying the number of groups he sees by 10 and then adding this product to his total count of individual spots. Results, however, vary greatly, since the measurement strongly depends on the observer's interpretation and experience and on the stability of the Earth's atmosphere above the observing site. The use of Earth as a platform from which to record these numbers additionally contributes to their variability because the Sun rotates and the evolving spot groups are distributed unevenly across solar longitudes. To compensate for these limitations, each daily international number is computed as a weighted average of measurements made from a network of cooperating observatories.

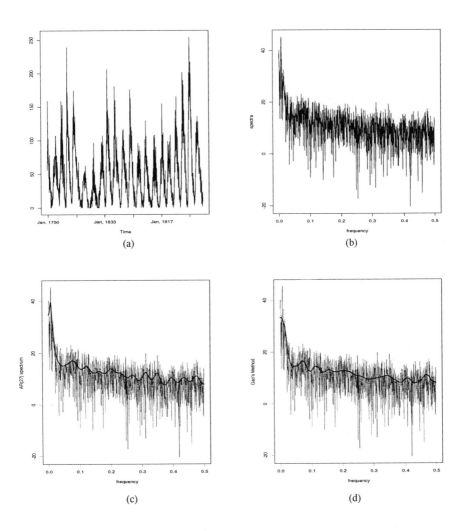

Fig. 9.2 (a) Wolf's sunspot numbers. (b) Raw periodogram. The units spectra are in decibels. (c) The estimator based on the Yule-Walker autoregressive fit. (d) Gao's method.

Fig. 9.2(a) depicts a time series of Wolf's sunspot numbers dating back to the 1750s. Fig. 9.2(b) gives a raw periodogram with the variable `spectra` measured in decibels (*dB*). The two estimators of the spectra, the Yule-Walker autoregressive fit and Gao's wavelet method, are given in Fig. 9.2, panels (c) and (d), respectively.

The large sample minimax optimality of wavelet-smoothed periodograms (not log-periodograms!) is discussed in Gao [164]. To obtain an estimator based on a sample $X_0, X_1, \ldots, X_{n-1}$, periodized Meyer wavelets and thresholds $\lambda_{j,n} = c_j \log n$ are utilized. The resulting estimators have near-optimal $\mathbb{L}_2$-convergence rates over scales of Besov spaces.

If the spectral density belongs to the Besov ball $\mathbb{B}_{pq}^\sigma(C)$, for $p \geq 1$ and $\sigma > \frac{1}{p}$, and if the threshold is

$$\lambda_n = \frac{2 \log n}{\sqrt{n}},$$

then the shrinkage estimator based on $\delta^s(\cdot, \lambda_n)$ satisfies

$$\sup_{f \in \mathbb{B}_{pq}^\sigma(C)} E\|\hat{f} - f\|_{\mathbb{L}_2}^2 \leq C_1 (\log n)^2 n^{-2\sigma/(2\sigma+1)}, \qquad (9.21)$$

where C_1 depends only on σ, p, q, and C.

The rate in (9.21) is nearly optimal, i.e., there is a constant C_2 such that

$$\inf_{\hat{f}} \sup_{f \in \mathbb{B}_{pq}^\sigma(C)} E\|\hat{f} - f\|_{\mathbb{L}_2}^2 \geq C_2 \cdot n^{-2\sigma/(2\sigma+1)} \cdot (1 + o(1)).$$

The classical methods of smoothing the periodogram (window method, AR approximation, etc.) can only achieve the rate $r' = \frac{\sigma + (1/(p \wedge 2) - 1/p)}{\sigma + 1/2 + (1/(p \wedge 2) - 1/p)}$, which is optimal for the linear procedure, see Donoho and Johnstone [130]. It is easy to check that $2\sigma/(2\sigma + 1) \geq r'$, with strict inequality when $p < 2$.

9.3.2 Non-Gaussian Stationary Processes

For possibly non-Gaussian wide-sense stationary processes, Neumann [313] considered wavelet smoothing a tapered periodogram,

$$I_T(\omega) = \frac{1}{2\pi H_{2,T}} \left| \sum_{t=0}^{T-1} h(t/T) X_t e^{-it\omega} \right|^2,$$

where $h(t) : [0,1] \mapsto [0,1]$ is a sufficiently smooth window function and $H_{k,T} = \sum_{t=0}^{T-1} h^k(t/T)$ is a normalizing factor, such that $H_{2,T} \sim T$. One example of an appropriate window function is the Hanning window, defined as $h(u) = \frac{1}{2}(1 - \cos(2\pi u))$, if $u \in [0, \frac{1}{2}]$, and $h(u) = h(1-u)$, if $u \in [\frac{1}{2}, 1]$.

The spectral density is defined on $[-\pi, \pi]$ and it is convenient to define a wavelet basis constrained to that interval, by periodizing an arbitrary wavelet basis (see Section 5.6). Define

$$\phi^*_{j_0,k}(t) = \sum_n \frac{1}{\sqrt{2\pi}} \phi_{j_0,k}\left(\frac{t}{2\pi} + n\right),$$

$$\psi^*_{jk}(t) = \sum_n \frac{1}{\sqrt{2\pi}} \psi_{jk}\left(\frac{t}{2\pi} + n\right).$$

The spectral density now has the representation

$$f(\omega) = \sum_k \alpha'_{j_0,k} \phi^*_{j_0,k}(t) + \sum_{j \geq j_0} \sum_k \alpha_{jk} \psi^*_{jk}(t).$$

As in density estimation, one defines the empirical counterparts of α_{jk} via

$$\widehat{\alpha}_{jk} = \int I_T(\omega) \psi^*_{jk}(\omega)\, d\omega,$$

with $\widehat{\alpha}'_{j_0,k}$ defined analogously.

Under mild regularity conditions on the wavelet basis, taper function and function f, Neumann proved that

(i) $\quad E\widehat{\alpha}_{jk} = \alpha_{jk} + O(2^{j/2} T^{-1} \log T),$

(ii) $\quad \mathrm{Var}\widehat{\alpha}_{jk} = \frac{2\pi H_{4,T}}{(H_{2,T})^2} \cdot \int_{-\pi}^{\pi} \psi_{jk}(\alpha)[\psi_{jk}(\alpha) + \psi_{jk}(-\alpha)] |f(\omega)|^2\, d\alpha$
$\quad + o(T^{-1}) + O(T^{-1} 2^{-j}).$

For a restricted set of indices $\{j,k\}$, $\mathcal{I} = \{(j,k), 2^j < CT^{1-\alpha}, 0 \leq k < 2^j\}$ where C and $0 < \alpha < \frac{1}{3}$ are fixed constants, the asymptotic normality of $\widehat{\alpha}_{jk}$ is established, which, in turn, provides a basis for threshold determination.

Based on the asymptotic normality of $\widehat{\alpha}_{jk}$, it can be demonstrated that hard or soft thresholding estimators $\widehat{f}$ achieve optimal, up to a logarithmic factor, minimax convergence rates for spectral densities of bounded variation belonging to $\mathbb{B}^\sigma_{p,q}$ ($p \geq 1, \sigma > \frac{1}{p}$ or $\sigma, p > 1$).

The rate obtained by Neumann is comparable to (9.21), but the price for non-normality is paid by the restriction to thresholding of certain levels only and by analytically more complicated asymptotic arguments.

Neumann and von Sachs [316] apply wavelet methods in estimating time-dependent spectrum in locally stationary time series as introduced by Dahlhaus [99]. The authors show that minimax rates are attained by the coordinate-wise thresholding of a bivariate wavelet expansion based on a tensor product. See also Donoho, Mallat, and von Sachs [134], Flandrin [152], Krim and Pesquet [247], and von Sachs [429].

9.4 WAVELET SPECTRUM

In this section, we provide some results about a wavelet decomposition of a stationary random process that parallel a time-localized Cramér (Fourier) spectral representation. We provide a time/scale instead of a time/frequency decomposition and, hence, instead of thinking of scale in terms of an "inverse frequency" we start from genuine time/scale building blocks or *atoms*.

We present several results of Chiann and Morettin [73] connected with stationary processes only. The best one can do for stationary processes is already contained in the classical Cramér spectral representation. However, the results of Chiann and Morettin provide another point of view to a frequency/scale analysis of a time series.

The wavelet-based counterparts of spectral tools are most useful in applications to non-stationary time series.

Using a class of locally stationary wavelet processes, a doubly indexed array of processes $\{X_{t,T}\}_{t=1,...,T}$, $T \geq 1$, the theory for the estimation of the "evolutionary wavelet spectrum" was developed in a series of papers by Nason, von Sachs, and Kroisandt [312], and von Sachs, Nason, and Kroisandt [430, 431]. This evolutionary wavelet spectrum measures the local power in the variance-covariance decomposition of the process $\{X_{t,T}\}$ at a certain scale and a (rescaled) time location. It is possible to estimate the evolutionary wavelet spectrum by means of a wavelet *periodogram* or *scalogram*; in other words the squared coefficients from a discrete or stationary wavelet transform.

A detailed treatment of these novel techniques for locally stationary processes is beyond the scope of this book.

9.4.1 Wavelet Spectrum of a Stationary Time Series

Let $\{X_t, t = 0, \pm 1, \pm 2, \dots\}$ be a discrete time, zero-mean, second-order stationary time series. Let $\gamma_h = EX_t X_{t+h}$ be the autocovariance function of X_t. Assume that we observed $X_0, X_1, \ldots, X_{T-1}$, where $T = 2^m$, for m integer.

Under the assumption

[A] $$\sum_{t \in \mathbb{Z}} (1+|t|)|\gamma(t)| < \infty,$$

Chiann and Morettin [73] propose the following definition of a wavelet-based spectrum:

Definition 9.4.1 *[73] The wavelet spectrum of X_t at the node (j, k), with respect to the wavelet ψ, is defined to be*

$$\eta_{jk}^{(\psi)} = \sum_{u \in \mathbb{Z}} \gamma(u) \Psi_{jk}(u), \quad j, k \in \mathbb{Z}, \tag{9.22}$$

where

$$\Psi_{jk}(u) = \sum_{t=0}^{\infty} \psi_{jk}(t) \psi_{jk}(t+|u|)$$

is the wavelet autocorrelation function at the node (j, k).

If assumption [A] holds, then $\eta_{jk}^{(\psi)}$ is a bounded, non-negative function. Let

$$I_{jk}^{(\psi)} = \left(\sum_{t=0}^{T-1} X_t \psi_{jk}(t) \right)^2$$

be the second-order wavelet periodogram based on the values $X_0, X_1, \ldots, X_{T-1}$. The expected value of $I_{jk}^{(\psi)}$ is

$$\begin{aligned} EI_{jk}^{(\psi)} &= \sum_{t=0}^{T-1} \sum_{s=0}^{T-1} EX_t X_s \psi_{jk}(t) \psi_{jk}(s) \\ &= \sum_{t=0}^{T-1} \sum_{s=0}^{T-1} \gamma(t-s) \psi_{jk}(t) \psi_{jk}(s) \\ &= \sum_{u=-(T-1)}^{T-1} \gamma(u) \sum_{t=0}^{T-1-|u|} \psi_{jk}(t) \psi_{jk}(t+|u|). \end{aligned}$$

Thus, when $T \to \infty$,

$$EI_{jk}^{(\psi)} \to \eta_{jk}^{(\psi)}, \tag{9.23}$$

and $I_{jk}^{(\psi)}$ is asymptotically unbiased.

Let $d_{jk}^{(\psi)} = \sum_{t=0}^{T-1} X_t \psi_{jk}$. Then $E d_{jk}^{(\psi)} = 0$ and

$$\begin{aligned} \mathrm{Var}(d_{jk}^{\psi}) &= \sum_{t=0}^{T-1} \sum_{s=0}^{T-1} \gamma(t-s) \psi_{jk}(t) \psi_{jk}(s) \\ &= \sum_{u=-(T-1)}^{T-1} \gamma(u) \sum_{t=0}^{T-1-|u|} \psi_{jk}(t) \psi_{jk}(t+|u|). \end{aligned}$$

If [A] holds, $\mathrm{Var}\, d_{jk}^{(\psi)} \to \eta_{jk}^{(\psi)}$, $T \to \infty$.

Remark 9.4.1 The multilinear and dependence description properties of cummulants allow an analytic derivation of various characteristics of empirical wavelet coefficients, particularly sample distributions. Brillinger [38] gives several applications of cummulants in wavelet asymptotic. See also Neumann [313], and Neumann and von Sachs [315]. We give the definition of multivariate cummulants needed for the exposition that follows:

Let $M(z_1, \ldots, z_n) = E \exp\{z_1 X_1 + \cdots + z_n X_n\}$ be the moment generating function of $(X_1, \ldots, X_n)$. Here

$$E X_1^{k_1} \ldots X_n^{k_n} = \left. \frac{\partial^{k_1}}{\partial z_1^{k_1}} \right|_{z_1=0} \cdots \left. \frac{\partial^{k_n}}{\partial z_n^{k_n}} \right|_{z_n=0} M(z_1, \ldots, z_n).$$

$$M(z_1, \ldots, z_n) = \exp\left\{ \sum_{i=1}^{\infty} \sum_{(m)} u_i(\underline{x}) \cdot \frac{z_1^{m_1} \ldots z_n^{m_n}}{m_1! \ldots m_n!} \right\},$$

where (m) indicates $\sum_{k=1}^{n} m_k = i$ and $\underline{x}$ in $u_i(\underline{x})$ consists of m_1 indices X_1, m_2 indices X_2, etc.

For example,

$$\begin{aligned} \mathrm{Cum}(X_1) &= u_1(X_1) = EX_1, \\ \mathrm{Cum}(X_1, X_2) &= u_2(X_1, X_2) = E(X_1 X_2) - EX_1 EX_2, \\ \mathrm{Cum}(X_1, X_2, X_3) &= u_3(X_1, X_2, X_3) = E(X_1 X_2 X_3) - EX_1 E(X_2 X_3) \\ &\quad - EX_2 E(X_1 X_3) - EX_3 E(X_1 X_2) + 2 EX_1 EX_2 EX_3, \end{aligned}$$

etc.

Let $u_1, u_2, \ldots, u_{k-1} = 0, \pm 1, \ldots$, and let $k > 1$. Let $C(u_1, \ldots, u_{k-1}) =$

$\mathrm{Cum}[X(t+u_1),\ldots,X(t+u_{k-1}),X(t)]$.
Assume

[B] $\quad \sum_{u_1}\cdots\sum_{u_{k-1}} |u_j|\,|C(u_1,\ldots,u_{k-1})| < \infty.$

Let

$$\eta_{(j,j'),(k,k')} = \sum_{u\in\mathbb{Z}}\sum_{t=0}^{\infty} \gamma(u)\psi_{j,k}(t+|u|\mathbf{1}(u>0))$$
$$\cdot \psi_{j',k'}(t+|u|\mathbf{1}(u\leq 0)),\quad j,j',k,k'\in\mathbb{Z},$$

be the *wavelet-cross spectrum* or asymptotic covariance of the wavelet transform (as in [73]).

If assumption [B] holds,

$$\mathrm{Cov}(I_{j,k}, I_{j',k'}) = 2[\eta_{(j,j'),(k,k')}]^2 + O(1), \tag{9.24}$$

when $T\to\infty$. In particular, when $j=j'$ and $k=k'$ we get the variance

$$\mathrm{Var}(I_{jk}) = 2\eta_{jk}^2 + O(1). \tag{9.25}$$

Relations (9.24) and (9.25) show that the wavelet periodogram is not consistent.

9.4.2 Scalogram and Periodicities

The wavelet periodogram gives the energy content at each node (j,k). We may be interested in the level-wise distribution of energies. Define the *scalogram* (or *scalegram*) at scale j as

$$S(j) = \sum_k d_{jk}^2 = \sum_k I_{jk},\ j = 0, 1, \ldots, m.$$

For some applications, see Scargle [368].

When assumptions [A] and [B] hold,

$$ES(j) = \eta_{j,.} + O(T^{-1}),$$
$$\text{Cov}(S(j), S(j')) = 2\sum_{k}\sum_{k'}[\eta_{(j,j'),(k,k')}]^2 + O(1),$$

where $\eta_{j,.} = \sum_k \eta_{jk}$ is the wavelet energy associated with the scale j. Recall that the Fourier periodogram $I(\omega) = \frac{1}{2\pi T}|\sum_{t=0}^{T-1} X(t)e^{-i\omega t}|^2$ has, for each frequency $-\pi < \omega < \pi$, $\omega \neq 0$, an asymptotic $f(\omega)\chi_2^2$ distribution.

The wavelet periodogram I_{jk} is an asymptotically $\eta_{jk}\chi_1^2$ random variable. Indeed, this follows from the fact that d_{jk} has an asymptotic $\mathcal{N}(0, \eta_{jk})$ distribution. Hence, the confidence intervals for η_{jk} can be constructed the same way as for $f(\omega)$.

Now $S(j)$ is an asymptotically $a\chi_b^2$ random variable with

$$a \approx \left[\sum_k \sum_{k'} \eta_{(j,j'),(k,k')}\right]^2 / \eta_{j,.},$$

$$b \approx (\eta_{j,.})^2 / \left[\sum_k \sum_{k'} \eta_{(j,j'),(k,k')}\right]^2.$$

Some other important interplays between wavelets and time series can be found in Carmona and Wang [65], Cohen and Ryan [87], Mallat [277], Morettin [298, 299], Percival [335], Percival and Walden [338], and Priestley [347].

9.5 LONG-MEMORY PROCESSES

Many economic and physical phenomena are well modeled by "long-memory" or "1/f" random processes. Such processes have power spectra that are represented by straight lines in the log-scale log-power coordinates, at least over several decades of frequency.

Examples of such a process are the *Fractional Brownian Motion (fBm)* and *Fractional Gaussian noise (fGn)* processes. Several researchers explored the connection between fBm, as an exemplary long-memory process, and wavelets. The reader interested in more detailed accounts on the topic and in techniques and proofs of the results cited below is directed to Abry, Gonçalvès, and Flandrin [5], Flandrin [154], McCoy [289], and Wornell [461].

9.5.1 Wavelets and Fractional Brownian Motion

Fractional Brownian motion processes offer a convenient tool for modeling nonstationary stochastic phenomena with long-term dependencies and 1/f-type spectral behavior over wide ranges of frequencies. Due to the nonstationary nature of these processes, the precise meaning of their spectra remains generally unclear.

Statistical self-similarity (self-affinity) is an essential feature of fBm, and wavelets are especially apt for both analysis and synthesis of such processes. The standard time-frequency analysis of fBm processes emphasizes time-averaged measurements as an answer to the nonstationary nature of the process. Wavelet methods are complementary. They emphasize the self-similar nature of fBm processes and partition such nonstationary process to stationary component-subprocesses associated with discretized levels of scaling.

We start with the definition of fBm and some associated notions.

Definition 9.5.1 *Fractional Brownian motion (fBm) is a Gaussian, zero-mean non-stationary process, $B_H(t)$, indexed by a parameter H (Hurst exponent, $0 < H < 1$), such that*

$$B_H(t) = 0$$
$$B_H(t+h) - B_H(t) \sim \mathcal{N}(0, \sigma_H^2 |h|^{2H}).$$

It follows that the autocovariance function is

$$E(B_H(t) \cdot B_H(s)) = \frac{\sigma_H^2}{2}(|t|^{2H} + |s|^{2H} - |t-s|^{2H}), \tag{9.26}$$

where

$$\sigma_H^2 = \text{Var} B_H(1) = \Gamma(1 - 2H)\frac{\cos(\pi H)}{\pi H}.$$

Since the process is non-stationary ($\gamma(t, s)$ is not a function of $|t - s|$ only), it has no spectrum in the usual sense. However, one can define an *average spectrum* (or pseudo-spectrum) of $B_H(t)$ as

$$f(\omega) = \frac{\sigma^2}{|\omega|^{2H+1}}.$$

The *fBm* process is self-similar, i.e., $B_H(at) \stackrel{d}{=} a^H B_H(t)$. Simply speaking, the graph $(t, B_H(t))$ remains statistically unchanged if the axes are simultaneously scaled by a and a^H, respectively.

The paths of the process are also "space-filling." Their Hausdorff dimension D is connected with the Hurst exponent, $D = 2 - H$.

Example 9.5.1 Standard Brownian motion corresponds to the fBm process with $H = \frac{1}{2}$.

We saw that the fBm process is not stationary; however, its increments form a zero-mean stationary Gaussian process known as *fractional Gaussian noise, fGn*.

Definition 9.5.2 *Let $B_H(t)$ be a fractional Brownian motion of exponent H. The fractional Gaussian noise is defined as*

$$G_{H,h}(t) = \frac{B_H(t+h) - B_H(t)}{h}. \tag{9.27}$$

For large lags, $\tau \gg h$, the autocovariance function $\gamma_{G_{H,h}}(\tau) = E(G_{H,h}(t+\tau) \cdot G_{H,h}(t))$ behaves like $\sigma^2 H(2H-1)|\tau|^{2H-2}$. The associated spectral density $f_{G_{H,h}}(\omega)$ behaves like $|\omega|^{1-2H}$, $0 < |\omega| \ll 1/h$. If t is an integer and $h = 1$ in (9.27) we obtain the discrete fractional Gaussian noise ($dfGn$). It is a zero mean, stationary sequence with autocorrelation

$$\gamma(k) = \frac{\sigma^2}{2}(|k+1|^{2H} + |k-1|^{2H} - 2|k|^{2H}).$$

The following result is from Kaplan and Kuo [226]. Let d_{jk} be the detail coefficients in Haar's wavelet transformation of $X(k) = B_H(k+1) - B_H(k)$. Then, for a fixed scale j,

(a) the variance of d_{jk} is

$$\text{Var}(d_{jk}) = \sigma^2 2^{\alpha(j-1)}(2 - 2^\alpha),$$

where $\alpha = 2H - 1$,

(b) the autocorrelation $\gamma(k-l)$ of d_{jk} decays like $O(|k-l|^{2(H-2)})$, for all k and l such that $|k-l| > 1$.

For an excellent general overview of fBm processes we direct the reader to Beran [26, 28]. Continuous wavelet transformations of fBm can be found in Flandrin [151], Wornell and Oppenheim [462], and Wornell [461]. See also Benassi [24] and Kaplan and Kuo [226] for a discussion of fGn.

9.5.2 Estimating Spectral Exponents in Self-Similar Processes

Phenomena in many fields (economics, geophysics, hydrology, medicine, etc.) can be successfully modeled by long-memory processes, and estimation of the associated spectral exponent α is crucial. For some phenomena, such as turbulence (see Section 11.5), theoretical power-laws hold and researchers are able to test the compliance of statistical models with the theory. Some general references on a spectral exponent estimation include Beran [26, 27, 28], Geweke and Porter-Hudak [174], and Smith [382].

We saw that the fBm process has a pseudo-spectrum of the form $f(\omega) \propto |\omega|^{-(2H+1)}$. The exponent $-(2H+1)$ can be associated with a negative slope in the log-log plot of $f(\omega)$. Fig. 9.3 compares the exponents for turbulence measurements and the matching ($H = \frac{1}{3}$) fBm process. Random processes for which the spectral density $f(\omega)$ behaves like $|\omega|^{-\alpha}$ for ω close to 0, can be classified as "blue noise" if $\alpha < 0$, or as "red noise" if $\alpha > 0$.

The time/scale approach allows an unbiased estimation of the spectral exponent α; and one can interpret this result in terms of matched tilings of the time-frequency plane.

Abry, Gonçalvès, and Flandrin [5] derive explicitly the probability density function of the estimated value of α. From this analysis they find that there exists an optimum number of scales to use in a discrete wavelet scheme for obtaining a minimum variance estimator, and that an improved procedure can be designed by making use of weighted least-squares in the estimation.

A standard estimator for the power spectrum density $f(\omega)$ is Welch's estimator, see Welch [454]. For a stationary process $\{X(t), 0 \leq t \leq T\}$, Welch's estimator is given by

$$\widehat{f}_1(\omega) = \frac{1}{N} \sum_{n=1}^{N} \left| \int_0^T X(t) w_\theta(t-\tau) \cdot e^{-i2\pi\omega t} dt \right|^2, \qquad (9.28)$$

with $w_\theta(t)$ as an arbitrary weight function (of "duration" θ) and with time instances $\{\tau_n, n = 1, \ldots, N\}$.

In analogy with the standard periodogram, Welch's estimator is asymptotically unbiased, inconsistent, and for $\omega_1 \neq \omega_2$, $\widehat{f}_1(\omega_1)$ and $\widehat{f}_1(\omega_2)$ are asymptotically uncorrelated.

The estimator in (9.28) is biased since $E\widehat{f}_1(\omega) = \int f(\varphi)|W_\theta(\varphi-\omega)|^2 d\varphi$, where W_θ is the Fourier transformation of w_θ, and the bias is convolutive. The variance $\text{Var}\widehat{f}_1(\omega)$ is approximately $\frac{1}{N}|f(\omega)|^2$, and its minimization increases the bias in $f_1(\omega)$.

The scalogram, $\widehat{f}_2(\omega_j) = \frac{1}{N_j} \sum_{k=1}^{N_j} d_{jk}^2$, as a wavelet counterpart of (9.28) is still biased, but the bias is no longer frequency-dependent.

For instance, for an fBm with spectral exponent α,

$$E\widehat{f}_2(\omega) = \sigma^2 |\omega|^{-\alpha} \cdot \int \left|\frac{\varphi}{\omega_\psi}\right|^{-\alpha} |\Psi(\varphi)|^2 d\varphi,$$

where ω_ψ and $\Psi(\varphi)$ depend on the wavelet basis and not on ω. Thus, the slope in the log-log plot of $E\widehat{f}_2(\omega)$ is exactly $-\alpha$.

The following result from Abry, Gonçavès, and Flandrin [5] describes some

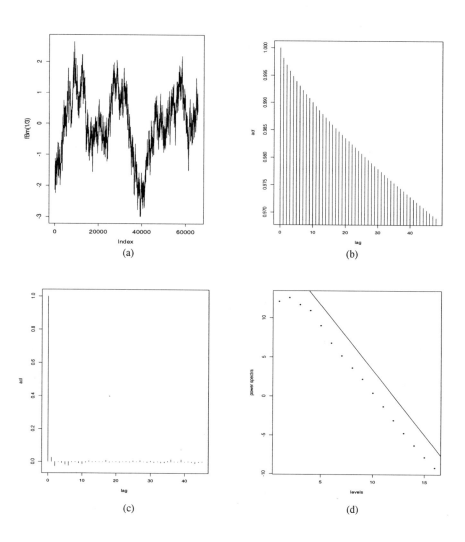

Fig. 9.3 (a) Simulated fractional Brownian motion with the exponent $H = 1/3$. (b) Autocorrelation function showing slow decay. (c) Autocorrelation function of the finest level in Haar's decomposition of the original $fBm(1/3)$ series. (d) The power law and the Kolmogorov-Obukhov slope -5/3.

asymptotic properties of the wavelet-based estimator of α.

Let N_j be the number of wavelet coefficients in the level j associated with time series $\{X(t),\ t=1,\ldots,N\}$, and let $\delta_j = \frac{1}{N_j}\sum_{k=1}^{N_j}\left(\frac{d_{jk}}{\sigma_j}\right)^2$, where $\sigma_j^2 = Ed_{jk}^2$. Let $\eta_j = \log_2 \delta_j$. Then,

(i) η_j is asymptotically normal, when $N \to \infty$.
(ii) $\hat{\alpha}(J)$ is linear fit of (j,η_j), $j = 1,\ldots,J$, i.e.,

$$\hat{\alpha}(J) = \frac{J\sum_{j=1}^{J} j\eta_j - \sum_{j=1}^{J} j \cdot \sum_{j=1}^{J} \eta_j}{J\sum_{j=1}^{J} j^2 - (\sum_{j=1}^{J} j)^2}. \tag{9.29}$$

The asymptotic distribution (when the N_j's are large and $N = 2^j N_j$) gives

$$E\hat{\alpha}(J)^2 = \frac{144}{N\ln^2 2} \cdot \frac{2^J \cdot (J^2 - 6J + 17) - (J^2 + 6J + 17)}{J^2(J^2-1)^2},$$

which provides a vehicle for selecting the number of levels J which minimizes the asymptotic variance.

Abry, Gonçavès, and Flandrin [5] also suggest a modification of $\hat{\alpha}(J)$ by a weighted estimator of the regression slope for which the variance approaches the Cramer-Rao bound. McCoy and Walden [290] suggest an iterative procedure for estimating α and σ_α^2 based on maximum likelihood. See also Percival and Bruce [336] for estimation of long memory processes with missing data.

Some pre-wavelet methods for estimating α for long-memory processes were suggested by Geweke and Porter-Hudak [174] and Smith [381].

9.5.3 Quantifying the Whitening Property of Wavelet Transformations for fBm Processes

The discrete wavelet transformation of an fBm process is defined in the standard way as

$$d_{jk} = \langle B_H(x), \psi_{jk}(x) \rangle, \tag{9.30}$$

and at each scale j the sequence d_{jk}, $k \in \mathbb{Z}$ forms a sequence of random variables. Consider $E(d_{jk}d_{jk'})$. Flandrin [153] shows that

$$E(d_{jk}d_{jk'}) = C \cdot 2^{-j(2H+1)} \int_{-\infty}^{\infty} \frac{|\Psi(\omega)|^2}{|\omega|^{2H+1}} e^{i2\pi(n-n')\omega}\, d\omega.$$

The asymptotic correlation structure in the detail coefficients is governed by the behavior of

$$\frac{|\Psi(\omega)|^2}{|\omega|^{2H+1}} \tag{9.31}$$

at the origin, see page 129 in Flandrin [153]. $\Psi(\omega)$ in (9.31) is the Fourier transformation of the wavelet ψ. When the decomposing wavelet possesses R vanishing moments, (9.31) behaves like $|\omega|^{2|R-H|-1}$. This corresponds to a perfect decorrelation if and only if $R = H + \frac{1}{2}$, which is the case for Brownian motion [$fBm(1/2)$] and the Haar's wavelet.

If $R \neq H + \frac{1}{2}$ the coefficients d_{jk} and $d_{jk'}$ are correlated, but their correlations decay like $O\left(|2^{-j}(k-k')|^{2(H-R)}\right)$ when $|2^{-j}(k-k')| \to \infty$. For related discussions see Tewfik and Kim [400] and Abry, Veitch, and Flandrin [7].

Applications of fBm processes to Internet traffic data can be found in Abry and Veitch [6]. The authors propose an estimator for the Hurst exponent, which is unbiased under very general conditions and efficient under the assumption of normality.

9.6 DISCUSSION AND REFERENCES

Basseville et al. [22] proposed stochastic models indexed by nodes on lattices or trees in which different depths in the tree or lattice correspond to different spatial scales in representing data. This perspective then leads directly to the proposal of several classes of dynamic models and related notions of multiscale stationarity in which scale plays the role of a time-like variable. Dijkerman and Mazumdar [118] propose multiresolution stochastic models of the discrete wavelet coefficients as approximations to the original time process. Their models are simple due to the strong decorrelation of the wavelet transform. Experiments show that these models significantly improve the approximation in comparison with the often-used assumption that the wavelet coefficients are completely uncorrelated. For a Bayesian approach in modeling $1/f$ processes by utilizing stationary wavelets, see Nowak [317]. Nonparametric estimation of variance of a diffusion is addressed in Genon-Catalot, Laredo, and Picard [170, 171]. Wavelet applications in threshold autoregressive models are addressed in Li and Xie [262]. For wavelet-based detection of hidden periodicities, see Li and Xie [261]. For additional readings on the interplay of wavelets and time series the reader is directed to monographs by Percival and Walden [338] and Carmona, Hwang and Torrésani [64].

9.7 EXERCISES

9.1. Derive the autocovariance functions of the MA(q) and AR(1) stationary time series, (9.2) and (9.3). [Hint: Express AR(1) as an MA(∞) process. What condition is needed on ϕ so that a stationary solution is $\mathbb{L}_2$ convergent and depends only on the "past", $\{Z_s, s \leq t\}$?].

9.2. Prove that $\theta_0(t)$ and $\theta^0(t)$ as defined by their Fourier transformations in (9.12) and (9.13), satisfy $\int \theta_0(t-k)\theta^0(t-l)\, dt = \delta_{k,l}$.

9.3. [439] A process $X(t)$ is called wide-sense *cyclostationary* if $EX(t+1) = EX(t)$ and $\gamma_X(t+1, s+1) = \gamma_X(t,s)$.

Let

$$X(t) = \sum_n c_n \phi(t-n), \qquad (9.32)$$

where c_n is a stationary random sequence and ϕ is a scaling function.

(a) Prove that the process $X(t)$ is cyclostationary.

(b) Let c_n in (9.32) be the white noise sequence. Show that $\gamma(t,s) = \mathbb{K}(t,s)$, where $\mathbb{K}(t,s)$ is the standard wavelet-based reproducing kernel of the multiresolution subspace V_0, as in (6.6).

(c) Consider a cyclostationary process $X(t)$ that randomly takes values 1 or -1 on intervals $n-1 \leq t < n$, $n \in \mathbb{Z}$.

Prove that $X(t)$ has the representation (9.32), identify ϕ, and demonstrate that the c_n are uncorrelated.

9.4. Can $\hat{\theta}_0(\omega)$ in (9.12) be defined simply as $\hat{\gamma}(\omega)^{-1/2}$? What happens in that case if $\gamma(h) = e^{-|h|}$?

9.5. If ω_j is any Fourier frequency, show that $I(\omega_j) = \frac{1}{2\pi}\sum_{|k|\leq T-1}\hat{\gamma}(k)e^{-ik\omega_j}$, where $\hat{\gamma}(k) = \frac{1}{T}\sum_{t=0}^{T-1-k}(X_{t+k}-m)(X_t-m)$, $k > 0$. Notice the similarity with (9.15).

9.6. Let B be the backward time-shift operator, defined as $B^j X_t = X_{t-j}$. Prove that the k-times differenced fractional Brownian motion $fBm(H)$ process,

$$Y_t = (1-B)^k X_t,$$

is a zero-mean process with autocovariance function

$$\gamma(h) = \frac{\sigma^2}{2} \sum_{i=0}^{2k}(-1)^{k+i-1}\binom{2k}{i}|h+(k-i)|^{2H}.$$

9.7. Let $X_t, t = 1, \ldots, n = 2^J$ be a stationary process and let $\underline{d}$ be its wavelet transformation.

(a) Prove that the finest level of detail (see the representation 4.6), $\underline{d}_{J-1} = \sum_{n \in \mathbb{Z}} g_n X_{2t+n}$, is a stationary sequence with autocovariance function

$$\gamma_{\underline{d}_{J-1}}(h) = \sum_{k,l=-\infty}^{\infty} g_k g_l \, \gamma_X(2h + k - l).$$

(b) Let f_X be the spectral density of X_t. Prove that the finest level of detail $\underline{d}_{J-1}$ has the spectral density

$$f_{\underline{d}_{J-1}}(\omega) = \left| m_1 \left(\frac{\omega}{2} \right) \right|^2 f_X \left(\frac{\omega}{2} \right) + \left| m_1 \left(\pi - \frac{\omega}{2} \right) \right|^2 f_X \left(\pi - \frac{\omega}{2} \right).$$

where $m_1(\omega)$ is the high-pass wavelet transfer function given by (3.27).

(c) Refer to Example 9.2.1. Find and plot the autocovariance functions and the spectral densities for the three finest levels of detail in Haar's transformation of an autoregressive AR(1) process with $\phi = 0.4$. Discuss the "whitening property" by inspecting the shapes of the obtained spectral densities.

9.8. Show that from the definition of fractional Gaussian noise, page 292, it follows that $G_{H,h}(t) \sim \mathcal{N}(0, \sigma^2 h^{2H-2})$.

9.9. Prove that if $X(t)$ is a Brownian bridge process, the random variables $Z_n = \int_0^1 X(t) \cdot \sqrt{2} \sin \pi n t \, dt$ (coefficients in K-L expansion), are uncorrelated.

9.10. [73] Let $X_t = \epsilon_t - \beta \epsilon_{t-1}$ be an MA(1) process, with $\epsilon_t \sim \mathcal{N}(0, \sigma^2)$. Prove that for Haar's filter:

$$\eta_{jk}^{(\psi)} = \{\gamma(0) 2^j \mathbf{1}[j > 0, k \equiv 0 \,(\text{modulo } 2^j)] + \gamma(0) \mathbf{1}(j \le 0) + 2\gamma(1)(1 - 3 \cdot 2^j) \mathbf{1}(j \le 0)\} \cdot \mathbf{1}(k \ge 0).$$

9.11. Simplify the estimator $\widehat{a}(J)$ given by (9.29).

9.12. Let $B(t)$ be a fractional Brownian motion with Hurst exponent H.

(a) Find $\text{Var}[B(k - p) - B(k)]$.

(b) Generate a discrete sample path of the fBm process with Hurst exponent $H = \frac{1}{3}$.

10
Wavelet-Based Random Variables and Densities

In this chapter, we give an overview of wavelet decompositions with random coefficients. We discuss two topics in greater detail: (i) random variables whose densities are squares of functions from an orthonormal wavelet basis, and (ii) the use of wavelet transformations to generate random densities. The first topic is based on the fact that elements of an orthonormal basis have $\mathbb{L}_2$-norm equal to 1, and the second topic relies on Parseval's identity, see Theorem 2.2.2 (iii).

10.1 SCALING FUNCTION AS A DENSITY

The following result adapted from Janssen [216] states that except in the Haar case, a scaling function cannot be a density, even though $\int \phi(x)\, dx = 1$.

Theorem 10.1.1 *[216] Let ϕ be a scaling function that corresponds to an orthogonal multiresolution analysis. Except for the Haar wavelet, ϕ can not be a probability density.*

Proof: Assume that ϕ is a probability density. Then, all h_n in $\underset{\sim}{h}$ are non-negative because

$$h_n = \int \sqrt{2}\phi(2x - n)\, \phi(x)\, dx = \sqrt{2}E\phi(2X - n) \geq 0.$$

From the relations

$$1 = m_0(0) = \frac{1}{\sqrt{2}} \sum h_n \text{ and}$$
$$0 = m_0(\pi) = \frac{1}{\sqrt{2}} \sum (-1)^n h_n,$$

we conclude

$$\sum_{n \text{ even}} h_n = \sum_{n \text{ odd}} h_n = \frac{1}{\sqrt{2}}.$$

Since $\sum_n h_n^2 = 1$ and $h_n \geq 0$, the sums $\sum_{n \text{ even}}$ and $\sum_{n \text{ odd}}$ must contain a single element each.

It is possible to construct biorthogonal bases whose primary scaling functions are densities. One such construction is applied in density estimation; see the discussion on page 229 and Walter and Shen [444].

10.2 WAVELET-BASED RANDOM VARIABLES

A class of probability densities can be obtained by squaring functions from any orthonormal basis of $\mathbb{L}_2$. For example, all squared basis functions from the Haar and Walsh bases are scaled uniform densities.

Let ϕ and ψ be the scaling and wavelet functions generated by an orthonormal multiresolution analysis. The functions $\phi^2(x)$ and $\psi^2(x)$ are probability densities, nonnegative with an integral of one. Panels (a) and (c) in Fig. 10.1 depict densities corresponding to the DAUB2 scaling and wavelet functions, respectively. Panels (b) and (d) give the corresponding cumulative distribution functions. The scaling equations (3.10) and (3.26) that recursively connect $\phi(x)$ and $\psi(x)$ with $\phi(2x - k)$, $k \in \mathbb{Z}$, are a basis for an algorithm for finding moments. Villemoes [427], and independently Shann and Yan [373], derived a recursive relation based on the scaling equations and their result is given in Theorem 10.2.1. Closely related results can be found in Dahmen and Micchelli [100]. We also give an *exact expression* for the first moment $\int x\psi^2(x)\,dx$ with a nice proof provided by Yanyuan Ma [270].

We first give the necessary definitions and the result of Villemoes. The exposition is similar to Shann and Yan [373].

The *generalized moments* of $\phi(x)$ are defined by

$$\mu_{k,t} = \int_{\mathbb{R}} x^k \phi(x)\phi(x-t)\,dx. \tag{10.1}$$

Theorem 10.2.1 *([427]) Let $T = 2N - 2$, where N is the number of vanishing* moments for the wavelet function under consideration.

The vector $\mu_k = (\mu_{k,t})$, $|t| \leq T$, is a solution of the system

$$(I - \frac{1}{2^k}A)\mu_k = b_k , \qquad (10.2)$$

where

$$A_{ij} = \sum_n h_n h_{n+i-2j} , \quad -T \leq i,j \leq T \qquad (10.3)$$

is the transition matrix (or Lawton's matrix, see [253] or Daubechies [104] page 189). The vector b_k has components

$$b_{k,t} = \frac{1}{2^k} \sum_n \sum_l h_n h_l \sum_{j=1}^{k} \binom{k}{j} n^j \mu_{k-j, l-n+2t} , \quad -T \leq t \leq T.$$

Proof:

$$\begin{aligned} \mu_{k,t} &= \int_{\mathbb{R}} x^k \phi(x)\phi(x-t)\,dx \\ &= \int_{\mathbb{R}} x^k \sqrt{2}\sum_n h_n \phi(2x-n) \sqrt{2}\sum_l h_l \phi(2x-2t-l)\,dx \\ &= 2\int_{\mathbb{R}} x^k \sum_n \sum_l h_n h_l \phi(2x-n)\phi(2x-2t-l)\,dx. \end{aligned}$$

If N is the number of vanishing moments, then the DAUBN scaling function is supported on $[0, 2N-1]$. Also, $-(2N-2) \leq t \leq 2N-2 = T$. Making the change of variables $2x - n = y$; $x = \frac{y+n}{2}$; $dx = \frac{dy}{2}$, one gets

$$\begin{aligned} \mu_{k,t} &= 2\int_{\mathbb{R}} \left(\frac{y+n}{2}\right)^k \sum_n \sum_l h_n h_l \phi(y)\phi(y+n-2t-l)\frac{dy}{2} \\ &= \sum_n \sum_l h_n h_l \int_{\mathbb{R}} \left(\frac{y+n}{2}\right)^k \phi(y)\phi(y-(2t+l-n))\,dy. \end{aligned}$$

The ranges of the indices n and l are

302 WAVELET-BASED RANDOM VARIABLES AND DENSITIES

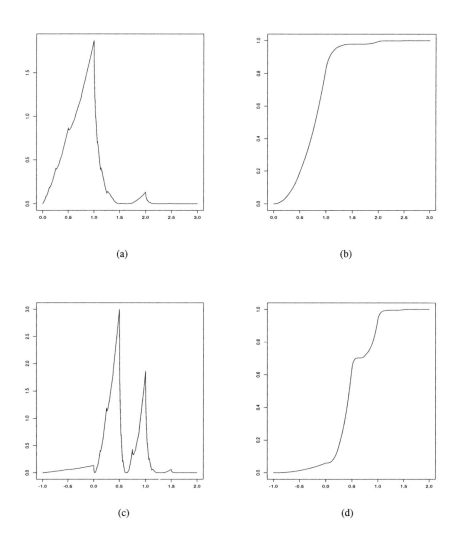

Fig. 10.1 Densities and corresponding cumulative distribution functions for squares of the DAUB2 scaling and wavelet functions.

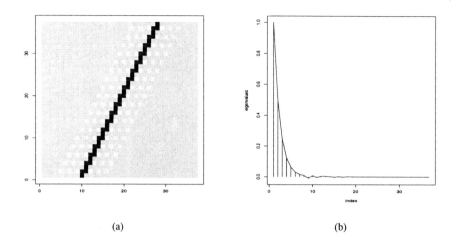

Fig. 10.2 Plots of the transition matrix [panel (a)] and its eigenvalues for $N = 10$ [panel (b)]. Asymptotically, the matrix A will have eigenvalues $1, \frac{1}{2}, \frac{1}{4}, \ldots, \frac{1}{2^n}, \ldots$.

$$0 \leq n \leq T+1$$
$$(n - T - 2t) \vee 0 \leq l \leq (n + T - 2t) \wedge (T+1),$$

where $a \vee b$ ($a \wedge b$) is the maximum (minimum) of a and b.

Finally,

$$\begin{aligned}
\mu_{k,t} &= \sum_n \sum_l h_n h_l \frac{1}{2^k} \sum_{i=0}^{k} \int_{\mathbb{R}} \binom{k}{i} y^i n^{k-i} \phi(y) \phi(y - (2t + l - n)) \, dy \\
&= \sum_n \sum_l h_n h_l \frac{1}{2^k} \sum_{i=0}^{k} n^{k-i} \binom{k}{k-i} \mu_{i, 2t+l-n} \\
&= \sum_n \sum_l h_n h_l \frac{1}{2^k} \mu_{k, l-n+2t} + \sum_n \sum_l h_n h_l \frac{1}{2^k} \sum_{j=1}^{k} n^j \binom{k}{j} \mu_{k-j, l-n+2t}.
\end{aligned}$$

The expression for $\mu_{k,t}$ can be represented as

$$\mu_{k,t} - \frac{1}{2^k} \sum_n \sum_l h_n h_l \mu_{k, l-n+2t} = \frac{1}{2^k} \sum_n \sum_l h_n h_l \sum_{j=1}^{k} n^j \binom{k}{j} \mu_{k-j, l-n+2t}.$$

One can write this equation in matrix form

$$\left(I - \frac{1}{2^k}A\right) \cdot \underset{\sim}{m} = \underset{\sim}{b},$$

where $\underset{\sim}{\mu}$ is a vector of dimension $2T+1$ given by

$$\underset{\sim}{\mu}_k = (\mu_{k,-T}, \mu_{k,-T+1}, \ldots, \mu_{k,0}, \ldots, \mu_{k,T})',$$

and

$$\underset{\sim}{b} = (b_{-T}, b_{-T+1}, \ldots, b_T)'$$

is a vector with components

$$b_t = \frac{1}{2^k} \sum_n \sum_l h_n h_l \sum_{j=1}^k n^j \binom{k}{j} \mu_{k-j, l-n+2t}, \quad -T \leq t \leq T;$$

also

$$A = (a_{ij})$$
$$a_{ij} = \sum_n h_n h_{n+i-2j}, \quad -T \leq i, j \leq T,$$

is the transition matrix. Panel (a) in Fig. 10.2 gives a plot of the transition matrix A corresponding to $N = 10$. Panel (b) gives the eigenvalues of A. Note that 1 is always the eigenvalue of A, which is one of Lawton's conditions for the orthogonality of $\{\phi(x-n), n \in \mathbb{Z}\}$. See also Strang and Nguyen [387].

The values $\mu_{k,0}$ are of interest to us since they represent the moments of the random variable with the density $\phi^2(x)$. The numbers in Table 10.1 are obtained by solving the matrix equation (10.2) for different values of N. The recursion starts with $\mu_{0,t} = \delta(t)$.

Once the generalized moments $\mu_{k,t}$ are known, one can find the corresponding generalized moments defined by the wavelet $\psi(x)$ by using the scaling equations (3.26). The following corollary makes this precise.

Corollary 10.2.1 *Let*

$$\xi_{k,t} = \int_{\mathbb{R}} x^k \psi(x) \psi(x-t)\, dx.$$

Table 10.1 Expectations and variances of random variables with densities $\phi^2(x)$ from Daubechies' family.

	MEAN $\mu_{1,0}$	VARIANCE
DAUB1	0.5	0.083333
DAUB2	0.770948	0.097718
DAUB3	1.022422	0.132056
DAUB4	1.266408	0.172921
DAUB5	1.506244	0.219200
DAUB6	1.743334	0.270688
DAUB7	1.978412	0.327253
DAUB8	2.211921	0.388751
DAUB9	2.444157	0.455041
DAUB10	2.675332	0.525993

Then,

$$\xi_{k,t} = \frac{1}{2^k} \sum_{n,l=-T}^{1} g_n g_l \sum_{r=0}^{k} n^{k-r} \binom{k}{r} \mu_{r,2t+l-n}. \tag{10.4}$$

Though the general relation (10.4) provides an effective way to calculate any moment $\xi_{k,0}$, it requires pre-calculation of many generalized moments $\mu_{k,t}$. A surprisingly simple result holds for the mean $\xi_{1,0}$.

The following result has been first proved by Villemoes [427], and independently proved by Ma, Strang and Vidakovic [270]. See also Exercise 10.3.

Theorem 10.2.2 *The mean* $\xi_{1,0} = \int x\psi^2(x)\,dx$ *is at the center of the support of* $\psi(x)$.

Proof: [270] We will place the support of $\phi(x)$ and $\psi(x)$ on $[0, 2N-1]$. This comes with choosing $g_k = (-1)^k h_{2N-1-k}$. The choice $g_k = (-1)^k h_{1-k}$ moves the support of ψ to $[1-N, N]$. Then the mean of ψ^2 moves to the new center point $\xi_{1,0} = \frac{1}{2}$.

Let $H_0(z) = \sum_{i=0}^{2N-1} h_i z^i$, where the h_i are the low-pass filter coefficients from (3.10). The polynomial $P(z) = H_0(z)H_0(z^{-1}) = \sum_{n=-2N+1}^{2N-1} p_n z^n$ is called *halfband* because of the requirements for orthogonality (see Strang and Nguyen [387] and the footnote on page 55).

The coefficients

$$p_i = \sum_k h_k h_{k+i}$$

satisfy

$$p_{2i} = \delta(i) \text{ and } p_i = p_{-i}. \tag{10.5}$$

We first prove that $\mu_{1,0} = \frac{1}{2}\sum_i p_i \mu_{1,i} + \frac{1}{2}\sum_i i h_i^2$. Indeed by using (3.10) and changing the order of integration and summation we obtain

$$\begin{aligned}
\int x\phi^2(x)\,dx &= \int x\, 2\sum_k \sum_s h_k h_s \phi(2x-k)\phi(2x-s)\,dx \\
&= \frac{1}{2}\sum_k \sum_s h_k h_s \int x\phi(x-k)\phi(x-s)\,dx \qquad (10.6) \\
&= \frac{1}{2}\sum_k \sum_i h_k h_{k-i} \int (x+k)\phi(x)\phi(x+i)\,dx \qquad (s=k-i) \\
&= \frac{1}{2}\sum_i p_i \mu_{1,i} + \frac{1}{2}\sum_i i h_i^2 \qquad \text{(by orthogonality)}.
\end{aligned}$$

Let $g_k = (-1)^k h_{2N-1-k}$ be the coefficients of the high-pass filter corresponding to the low-pass filter $\underline{h}$. Then,

$$p_i = \sum_k h_k h_{k+i} = (-1)^i \sum_k g_k g_{k+i}. \tag{10.7}$$

The relation of $\sum_i i g_i^2$ to $\sum_i i h_i^2$ is straightforward:

$$\begin{aligned}
\sum_i i g_i^2 &= \sum_i i h_{2N-1-i}^2 \\
&= (2N-1)\sum_i h_{2N-1-i}^2 - \sum_i (2N-1-i) h_{2N-1-i}^2 \\
&= 2N - 1 - \sum_i i h_i^2. \tag{10.8}
\end{aligned}$$

By imitating the steps in (10.6), we obtain

$$\xi_{1,0} = \frac{1}{2}\sum_i (-1)^i p_i \mu_{1,i} + \frac{1}{2}\sum_i i g_i^2.$$

From (10.6) we express $\frac{1}{2}\sum_i i h_i^2$ as $\mu_{1,0} - \frac{1}{2}\sum_i p_i \mu_{1,i}$. Then, (10.5), (10.7), and (10.8) imply that the first moment falls halfway along the support:

$$\begin{aligned}
\xi_{1,0} &= \frac{1}{2}\sum_i (-1)^i p_i \mu_{1,i} + \frac{2N-1}{2} - \frac{1}{2}\sum_i ih_i^2 & \text{by (10.8)}\\
&= \frac{1}{2}\sum_i (-1)^i p_i \mu_{1,i} + \frac{2N-1}{2} - \left(\mu_{1,0} - \frac{1}{2}\sum_i p_i \mu_{1,i}\right) & \text{by (10.6)}\\
&= \sum_{i \text{ even}} p_i \mu_{1,i} + \frac{2N-1}{2} - \mu_{1,0}\\
&= \mu_{1,0} + \frac{2N-1}{2} - \mu_{1,0} & [\text{because } p_{2k} = \delta(k)]\\
&= \frac{2N-1}{2}.
\end{aligned}$$

Example 10.2.1 Let f be a density and let $g = \sqrt{f}$ have the representation

$$g = \sum_{j,k} d_{jk} \psi_{jk}(x), \qquad (10.9)$$

in some fixed wavelet basis. Then, any density $f(x)$ can be decomposed into two components, $f_1(x) + f_2(x)$, where $f_1(x)$ is a mixture of wavelet densities and is a density itself, and $f_2(x)$ is a "detail function" that integrates to 0.

The above statement follows from the wavelet representation of the square root of the density f,

$$f(x) = f_1(x) + f_2(x) = \sum_{j,k} d_{jk}^2 \psi_{jk}^2(x) + \sum_{j,k}\sum_{j',k'} d_{jk}\, d_{j'k'} \psi_{jk}(x)\psi_{j'k'}(x).$$

If (J, K) is a pair of integer-valued random variables with joint distribution

$$P(J = j,\, K = k) = d_{jk}^2,\ j, k \in \mathbb{Z}, \qquad (10.10)$$

where the d_{jk}s are as in (10.9), and if X_{jk} has the density $\psi_{jk}^2(x)$, then X_{JK} has the density $f_1(x)$.

Also,

$$EX_{JK} = E(E(X_{JK}|J,K)) = \sum_{jk} EX_{jk} d_{jk}^2 = \sum_{jk} 2^{-j}\left(k + \frac{1}{2}\right) d_{jk}^2.$$

10.3 RANDOM DENSITIES VIA WAVELETS

In this section, we first give some background on random densities represented as orthogonal series, and provide an algorithm for generating wavelet-based random densities. We discuss the properties of random densities and explain how to generate random densities from some common families of densities. The idea of representing a probability density function as an orthogonal series dates back to 1962 when Čencov published his seminal work [66]. Rubin and Chen [68, 355] first used Čencov's orthogonal decompositions to generate random densities. Their idea relies on Parseval's identity.

Let $\{\psi_i, i \in I\}$ be a complete orthonormal basis for $\mathbb{L}_2(D), D \subseteq \mathbb{R}$. Then, any $\mathbb{L}_2(D)$ function g may be uniquely represented as $g(x) = \sum_{i \in I} a_i \psi_i(x)$. Moreover, Parseval's identity states that the $\mathbb{L}_2$-norm of g coincides with the ℓ_2-norm of its Fourier coefficients. Assume now that this ℓ_2 norm is 1. Then

$$1 = \sum_i a_i^2 = \int g^2(x)\, dx.$$

Because $f = g^2$ is non-negative and integrates to 1, f is a density. We next provide a formal definition of a wavelet-based random density.

Definition 10.3.1 *A wavelet based random density* $\mathbf{f}(x)$ *generated by the wavelet* ψ *is defined by* $\mathbf{f}(x) = \left[\sum_{j,k \in \mathbb{Z}} d_{jk} \psi_{jk}(x) \right]^2$, *where the coefficients* d_{jk} *are random variables satisfying the constraint* $\sum_{jk} d_{jk}^2 = 1$.

We restrict our attention throughout the rest of the chapter to generating random densities with compact support, which, without loss of generality, will be the interval [0,1]. To this end, one may choose as an appropriate basis the family of periodized wavelets $\{\phi_{00}^{per}, \psi_{j,k}^{per}, j \geq 0, 0 \leq k \leq 2^j - 1\}$ as discussed in the previous chapter. Related constructions can be found in Basseville et al. [22], in the context of stochastic processes indexed by nodes on lattices or trees in wavelet decompositions of signals and images.

10.3.1 Tree Algorithm

The following algorithm, which generates a random density is a modification of an algorithm proposed by Vidakovic [421] and Rios and Vidakovic [209].

Let p be a fixed number between 0 and 1 and let X_{jk} be a family of i.i.d. Bernoulli(p) random variables. Let $r_{jk}(p)$ be a *random sign* defined as

$$r_{jk}(p) = 2X_{jk} - 1,$$

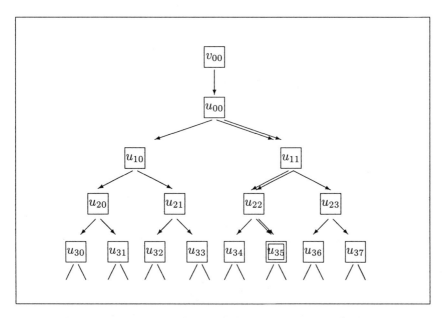

Fig. 10.3 Tree with $\{v_{00}, u_{jk}, j \geq 0, k = 0, \ldots, 2^j - 1\}$.

$j = 0, 1, \ldots;\ 0 \leq k \leq 2^j - 1$.

Let $\mathcal{T} = \{v_{00}, u_{jk}, j = 0, 1, \ldots;\ k = 0, 1, \ldots, 2^j - 1\}$ be a family of i.i.d. random variables on $[0, 1]$. We additionally assume that $P(u = 1) < 1$, for $u \in \mathcal{T}$. The family $\mathcal{T}$ will be called a *tree*. Indices j in the tree $\mathcal{T}$ correspond to levels of detail in the wavelet representation. The random variable v_{00} is in the *root* of the tree; u_{00} is on level zero, u_{10} and u_{11} are on level one, and so on. At each node (j, k), we identify a unique path (set of nodes) leading from v_{00} to u_{jk}, which we designate as $path(j, k)$. For example, $path(3, 5)$ emphasized in Figure 10.3 is (00), (11), and (22).

Lemma 10.3.1 *Given the tree $\mathcal{T}$ and random signs r_{jk}, the random variables*

$$\begin{aligned}
c_{00} &= r'_{00}(p)\sqrt{v_{00}}, \\
d_{00} &= r''_{00}(p)\sqrt{(1-v_{00})(1-u_{00})} \\
d_{jk} &= r_{jk}(p)\sqrt{\frac{(1-v_{00})(1-u_{jk})}{2^j} \prod_{j'k' \in path(j,k)} u_{j'k'}}
\end{aligned} \quad (10.11)$$

satisfy

$$c_{00}^2 + \sum_{j \geq 0} \sum_{k=0}^{2^j-1} d_{jk}^2 = 1. \quad \text{a.s.} \tag{10.12}$$

All finite products in the left-hand side of (10.12) cancel. The proof is completed by utilizing this result: Let $\{u_n\}$ be a sequence of i.i.d. random variables on $[0, 1]$ such that $P(u_1 = 1) < 1$. Then,

$$\prod_{i=1}^{n} u_i \to 0, \quad \text{a.s.}$$

as $n \to \infty$. Indeed, the product $\prod_{i=1}^{n} u_i$ is a nonnegative supermartingale. From the Chebyshev inequality, it converges to 0 in probability and therefore it converges to 0 a.s.

10.4 PROPERTIES OF WAVELET-BASED RANDOM DENSITIES

Once the wavelet basis is selected, the algorithm in Lemma 10.3.1 describes the construction of a random density.

Theorem 10.4.1 *Let* $\mathbf{f}(x) = \left[c_{00} + \sum_{j \geq 0} \sum_{0 \leq k \leq 2^j - 1} d_{jk} \psi_{j,k}^{per}(x)\right]^2$, *with coefficients defined as in Lemma 10.3.1. Then*

$$\int \mathbf{f}(x)\, dx = 1, \quad \text{a.s.}$$

The proof of this result, which justifies the term *random density*, follows directly from Lemma 10.3.1.

Fig. 10.4 depicts a random density with the support $[0,1]$, generated by uniform $\mathcal{U}[0, 0.7]$ random variables from $\mathcal{T}$, utilizing the DAUB10 family of wavelets.

We now compute the expectation of the constructed random density.

Theorem 10.4.2 *Let* λ *be the expectation of the random variables* u_{jk}. *For any periodized wavelet basis* $\{\mathbf{1}(0 \leq x \leq 1), \psi_{j,k}(x), j \geq 0, 0 \leq k \leq 2^j - 1\}$ *on* $[0, 1]$ *and for* $p = \frac{1}{2}$, *one has*

$$E\mathbf{f}(x) = \lambda + (1-\lambda)^2 \sum_{j \geq 0} \left(\frac{\lambda}{2}\right)^j \sum_k \psi_{j,k}^2(x). \tag{10.13}$$

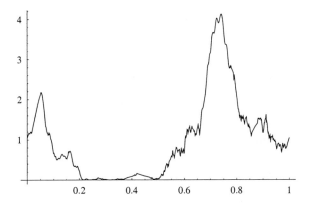

Fig. 10.4 An example of a random density on $[0, 1]$. The random variables from $\mathcal{T}$ are uniform $[0, 0.7]$. The wavelet basis used is DAUB10.

Proof: The theorem follows by Fubini's theorem, independence of random signs, the identities $Ec_{00}^2 = \lambda$; $Ed_{jk}^2 = (1 - \lambda)^2 \left(\frac{\lambda}{2}\right)^j$, and the zero-expectation property of random signs.

In general, the function $\sum_k \psi_{jk}^2(x)$ has no a finite form and further simplification of expression (10.13) is impossible. The only exception is the Haar basis for which we can find the expectation of **f** explicitly.

Corollary 10.4.1 *For the Haar basis and* $p = \frac{1}{2}$,

$$E\mathbf{f}(x) = \mathbf{1}(x \in [0, 1]).$$

For the Haar wavelet, we have $\sum_{k=0}^{2^j - 1} \psi_{j,k}^2(x) = 2^j \mathbf{1}(x \in [0, 1])$. Then,

$$E\mathbf{f}(x) = \left[\lambda + (1 - \lambda)^2 \sum_{j \geq 0} \lambda^j\right] \mathbf{1}(x \in [0, 1]) = \mathbf{1}(x \in [0, 1]). \quad (10.14)$$

In other words, the expected value of the random density **f**, based on Haar's basis, is uniform on $[0,1]$ and is independent of the distribution of the random variables in $\mathcal{T}$.

Next we consider the moments of random variables with the random densities. Denote by

$$\xi_l^{jk} = \int_0^1 x^l \psi_{jk}^2(x)\, dx,$$

the lth moment of a random variable with density $\psi_{j,k}^2(x)$.

Let $M_l = \int_0^1 x^l \mathbf{f}(x)\,dx$. Then, from (10.13) and (10.14)

$$EM_l = \frac{\lambda}{l+1} + (1-\lambda)^2 \sum_{j\geq 0} \left(\frac{\lambda}{2}\right)^j \sum_{k=0}^{2^j-1} \xi_l^{jk},$$

and for the Haar basis,

$$EM_l = \frac{1}{l+1}.$$

10.5 RANDOM DENSITIES WITH CONSTRAINTS

For modeling purposes, it is of interest to generate random densities that satisfy some constraints. Most typical are requirements on the smoothness, symmetry, unimodality, and skewness. We give procedures that transform general random densities to constrained ones. We emphasize that such procedures are not unique and the reader may invent his own transformation leading to a class of constrained densities.

10.5.1 Smoothness Constraints

As unconditional bases for important smoothness spaces, wavelets provide natural building blocks in describing smooth functions.

According to Meyer's results, the magnitudes of random variables $\{c_{00}, d_{jk}\}$ from Lemma 10.3.1 affect the global regularity of random densities. The following result connects the expectation of random variables in $\mathcal{T}$ with the regularity of the generated density. We assume that the wavelet ψ comes from a β-regular multiresolution analysis ($\beta \geq \alpha$).

Theorem 10.5.1 [421] *Let the random density* $\mathbf{f}$ *be generated (in the sense of Lemma 10.3.1) by a sequence of i.i.d. random variables* $\{u_{jk}\}$ *such that* $Eu_{jk} = \lambda$. *Then,* $\mathbf{f} \in \mathcal{C}^\alpha([0,1]), \alpha = \frac{1}{2}\log_2 \frac{1}{\lambda}$, *with probability 1.*

10.5.2 Constraints on Symmetry

To generate random symmetric densities on $[-1, 1]$, one can mix an already generated random density on $[0, 1]$ with the random sign $r(1/2)$.

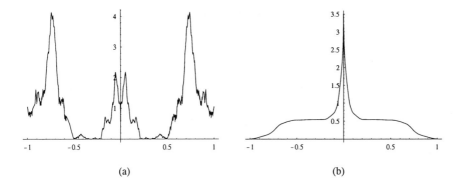

Fig. 10.5 (a) Symmetric density on [-1,1]. (b) Unimodal random density.

Lemma 10.5.1 *Let X be a random variable with a random density. Then, the random variable $S = r(0.5) \cdot X$ has a symmetric random density.*

Fig. 10.5(a) gives a symmetric density obtained by symmetrizing the random density from Fig. 10.4.

10.5.3 Constraints on Modality

Any symmetric, unimodal distribution can be represented as a mixture of uniformly distributed random variables. The result (attributed to Feller, the proof can be found in [116]) states

Proposition 1 *Let X be a symmetric unimodal random variable on $(-m, m)$, with $0 < m \leq \infty$. Then, X can be represented as the product UZ, where U is uniform on $[-1, 1]$ and Z is a random variable on $[0, m)$. The mixing random variable Z is uniquely defined (up to $\stackrel{d}{=}$) and independent of U. If Z has density $g(z)$ and X has density $f(x)$, then*

$$f(x) = \int_{|x|}^{\infty} \frac{g(z)}{2z} dz. \qquad (10.15)$$

The relation (10.15) suggests a procedure for generating symmetric unimodal random densities. One first generates unconstrained random densities and then mixes the uniforms according to them, as in (10.15) [See Fig. 10.5(b)].

10.5.4 Skewed Random Densities

In many cases of Bayesian inference, prior knowledge about the parameter of interest suggests skewed densities. For example, suppose that the parameter space is truncated (it is of the form $\Theta = [\theta_0, 1]$, say), but the truncation point $-1 \le \theta_0 < 1$ is unknown to the statistician. One way to incorporate this prior information about the parameter in the simulation procedure is by generating random densities on [-1,1] that are skewed to the right. Other issues concerning the use of skewed densities are discussed in O'Hagan and Leonard [327], and Azzalini [21], among others.

Lemma 10.5.2 *Let $f(x)$ and $G(y)$ be the density and the cumulative distribution function of independent symmetric random variables X and Y, respectively. Then, for any λ,*

$$h(x) = 2f(x)G(\lambda x) \tag{10.16}$$

is a density. The density h is symmetric only if $\lambda = 0$. For $\lambda > 0(< 0)$ the density is skewed to the right (left).

Proof: Since X and Y are symmetric it follows that $1 = 2 \cdot P(Y - \lambda X \le 0)$. The conditioning of $2P(Y - \lambda X \le 0)$ on X results in a non-negative function $2f(x)G(\lambda x)$ that integrates to 1.

Fig. 10.6 gives four different skewed densities. The plots are obtained by choosing G in (10.16) to be a standard normal distribution for $\lambda = 0.2, 1, 5, 20$.

In Example 10.5.1 we provide an application of random densities in assessing Γ-minimax performance of linear estimators.

Example 10.5.1 Minimax and Γ-minimax rules are often criticized as being overly-conservative, see Robert [353]. By generating priors from some fixed class of densities, we will compare Bayes risks with the Γ-minimax risk. In the ensuing example, we will see that Bayes risks are not substantially smaller and that the price for robustness induced by minimaxity is not unduly high.

Let $X|\theta \sim \mathcal{N}(\theta, 1)$ and let $\Gamma_{SU}[-1, 1]$ be the class of all symmetric unimodal distributions on $[-1, 1]$. The linear Γ-minimax rule for θ is $\delta^*(x) = \frac{x}{4}$ and the least favorable distribution π^* is the uniform [-1,1]. The Γ-minimax risk is $r(\pi^*, \delta^*) = \frac{1}{4}$, see Vidakovic and DasGupta [424].

The linear Γ-minimax rule gives a slightly more conservative risk when compared to the Bayes risk of linear Bayes rules evaluated for randomly selected priors from the class $\Gamma_{SU}[-1, 1]$. In order to illustrate this, we generated thirty symmetric unimodal priors and calculated the Bayes risks of the corresponding linear Bayes rules

$$\delta_\pi(x) = \frac{E^\pi \theta^2}{1 + E^\pi \theta^2} x.$$

The risks are

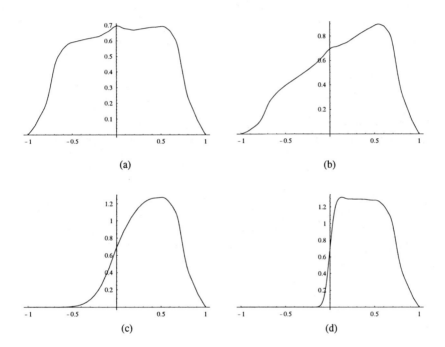

Fig. 10.6 Skewed densities illustrating (10.16). The f is as in Fig. 10.7(b) and G is the standard normal. (a) $\lambda = 0.2$. (b) $\lambda = 1$. (c) $\lambda = 5$. (d) $\lambda = 20$.

316 WAVELET-BASED RANDOM VARIABLES AND DENSITIES

$$r_\pi = \frac{E^\pi \theta^2}{1 + E^\pi \theta^2}, \qquad (10.17)$$

where, given ξ, $E^\pi \theta^2$ is a linear combination:

$$E^\pi \theta^2 = \xi * E^\mathbf{f} \theta^2 + \frac{1}{4}(1-\xi).$$

The expectation $E^\mathbf{f}$ was taken with respect to a symmetric unimodal density generated by (10.15) with smoothness constraints, as in Exercise 10.8.

Simulation suggested that the Bayes risks of the linear Bayes rules are below the "Γ-minimax line" of $\frac{1}{4}$. The explanation is simple. The least favorable distribution is uniform on [-1,1] and it maximizes the second moment in the class $\Gamma_{SU}[-1,1]$. Any other random density from $\Gamma_{SU}[-1,1]$ will produce a smaller $E\theta^2$ and consequently smaller risk (10.17).

However, our simulation shows that the Γ-minimax risk is not overly conservative and that users who use the simple linear rule δ^* are not penalized much more than the users of δ_π. The Bayes risks are, on average, about 20% smaller than the Γ-minimax risk and often very close to it.

Conti and Lasinio [96] propose the use of wavelet-based random prior measures in regression problems. See also the related Exercise 10.4. Nason, von-Sachs, and Kroisandt [312] introduced *wavelet processes*, double-indexed sequences $X_{t,T} = \sum_{j,k} w^0_{j,k,T} \xi_{jk} \psi_{jk}(t)$, where the ξ_{jk} are zero-mean uncorrelated random variables and the $w^0_{j,k,T}$ are appropriate weights.

10.6 EXERCISES

10.1. Consider the standard deviations of the wavelet random variables, $\sigma_N(X)$. Assume that the density is the square of the scaling function from the Daubechies' family and that the corresponding wavelet has N vanishing moments. Find the linear regression of $\sigma_N(X)$ on N, for $2 \le N \le 10$ (You can use the data in Table 10.12). Calculate the residuals and comment on your findings.

10.2. By using the definition of $\mu_{k,t}$ and the orthogonality of the $\phi(x-k)$, prove $\mu_{1,t} = \mu_{1,-t}$.

10.3. [Shann, Yan, and Tzeng: personal communication] By following steps (i)-(iii), give an alternative proof of Theorem 10.2.2.

   (i) Prove that $\langle x\psi(x), \psi(x) \rangle = \frac{i}{2\pi} \langle \Psi'(\omega), \Psi(\omega) \rangle$.

   (ii) Show that

$$\langle \Psi'(\omega), \Psi(\omega) \rangle = \int_0^{2\pi} m_1'(\omega)\overline{m_1(\omega)}\, d\omega + \int |m_1(\omega)|^2 \Phi'(\omega)\overline{\Phi(\omega)}\, d\omega,$$

$$\langle \Phi'(\omega), \Phi(\omega) \rangle = \int_0^{2\pi} m_0'(\omega)\overline{m_0(\omega)}\, d\omega + \int |m_0(\omega)|^2 \Phi'(\omega)\overline{\Phi(\omega)}\, d\omega,$$

and thus, because of (3.33),

$$\langle \Psi'(\omega), \Psi(\omega) \rangle = \int_0^{2\pi} m_1'(\omega)\overline{m_1(\omega)}\, d\omega + \int_0^{2\pi} m_0'(\omega)\overline{m_0(\omega)}\, d\omega.$$

(iii) By using (ii) and the definition of m_1, show that

$$\langle \Psi'(\omega), \Psi(\omega) \rangle = -i \int_0^{2\pi} |m_0(\omega)|^2\, d\omega = -i\pi.$$

10.4. Let $\pi(\theta) = (\sum_{jk} d_{jk}\psi_{jk}(\theta))^2$ be a random prior, and let $l(\theta)$ be a likelihood. Then the posterior $\pi^*(\theta)$ is given by

$$\pi^*(\theta) = \frac{\pi(\theta)\, l(\theta)}{m(x)} = \left(\sum p_{jk}\psi_{jk}(\theta)\right)^2, \tag{10.18}$$

where $m(x)$ is the corresponding marginal. Show that

$$p_{jk} = \left\langle \frac{S(\theta)}{C} \cdot \sum_{j'k'} d_{j'k'}\psi_{j'k'}(\theta),\ \psi_{jk}(\theta) \right\rangle$$

$$= \frac{1}{C} \cdot \sum_{j'k'} [d_{j'k'}\, \langle S(\theta)\psi_{j'k'}(\theta),\ \psi_{jk}(\theta) \rangle],$$

where $S(\theta) = \sqrt{l(\theta)}$ and $C = \sqrt{m(x)}$.

10.5. *Random sample from a random density.* Suggest an approximate procedure for generating a random sample from a random density.

10.6. Let ψ be a Daubechies' wavelet and let $\psi_{jk}(x) = 2^{j/2}\psi(2^j x - k)$. For $X_{jk} \sim \psi_{jk}^2$, show

$$EX_{jk} = 2^{-j}(\xi_{1,0} + k).$$

10.7. *Project.* By mimicking the algorithm for generating univariate random densities (Section 10.3.1) propose an algorithm for generating bivariate random densities on $[0, 1] \times [0, 1]$.

10.8. *Project.* The symmetric unimodal random densities obtained by (10.15) usually have a spike at zero [Fig. 10.5(b)]. For instance, for the Haar basis the expected symmetric unimodal density is $-\frac{1}{2} \log |x| \mathbf{1}(x \in [-1, 1])$, which is infinite at zero. This unbounded spike at the origin is a consequence of the fact that in a neighborhood of zero the mixing random density is not close to zero.

(a) Generate a random symmetric unimodal density that is smooth at zero.

[Hint: The mixing random density should be "tied" to 0. Let the function $g(x)$ (the square root of the random density) be multiplied by x^k, $k > \frac{1}{2}$. Let $\mathbf{a} = \{a_n\}$ be the wavelet coefficients of the product $x^k g(x)$. Define the mixing distribution h as the square of the function obtained from the coefficients $\left\{\frac{a_n}{\|\mathbf{a}\|}\right\}$. The normalized coefficients ensure that h is a density. One such "tied" random density and resulting mixture are given in Fig. 10.7.]

(b) Even though the uniform density is symmetric and unimodal, it can never be generated with a procedure from (a). In fact, no symmetric unimodal density for which $\mathbf{f}(-1) = \mathbf{f}(1) = C$, $0 < C \leq 1$, can be generated in (a).

Suggest a modification of (a) so that *all* symmetric unimodal densities can be generated.

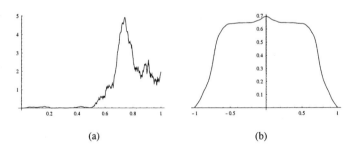

(a) (b)

Fig. 10.7 (a) Random density with the constraint $f(0) = 0$. (b) Unimodal density smooth at zero, generated by mixing uniforms by the density from panel (a).

11

Miscellaneous Statistical Applications

Wavelets in statistics are used mostly in regression and density estimation problems. In many cases, an application can be classified simultaneously in several categories. Spectral density smoothing or wavelet-based random priors are examples.

Wavelets can be successfully used in a range of statistical areas. Virtually any procedure involving orthogonal series decompositions can be "waveletized" with a hope that wavelets will bring locality and calculational efficiency. Some applications lie on the cross-boundaries between disciplines. For instance, recent applications of wavelets in the shape analysis are bridging the areas of statistical modeling, statistical theory of shapes, computational geometry, and image processing.

In this chapter, we discuss four miscellaneous applications in some detail (deconvolution, pursuit regression, order statistics calculation, and statistical turbulence), and provide pointers to some popular software for wavelet analysis and www resources.

11.1 DECONVOLUTION PROBLEMS

Deconvolution problems arise in many application areas and are extensively studied. The most popular approach to the problem utilizes kernel density estimation and Fourier transformation; see, for example, Diggle and Hall [117], Fan [142, 141], Stefansky and Carrol [384], and Masry [287].

The problem can be formulated as follows. Let θ and ϵ be independent random variables with densities π and f, where π is unknown and f is known. Let $X_1, \ldots, X_n$ be a sample such that $X_i = \theta_i + \epsilon_i$, $i = 1, \ldots, n$. The goal is to estimate the density

π on the basis of $X_1, \ldots, X_n$, in which each X_i has a marginal distribution

$$m(x) = \int_{\mathbb{R}} f(x - \theta)\pi(\theta)\, d\theta.$$

Two separate cases are usually considered: the case when the distribution of the error ϵ is supersmooth [i.e. the Fourier transform $\hat{f}(\omega)$ of f has an exponential descent], and the case when $\hat{f}(\omega)$ has a polynomial descent. In the first case, even when the degree of smoothness of π is unknown, the linear wavelet estimator can be adaptive and achieves the optimal convergence rate. In the case when $\hat{f}$ has a polynomial descent, the linear wavelet estimator fails to provide the optimal convergence rate if the degree of smoothness of π is unspecified. In this case, a nonlinear wavelet estimator can be constructed that is adaptive and achieves the optimal convergence rate.

The estimators obtained by Pensky and Vidakovic [333] are asymptotically optimal in the sense that the rates of convergence of the MISE cannot be improved; see Fan [143]. They are based on Meyer-type wavelets rather than wavelets with bounded support since this is the only type of wavelets that allows the construction of an estimator in the case of a supersmooth f.

Suppose that $\pi \in \mathbb{L}_2(\mathbb{R})$ and $\hat{f}(\omega)$ does not vanish on $(-\infty, \infty)$. Let

$$\pi(\theta) = \sum_{k \in \mathbb{Z}} a_{mk} \phi_{mk}(\theta) + \sum_{j \geq m} \sum_{k \in \mathbb{Z}} b_{jk} \psi_{jk}(\theta),$$

where

$$a_{mk} = \int_{\mathbb{R}} \phi_{mk}(\theta)\pi(\theta)d\theta \text{ and } b_{jk} = \int_{\mathbb{R}} \psi_{jk}(\theta)\pi(\theta)d\theta.$$

The coefficients a_{mk} and b_{jk} can be viewed as expectations, with respect to the marginal probability distribution function $m(x)$, of the functions u_{mk} and v_{jk},

$$a_{mk} = \int_{\mathbb{R}} u_{mk}(x)m(x)\, dx \text{ and } b_{jk} = \int_{\mathbb{R}} v_{jk}(x)m(x)\, dx,$$

provided that

$$\int_{\mathbb{R}} f(x - \theta)u_{mk}(x)\, dx = \phi_{mk}(\theta) \text{ and}$$

$$\int_{\mathbb{R}} f(x - \theta)v_{jk}(x)\, dx = \psi_{jk}(\theta). \tag{11.1}$$

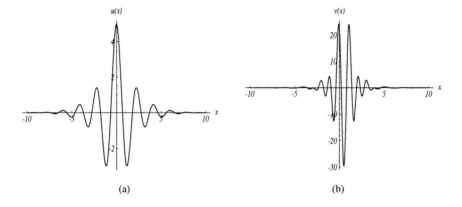

Fig. 11.1 Deconwavelets: (a) u_{00} and (b) v_{00}.

By taking the Fourier transformation of (11.1), one obtains

$$u_{mk}(x) = 2^{m/2}\mathfrak{U}_m(2^m x - k), \text{ and}$$
$$v_{jk}(x) = 2^{j/2}\mathfrak{V}_j(2^j x - k),$$

where $\mathfrak{U}_m(x)$ and $\mathfrak{V}_j(x)$ are the inverse Fourier transformations of

$$\hat{\mathfrak{U}}_m(\omega) = \frac{\Phi(\omega)}{\hat{f}(-2^{-m}\omega)} \text{ and } \hat{\mathfrak{V}}_j(\omega) = \frac{\Psi(\omega)}{\hat{f}(-2^{-j}\omega)}, \text{ respectively.}$$

Therefore, we estimate a_{mk} and b_{jk} directly by

$$\hat{a}_{mk} = \frac{1}{n}\sum_{i=1}^{n} 2^{m/2}\mathfrak{U}_m(2^m X_i - k), \text{ and}$$
$$\hat{b}_{jk} = \frac{1}{n}\sum_{i=1}^{n} 2^{j/2}\mathfrak{V}_m(2^j X_i - k).$$

Notice that the functions $u_{mk}(x) = 2^{m/2}\mathfrak{U}_m(2^m x - k)$ fail to generate a standard MRA since their analytic forms change from level-to-level. "Deconwavelet" functions u_{00} and v_{00} are given in Fig. 11.1.

The linear and nonlinear estimators are defined as

$$\hat{\pi}_n^L = \sum_{k \in \mathbb{Z}} \hat{a}_{mk}\phi_{mk}(\theta), \qquad (11.2)$$

and

$$\hat{\pi}_n^N = \sum_{k\in\mathbb{Z}} \hat{a}_{mk}\phi_{mk}(\theta) + \sum_{j=m}^{\infty} \left[\sum_{k\in\mathbb{Z}} \hat{b}_{jk}\psi_{jk}(\theta)\right] \cdot \mathbf{1}\left(\sum_{k\in\mathbb{Z}} \hat{b}_{jk}^2 > \lambda_{j,n}^2\right), \quad (11.3)$$

respectively. Notice that the nonlinear estimator is "block-thresholding" with complete levels as blocks.

The following result establishes the convergence rates for the linear estimator in the cases of both exponential and polynomial decay of $\hat{f}$.

Theorem 11.1.1 *Let π be a density belonging to the Sobolev space $\mathbb{W}^\alpha$ and let the Fourier transformation of f decay exponentially, $|\hat{f}(\omega)| \geq A_0(\omega^2 + 1)^{-\gamma/2} \cdot \exp\{-B|\omega|^\beta\}$. Select an integer m such that:*

$$2^m \approx 3(4\pi)^{-1}(2B)^{-1/\beta}[\ln n - (\ln \ln n)^2]^{1/\beta}.$$

Then,

$$E\int_{\mathbb{R}} [\hat{\pi}_n^L(\theta) - \pi(\theta)]^2 \, d\theta = O((\ln n)^{-2\alpha/\beta}).$$

In the case of polynomial decay of $\hat{f}$, $|\hat{f}(\omega)| \geq A_0(\omega^2 + 1)^{-\gamma/2}$, when m is given by $2^m \approx n^{-1/(2\alpha+2\gamma+1)}$, one has

$$E\int_{\mathbb{R}} [\hat{\pi}_n^L(\theta) - \pi(\theta)]^2 \, d\theta = O\left(n^{-\frac{2\alpha}{2\alpha+2\gamma+1}}\right).$$

Observe that the rates of convergence coincide with the optimal rate of convergence and that in the case of exponential decay the linear estimator is adaptive [choice of the parameter m does not depend on the unknown smoothness α of $\pi(\theta)$].

In the case of polynomial decay, the rate is not optimal and a non-linear estimator is needed.

Theorem 11.1.2 *Suppose $\pi \in \mathbb{W}^\alpha$ and $|\hat{f}(\omega)| \sim A_0(\omega^2 + 1)^{-\gamma/2}$ as $|\omega| \to \infty$. Then, for the nonlinear estimator in (11.3) with $m = (2+\epsilon)\log_2(\ln n)$, $m + r = (2\gamma + 1)^{-1}\log_2 n$ and $\lambda_{j,n} = 2^{j(\gamma+0.5)} \cdot n^{-1/2}\delta_0$, with $\delta_0 \geq 2\sqrt{2}K$, the following relation is valid*

$$E\int_{\mathbb{R}} [\hat{\pi}_n^N(\theta) - \pi(\theta)]^2 \, d\theta = O\left(n^{-\frac{2\alpha}{2\alpha+2\gamma+1}}\right).$$

Here K satisfies certain conditions and ϵ is an arbitrary constant.

For the rationale behind selecting the levels between m and $m+r$ in (11.3), and details and proofs of Theorems 11.1.1 and 11.1.2, see Pensky and Vidakovic [333]. Some complementary results can be found in Walter [440] and Walter and Shen [445]. See also Walter [439] pages 206-210, for results on the consistency of estimators defined in (11.2).

Since Meyer wavelets used in the construction have unbounded support, the range of summation indices k in (11.2) and (11.3) is infinite. The range of summation can be restricted to a finite set of integers without affecting asymptotic properties of the estimators.

11.2 WAVELET-VAGUELETTE DECOMPOSITIONS

Donoho [119] proposed a solution to linear inverse problems by building blocks, *vaguelettes* (wavelet-like), introduced previously by Meyer [292]. Suppose

$$y(u) = (Kf)(u) + \epsilon(u), \; u \in \mathcal{U},$$

where $\mathcal{U}$ is an index set (possibly continuous), ϵ the error, and K a linear operator (Radon transformation, convolution, Abel transformation, etc.). A natural attempt to estimate f is via $\hat{f} = K^{-1}y$, but in most interesting cases the inverse operator K^{-1} does not exist (as a bounded linear operator).

Donoho's solution waveletizes the standard singular value decomposition (SVD) of the operator K by the wavelets-vaguelette pairs. We give a brief overview; the interested reader is directed to [119] for details.

The standard SVD of K is

$$\begin{aligned} Kf &= \sum [Kf, f_\nu] \lambda_\nu f_\nu \text{ and} \\ f &= \sum [Kf, f_\nu] \lambda_\nu^{-1} e_\nu, \end{aligned} \quad (11.4)$$

where $[\,\cdot\,,\,\cdot\,]$ is the inner product, e_ν are eigenfunctions of the operator K^*K (K^* is an adjoint operator for K), and $f_\nu(u) = (Ke_\nu)(u)/\|Ke_\nu\|$. The wavelet-vaguelette counterpart of (11.4) is

$$\begin{aligned} Kf &= \sum [Kf, u_{jk}] \kappa_j v_{jk} \text{ and} \\ f &= \sum [Kf, u_{jk}] \kappa_j^{-1} \psi_{jk}, \end{aligned}$$

where $u_{jk} = \gamma_{jk} \cdot \kappa_j$ and $v_{jk} = \xi_{jk}/\kappa_j = K\psi_{jk}/\kappa_j$. The values γ_{jk} (Riesz representers of the functionals c_{jk} that solve $c_{jk}(Kf) = \langle \psi_{jk}, f \rangle$) are solutions of $[\gamma_{jk}, Kf] = \langle \psi_{jk}, f \rangle$.

The κ_j's are the WVD analogue of the SVD singular values. Donoho refers to them as quasi-singular values. As the scale j increases, the κ_j usually tend towards zero – just as the singular values tend towards zero with increasing frequency in the SVD.

The reconstructed function $\hat{f}$ is defined as

$$\hat{f} = \sum_{j,k} \delta^s([\underline{y}, u_{jk}]\kappa_j^{-1}, \lambda_j)\psi_{jk}, \qquad (11.5)$$

where δ^s is soft-thresholding and λ_j are level-dependent thresholds.

The vaguelettes are biorthogonal,

$$[u_{jk}, v_{j'k'}] = \delta_{j-j'}\delta_{k-k'},$$

and near orthogonal,

$$\|\sum_{j,k} a_{jk} u_{jk}\|_{\mathrm{L}_2} \approx \|(a_{jk})\|_{\ell_2}, \text{ and}$$

$$\|\sum_{j,k} b_{jk} v_{jk}\|_{\mathrm{L}_2} \approx \|(b_{jk})\|_{\ell_2}.$$

Example 11.2.1 Let K be the integration operator, i.e., $(Kf)(u) = \int_{-\infty}^{u} f(t)dt$. Suppose that the wavelet function ψ is of compact support, integral 0, and $\mathbb{C}^1$ regularity. Then the functions ψ' and $\psi^{(-1)}(u) = \int_{-\infty}^{u} \psi(x)dx$ are both continuous and of compact support. Set $\xi_{jk}(x) = (K\psi_{jk})(x)$ and $\gamma_{jk}(x) = -[\psi_{jk}(x)]'$. In this case $\kappa_j = 2^{-j}$.

For precise definitions, details and minimax optimality results for estimators in (11.5), see Donoho [119, 121]. For applications in tomography and software implementations in MATLAB, see Kolaczyk [238, 239]. Lee [254] connected the WVD-based method and variational problems for solving a homogeneous equation. The algorithms in [254] are derived as exact minimizers of some variational problems. Abramovich and Silverman [4] gave exact risk calculations, in the context of the estimation of the derivative of the observed noisy function. See also related results by Johnstone [221].

In addition to extracting global information about $f(x)$ by observing $(Kf)(x)$, the WVD can also characterize localized features of $f(x)$, near a point. This suitability of the WVD for studying change-points for indirect data was explored by Wang [452]. Some asymptotic results for detection and estimation were established. See also Wang [450].

11.3 PURSUIT METHODS

Pursuit methods were introduced in regression problems by Friedman and Stuetzle [158] in 1981. The paper by Friedman, Stuetzle, and Schroeder [159] is an early reference on pursuit methods in the context of density estimation. We saw that any $\mathbb{L}_2$ function can be represented exactly by functions belonging to an appropriate basis or frame. However, the bases are, in some sense, *minimal* collections and the representations may be very complicated. Problems requiring simple decompositions often arise in the analysis of measurements connected with sound or with image processing. Such need for the parsimony motivates decompositions over rich and redundant families of functions – *dictionaries*. A dictionary is an over-complete family that contains functions able to parsimoniously describe a variety of local features of an object that is analyzed. The price for parsimony is paid by high calculational complexity of the pursuit algorithm. An optimal decomposition is an NP-hard problem.

Several methods for finding over-complete representations have been proposed in the literature; we list a few: *Method of frames* (Daubechies [103]), *Matching pursuit* (Mallat and Zhang [279]), *Basis pursuit* (Chen and Donoho [70, 71] and Chen [69]), and *Best orthogonal basis* (Coifman and Wickerhauser [94]). Next, we describe the matching pursuit method and give two examples.

In the rest of this section we focus only on the matching pursuit method. Before explaining the pursuit algorithm, we give a more formal description of dictionaries. A dictionary is a family of elements (atoms) $\mathcal{D} = \{g_\gamma\}_{\gamma \in \Gamma}$ from a Hilbert space $\mathcal{H}$. It is assumed that (i) $||g_\gamma|| = 1$, and (ii) linear combinations of vectors from $\mathcal{D}$ are dense in $\mathcal{H}$ (completeness property). The minimal dictionaries are bases. We have already seen that wavelet packet tables constitute over-complete sets of functions; they are often used as dictionaries in pursuit methods. Many other dictionaries are possible (stationary wavelets, cosine packets, chirplets, steerable and segmented wavelets, Gabor bases, etc.).

A decomposition of a function in an over-complete system is not unique and several methods are proposed that reflect different goals to which the over-complete representations are aimed. Since elements of a dictionary (atoms) can possess different properties (trigonometric, bumps, chirps, etc.), the corresponding decompositions are capable of providing much better resolution. Also, it is often easier to separate signals made of a few disparate phenomena, like impulses and sinusoids.

Matching pursuit (MP) is a method that originated from the *pursuit regression* method of Friedman and Stuetzle [158]. Next is a brief description.

Let $\mathcal{D} = \{g_\gamma\}_{\gamma \in \Gamma}$ be a dictionary and let $f \in \mathcal{H}$. One selects $g_{\gamma_0} \in \mathcal{D}$ and projects f onto it. Then,

$$f = \langle f, g_{\gamma_0} \rangle g_{\gamma_0} + R(f).$$

The residual $R(f)$ is orthogonal to g_{γ_0}, and by the Pythagorean theorem,

$$\|f\|^2 = |\langle f, g_{\gamma_0}\rangle|^2 + \|R(f)\|^2. \tag{11.6}$$

Since $\|f\|^2$ in (11.6) is fixed, minimizing $\|R(f)\|^2$ amounts to maximizing $|\langle f, g_{\gamma_0}\rangle|^2$. The atom maximizing $|\langle f, g_{\gamma_0}\rangle|^2$ will be optimal. It is often computationally beneficial to adopt a suboptimal choice for g_{γ_0}, close to optimal in the sense that,

$$|\langle f, g_{\gamma_0}\rangle| > \alpha \sup_{\gamma \in \Gamma} |\langle f, g_{\gamma}\rangle|, \tag{11.7}$$

where $\alpha \in (0, 1]$ is an optimality factor. The selection condition (11.7) softens a greedy algorithm that is the cause of exponential computational complexity. By looking only one step ahead, greedy algorithms often miss the optimal decompositions. For example, Fig. 11.3 (b) depicts a function that is the sum of two normal densities, $\phi(x-2) + \phi(x+2)$. When the dictionary contains normal densities as atoms, $\{\frac{1}{\sigma}\phi(\frac{x-\mu}{\sigma}), \mu \in \mathbb{R}, \sigma \in \mathbb{R}^+\}$, the optimal and exact decomposition has only two functions, $\phi(x-2)$ and $\phi(x+2)$. A greedy algorithm approximates $\phi(x-2) + \phi(x+2)$ by a single function [e.g., $\frac{2}{3}\phi(\frac{x}{3})$ is better than any of $\phi(x-2)$ and $\phi(x+2)$] and compensates for the error in the subsequent steps. The resulting decomposition is infinite.

The matching pursuit is the following iterative process:

Assume that $R^0(f) = f$. Let the pursuit be at the iterative step m, i.e., let $R^m(f)$ be defined. To define the residual of order $m+1$, select $g_{\gamma_m} \in \mathcal{D}$ such that

$$|\langle R^m(f), g_{\gamma_m}\rangle| > \alpha \sup_{\gamma \in \Gamma} |\langle R^m(f), g_{\gamma}\rangle|,$$

for some α close to 1, and project $R^m(f)$ on g_{γ_m}

$$R^m(f) = \langle R^m(f), g_{\gamma_m}\rangle g_{\gamma_m} + R^{m+1}(f).$$

Since $R^{m+1}(f)$ and g_{γ_m} are orthogonal,

$$\|R^m(f)\|^2 = |\langle R^m(f), g_{\gamma_m}\rangle|^2 + \|R^{m+1}(f)\|^2.$$

By iterating the process up to order M we obtain a decomposition of f,

$$f = \sum_{m=0}^{M-1} \langle R^m(f), g_{\gamma_m}\rangle \cdot g_{\gamma_m} + R^M(f).$$

Similarly, as in (11.6), an "energy preservation" equation holds for $||f||^2$,

$$||f||^2 = \sum_{m=0}^{M-1} |\langle R^m(f), g_{\gamma_m}\rangle|^2 + ||R^M(f)||^2.$$

The following theorem ensures that the matching pursuit procedure converges.

Theorem 11.3.1 *(Jones [224]) There exists $\mu > 0$ such that for all $M \geq 0$,*

$$||R^M(f)|| \leq 2^{-\lambda M}||f||.$$

Thus,

$$f = \sum_{m=0}^{\infty} \langle R^m(f), g_{\gamma_m}\rangle \cdot g_{\gamma_m} \quad \text{[In $\mathbb{L}_2$ sense]}$$

and

$$||f||^2 = \sum_{m=0}^{\infty} |\langle R^m(f), g_{\gamma_m}\rangle|^2.$$

The identity

$$\langle R^{m+1}(f), g_{\gamma_m}\rangle = \langle R^m(f), g_\gamma\rangle - \langle R^m(f), g_{\gamma_m}\rangle \cdot \langle g_{\gamma_m}, g_\gamma\rangle, \quad (11.8)$$

can be incorporated into the algorithm. The procedure now can be described as follows:

- **STEP 1.** Set $m = 0$ and compute $\langle R^m(f), g_\gamma\rangle$.
- **STEP 2.** Find a match, $|\langle R^m(f), g_\gamma\rangle| \geq \alpha \sup |\langle R^m(f), g_\gamma\rangle|$.
- **STEP 3.** Update the error by (11.8).
- **STEP 4.** If $||R^{m+1}f||^2 \leq \epsilon^2 ||f||^2$, **STOP.** Otherwise increase m and return to **STEP 2.**

We conclude this section with two illustrative examples.

Example 11.3.1 In a no-noise situation and an efficient representation of the `doppler` signal is needed. Assume that the dictionary is a wavelet packet table generated by the SYMM4 wavelet. Four panels in Fig. 11.2 represent successive pursuit approximations with $M = 1, 3, 10$ and 50 best atoms from the dictionary. Table 11.1 compares

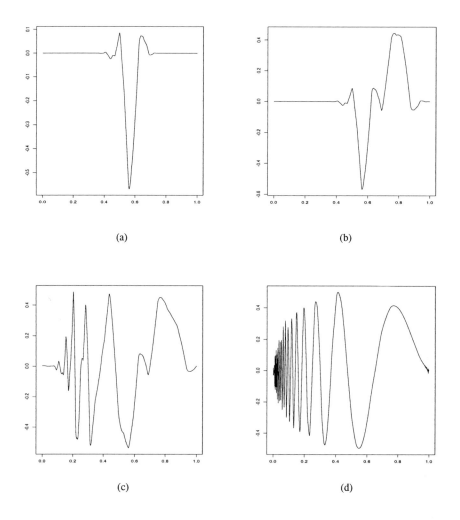

Fig. 11.2 Matching pursuit approximations of doppler signal with: (a) $M = 1$; (b) $M = 3$; (c) $M = 10$; and (d) $M = 50$ atoms.

Table 11.1 Energy explained and errors by approximating the doppler signal with $M = 1, 3, 10$ and 50 atoms from the wavelet packet library.

Number of coefficients (M)	Energy explained	Error
1	14.11492	73.80427
3	33.35139	54.56780
10	70.14615	17.77304
50	87.75214	0.16752

the energies explained and the error for different approximations.

Example 11.3.2 In this example we demonstrate that for a fixed number of "best atoms," matching pursuit gives the best reconstruction of the original signal. A discrete version of the doppler signal is generated and normal noise is added so that SNR $= 3$. This noisy signal is decomposed with the SYMM4 and the wavelet packet table is formed. We selected 17 coefficients with maximum energy from the table subject to the following constraints: (a) the coefficients are from the discrete wavelet transformation; (b) the coefficients are from the best basis (basis minimizing the entropy cost in the table); and (c) without restrictions (pursuit).

The reconstructions are compared in Fig. 11.3. It is evident that the matching pursuit method dominates the other two decompositions.

Approximation in matching pursuit can be improved by ortogonalizing the directions of projection with the Gram-Schmidt procedure. This approach was proposed by Pati, Rezaifar, and Krishnaprasad [328]. The price for the improved convergence and parsimony is paid by increased calculational complexity of the pursuit algorithm. See also Davis [109], Section 9.4 in Mallat [277], and the manuscripts at ftp://cs.nyu.edu/pub/wave/report/. Walden and Cristan [436] discuss matching pursuit in the library of nondecimated wavelets.

11.4 MOMENTS OF ORDER STATISTICS

In this section, we illustrate that wavelets can successfully replace standard orthonormal bases and enhance some of the existing methods in which orthogonal decompositions are used. We give an application in the efficient calculation of moments of order statistics.

Siguira [376, 377] developed a method for the approximate calculation of moments of order statistics by using orthogonal series. He utilized Legendre polynomials and proposed approximations for moments of order statistics from normal distribution on the basis of only several first coefficients in appropriate decompositions. The errors

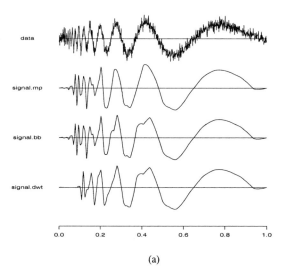

(a)

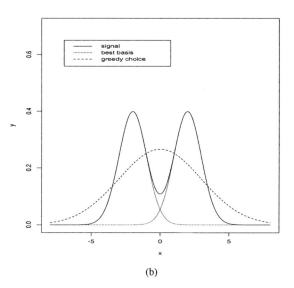

(b)

Fig. 11.3 (a) Comparison between matching pursuit, best basis and discrete wavelet transformation. Wavelet SYMM8, n=1024, SNR=3, n.top=17. (b) An illustration of MP greedy algorithm. The sum of two normal densities $f(x) = \phi(x-2) + \phi(x+2)$ is the function to be decomposed by matching pursuit. The atom $\frac{2}{3}\phi(\frac{x}{3})$ has smaller MSE in estimating $f(x)$ than either of $\phi(x-2)$ and $\phi(x+2)$.

reported were of the order 10^{-4}. Joshi [225] generalized Siguira's method to other distributions and orthogonal systems. Next, we will demonstrate how wavelet-based orthogonal systems can improve Siguira's procedure in both speed and precision.

Let $\{\psi_i, i \in \mathbb{N}\}$ be an arbitrary orthonormal basis of $\mathbb{L}_2$. Assume $f, g \in \mathbb{L}_2$ and let

$$f(x) = \sum_i d_i \psi_i(x) \quad \text{and} \quad g(x) = \sum_i e_i \psi_i(x),$$

be expansions of $f(x)$ and $g(x)$, where $\underline{d} = \{d_i, i \in \mathbb{N}\}$ and $\underline{e} = \{e_i, i \in \mathbb{N}\}$ are vectors of decomposition coefficients. Denote by K (keep) and R (reject) any partition of $\mathbb{N}$ such that $K \cup R = \mathbb{N}$, and $K \cap R = \emptyset$. Let $\mathbf{1}_A$ be the indicator of a set $A \subset \mathbb{N}$, i.e. $(\mathbf{1}_A)_i = \mathbf{1}(i \in A)$, $i \in \mathbb{N}$. Let $\underline{a} \star \underline{b} = \{a_1 b_1, a_2 b_2, \dots\}$ be the Hadamard product of the vectors $\underline{a} = \{a_1, a_2, \dots\}$ and $\underline{b} = \{b_1, b_2, \dots\}$.

Then, as easy corollary of Parseval's identity $\langle f, g \rangle_{\mathbb{L}_2} = \langle \underline{d}, \underline{e} \rangle_{\ell_2}$,

$$|\langle f, g \rangle_{\mathbb{L}_2} - \langle \underline{d}, \underline{e} \star \mathbf{1}_K \rangle_{\ell_2}| \leq \|\underline{d} \star \mathbf{1}_R\| \, \|\underline{e} \star \mathbf{1}_R\|. \tag{11.9}$$

Relation (11.9) provides a basis for an efficient evaluation of $\langle f, g \rangle$. The method is convenient when the functions f and g are given by their equidistant sampled values. Of course, the standard methods of numerical integration will give better precision, but an interesting feature of methods based on (11.9) is that only a few cross-product coefficients in the wavelet domain approximate the integral of the product with high precision. This feature is due to the "energy-packing" ability of wavelets.

We apply (11.9) in calculating the moments order statistics, and give an illustration for the normal distribution.

Let $X_1, X_2, \dots, X_n$ be a sample from a population with an absolutely continuous distribution F, with finite second moment. Let $X_{1:n} \leq X_{2:n} \leq \dots \leq X_{n:n}$ be the ordered sample (order statistic). The distribution of $X_{r:n}$ is given by

$$F_{r:n}(x) = \mathbf{I}_{F(x)}(r, n-r+1),$$

where $\mathbf{I}_p(a,b) = \frac{1}{B(a,b)} \int_0^p t^{a-1} (1-t)^{b-1} dt$, $a > 0, b > 0$, $0 \leq p \leq 1$, is the *incomplete Beta function*, (see, for example, David [108]).

Let $u = F(x)$ and $x(u) = F^{-1}(u)$. Then,

$$x(u) \in \mathbb{L}_2([0,1]).$$

Indeed, $\int_0^1 x^2(u) du = \int_{\mathbb{R}} x^2 \frac{du}{dx} dx < \infty$ by the assumption that the second moment is finite.

Let $\{\phi_{00}, \psi_{jk}, j \geq 0, 0 \leq k \leq 2^j - 1\}$ be periodized wavelets on $[0, 1]$. Let

$$d_{jk} = \langle x(u), \psi_{jk}(u) \rangle$$
$$e_{jk} = \left\langle \frac{u^{r-1}(1-u)^{n-r}}{B(r, n-r+1)}, \psi_{jk}(u) \right\rangle$$

be the wavelet coefficients of $x(u)$ and the density of $X_{r:n}$, respectively, which are both $\mathbb{L}_2[0, 1]$ functions. Then, for any "keep" subset K of indices,

$$|EX_{r:n} - \langle \underline{d}, \underline{e} \star \mathbf{1}_K \rangle| \leq \|\underline{d} \star \mathbf{1}_R\| \, \|\underline{e} \star \mathbf{1}_R\|, \qquad (11.10)$$

where $R = \mathbb{N} \setminus K$.

Example 11.4.1 It is well-known that the only exact calculations of moments for order statistics from a normal distribution are possible if $n \leq 5$. For example,

$$EX_{2:5} = \frac{30}{\pi\sqrt{\pi}} \arctan(\sqrt{2}) - \frac{10}{\sqrt{\pi}}.$$

A simple S+Plus program implements the wavelet approximation,

```
> a <- Exp(n=5, r=2, N=2^19)
[1] -0.4950189704756
```

and the error of approximation is of the order 10^{-11}.

Interestingly, the waveletized Siguira's method can approximate moments of order statistics for extremely large sample sizes. For example,

```
> a <- Exp(n=10000000000, r=4000000000, N=2^19)
[1] -0.2533471032
```

It is also possible to approximate mixed moments, $EX_{s:n}X_{r:n}$, under the conditions required for (11.10) to hold. The idea is similar, but two-dimensional wavelets and distributions have to be used. For details, see Siguira [377] and Joshi [225]. S-Plus programs are available at

http://www.isds.duke.edu/~brani/wiley.html,

and require the S+Wavelets module.

11.5 WAVELETS AND STATISTICAL TURBULENCE

Turbulence has become one of the most profound and yet most elusive problems in physics in the last century. Several different approaches taken by researchers failed to fully describe, explain, and predict turbulent motion. The turbulence usually carries the prefix *stochastic* and a large portion of turbulence research involves both descriptive and inferential statistics. The two monographs on statistical fluid mechanics by Monin and Yaglom [296, 297] illustrate the use of statistical concepts in describing turbulence.

The McGraw-Hill Encyclopedia of Science and Technology quotes: *Turbulence: motion of fluids in which the local velocities and pressures fluctuate irregularly: Most flows observed in nature such as rivers and wind are turbulent. ... The essential characteristic of turbulent flow is that the fluctuations are unpredictable ...*

A turbulent flow is characterized by the value of its Reynolds number, a dimensionless parameter that quantifies the ratio of inertial to viscous forces. If the Reynolds number is small, the flow is *laminar*, i.e., the flow is fully predictable in both space and time. With an increase of the Reynolds number, the flow becomes unstable, and when the Reynolds number exceeds a given threshold it becomes fully turbulent. An example of an extremely turbulent flow is our atmosphere with Reynolds numbers of up to $3 \cdot 10^8$.

An example measurement of turbulence velocity and temperature time series is shown next. These measurements illustrate the stochastic nature of turbulent flows at such high Reynolds numbers.

Example 11.5.1 (Duke Forest turbulence data set)[1]

> The velocity and air temperature measurements were carried out on July 12-16, 1995, at 5.2 m above the ground surface over an *Alta Fescue* grass site at the Blackwood division of the Duke Forest in Durham, North Carolina. During this time period, a heat wave resided in North Carolina for several days after it swept from the midwest to the east coast. During the experiment, maximum mean air temperature up to $38°C$ was measured in Durham. The sky condition during these five days was clear with low to moderate winds. The site is a 480 m by 305 m grass-covered forest clearing ($36°2'N$ $79°8'W$, elevation = 163 m), and a mast, situated at 250 m and 160 m from the north-end and west-end portions of a 12 m tall Loblolly pine forest edge, respectively, was used to mount a triaxial sonic anemometer. The three velocity components (U, V, W) and air temperature T were measured using a triaxial ultrasonic anemometer (Gill Instruments/1012R2).
>
> The sampling frequency (f_s) and period (T_p) were 56 Hz and 19.5 minutes, respectively, resulting in $N = 65,536$ measurements per velocity component per run.

The graphs of the measurements are given in Fig. 11.4.

---
[1] Data courtesy of Gabriel Katul, SOE, Duke University.

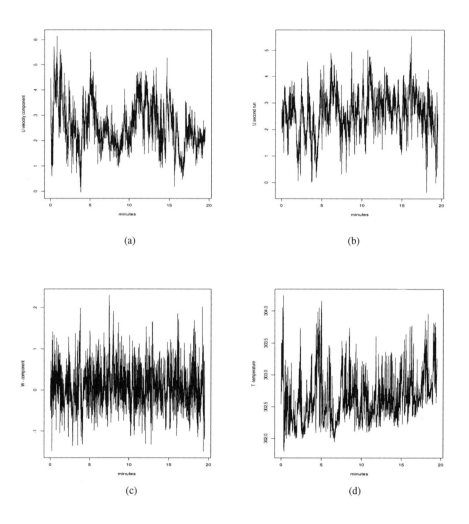

Fig. 11.4 Duke Forest measurements: U, V, and W velocity measurements in panels (a)-(c), and temperature T in panel (d).

A turbulent flow field is fully described by the well-known set of partial differential equations of hydrodynamics (Navier-Stokes equations): the conservation equations for momentum, and the balance equation for dissipative entropy production. Due to their complexity, these equations cannot be solved analytically. In fact, the existence and uniqueness of solutions to the Navier-Stokes equations is still in debate. Direct numerical simulation, which would involve numerically solving the full unsteady Navier-Stokes equations is currently limited to only the simplest flow geometries and low to intermediate Reynolds numbers. These limitations gave rise to the development of many phenomenological models to describe flow statistics. Perhaps the most universal phenomenological model put forth so far which has survived the "test of time" for more than half a century is due to Kolmogorov and is considered next.

11.5.1 K41 Theory

In 1941, Kolmogorov introduced his theory for locally isotropic, fully developed turbulence based on earlier ideas of energy cascading by Richardson. Several models based on energy cascades have been proposed since then. Such models consider energy dissipation ϵ [*energy per unit of fluid mass per unit time*] as a key parameter. It depends on the kinematic viscosity, velocity components and the position of the moving fluid. The parameter ϵ describes the energy transmission from large eddies, where the energy is injected, to small eddies where the energy is converted to heat by viscosity.

Kolmogorov [242] developed his theory, often referred to as **K41** theory, for *locally isotropic* turbulence. Let $\underline{x} = (x_1, x_2, x_3)$ be the position vector and $\underline{u} = (u_1(\underline{x}), u_2(\underline{x}), u_3(\underline{x}))$ be the velocity components. Locally isotropic turbulence describes the flow in which the probability distribution of the relative velocity differences

$$\Delta \underline{u}(\underline{r}) = \underline{u}(\underline{x} + \underline{r}) - u(\underline{x}),$$

is independent of time, and invariant under translations, reflections, and rotations. The fundamental objects in **K41** theory are *structure functions*. Structure functions are closely related to correlations of two-point velocity differences,

$$\langle \Delta \underline{u}(\underline{r})^2 \rangle = 2\sigma_u^2 (1 - \rho_u(r)).$$

A (longitudinal) structure function of order p is defined as

$$D_p(r) = \langle ||\Delta \underline{u}(\underline{r})||^p \rangle$$

where the angular brackets denote time averaging.

A functional description for the moments of velocity differences and thus for the structure functions can be derived using *dimensional analysis* and leads to

$$D_p(r) = C_p \langle \epsilon \rangle [r]^{\frac{p}{3}}, \qquad (11.11)$$

where C_p is a universal constant. For the third-order structure function, it can be inferred directly from the Navier-Stokes equations that $C_3 = -\frac{4}{5}$. From (11.11) it follows that structure functions possess scaling behavior. Let the symbol $\propto$ denote "proportional to." Then,

$$D_p(r) \propto r^{\zeta_p}.$$

The exponent ζ_p is called the *scaling exponent*. The **K41** theory gives the simple model $\zeta_p = \frac{p}{3}$.

Similarly, as for the structure functions, a description of the energy of the turbulent fluctuations per unit of mass of fluid in scales **r** can be derived from the hypotheses and by dimensional analysis,

$$E_r \propto (r)^{\frac{2}{3}}. \qquad (11.12)$$

Via the Fourier transform of (11.12), which results in the spectral density $\phi(k)$ of the energy, the celebrated "$-\frac{5}{3}$ law" for the power spectrum, as shown in Fig. 1.9 is obtained,

$$\begin{aligned} E_k &= 2R^{-1}k^2 \phi(k) \\ &\propto r^{\frac{2}{3}} k^{-\frac{5}{3}}. \end{aligned}$$

11.5.2 Townsend's Decompositions

There are several applications of wavelets in turbulence research. Some references are Yamada and Ohkitani [470], Hudgins, Friehe, and Mayer [206], Katul and Paralange [229], Katul and Vidakovic [230], and Wickerhauser et al. [458], among others. We focus here only on the method of separating an organized (attached, energetic) part of a turbulence signal from the detached (non-energetic) part. We demonstrate how this separation resembles a denoising procedure but the criteria for separation are different.

Townsend's [407] attached eddy hypothesis states that the turbulent structure in the constant stress layer (i.e., the layer in which the mean turbulent fluxes do not vary by more than 10%) can be decomposed into attached and detached eddy motion. One of the difficulties in separating the organized and less organized eddies from time series measurements is the locality and non-periodicity of the organized events. Traditional Fourier and eigenvector decompositions are less effective in separating these two different modes of turbulence properties. The *footprint* of such organized eddy motion from time series measurements is usually characterized by sharp edges (e.g., ramp-like structures in temperature measurements) that result in large local gradients. These sharp edges are somewhat ambiguous in Fourier space because these sharp edges contribute considerable spectral and co-spectral energy at scales much smaller than the structure itself. This limitation motivated the use of wavelet shrinkage approaches.

The wavelet shrinkage methodology is capable of extracting low-dimensional organized perturbations $(U^{(o)}, W^{(o)}, T_a^{(o)})$ from velocity (U, W) and temperature (T_a) time series measurements, given by

$$U = U^{(o)} + U^{(r)}$$
$$W = W^{(o)} + W^{(r)}$$
$$T_a = T_a^{(o)} + T_a^{(r)}$$

and filters out the high-dimensional part $(U^{(r)}, W^{(r)}, T_a^{(r)})$. As a graphical illustration, Fig. 11.5(a) and (b) compares the original (U) and the thresholded $(U^{(o)})$ longitudinal velocity time series for Lorentz-curve-based and Fourier ranking thresholding procedures. Fourier ranking is in fact hard thresholding of discrete Fourier transformations and, for turbulence signals, appears to give satisfactory results despite the limitations of the Fourier kernel described above.

Whether the thresholded time series represents organized structures is investigated via "optimality" criteria derived from immediate consequences of Townsend's hypothesis. The optimality criteria include not only the conservation of the energies but also conservation of the fluxes,

$$\langle uw \rangle \approx \langle u^{(o)} w^{(o)} \rangle$$
$$\langle wT_a \rangle \approx \langle w^{(o)} T_a^{(o)} \rangle$$

as required in Townsend's [407] attached eddy hypothesis. In simple terms, we are interested whether a small portion of coefficients from each time series in the wavelet domain is capable of conserving the energy and turbulent fluxes. Additional verification for two-point statistics can be indirectly performed via **K41** laws.

In **K41** theory, the second- $[D_2(r)]$ and third-order $[D_3(r)]$ structure functions are given by

$$D_2(r) = C_2 \langle \epsilon \rangle^{2/3} r^{2/3} \text{ and}$$
$$D_3(r) = -\frac{4}{5} \langle \epsilon \rangle r,$$

338 MISCELLANEOUS STATISTICAL APPLICATIONS

where r is the separation distance, $C_2 (= 0.55)$ is the Kolmogorov constant, and $\langle \epsilon \rangle$ is the mean turbulent kinetic energy dissipation rate (Monin and Yaglom, [297]) per unit mass. In Fig. 11.5(c) and (d), $D_2(r)$ and $D_3(r)$ are shown for the original ($N = 65,536$) velocity component, and thresholded time series for three thresholding models (Lorentz, universal and Fourier ranking). This analysis further supports the hypothesis that the high-dimensional component is associated with detached eddies.

Some additional references on related problems include Collineau and Brunet [95], Farge et al. [149], Katul and Albertson [228], and Yamada and Ohkitani [469]. Berliner, Wikle, and Milliff [30] and Vidakovic and Katul [425] propose hierarchical Bayesian models in the wavelet domain and address problems of multiscale analysis and de-noising of turbulent signals.

11.6 SOFTWARE AND WWW RESOURCES FOR WAVELET ANALYSIS

In this section, we give a short overview of some popular wavelet software and wavelet-related sources on the world wide web (www). We discuss both commercial and free software and give pointers to home pages giving useful information about the theory and practice of wavelets.

11.6.1 Commercial Wavelet Software

- S+WAVELETS
  http://www.mathsoft.com/splus/splsprod/wavelets.html
  is a module working under the S-PLUS programming language. It was created by Bruce and Gao, ([42]). The advantage of this software for statisticians is that the wavelet module is interfaced with the superb graphical and statistical features of the S-Plus software. The complete S+Wavelets toolkit includes an extensive set of over 500 functions embedded in an object-oriented environment. It contains functions performing the discrete wavelet transform, wavelet optimal signal estimation, wavelet packet analysis, local cosine analysis, "best basis" selection, matching pursuit analysis, robust wavelets analysis, and more. A comprehensive manual accompanying the module is well-written.

- WAVELET EXPLORER [http://store.wolfram.com/view/wavelet/] is a package for Wavelets in Mathematica. It is a collection of Mathematica programs that comprise a Mathematica tool-box for basic wavelet analysis tasks.

  Wavelet Explorer's functions and utilities let the user apply a variety of wavelet transforms. It includes common filters such as the Daubechies' extremal phase and least asymmetric filters, coiflets, spline filters, and more. Basic data compression and denoising routines are available as well. It comprises several interesting data sets and puts emphasis on basic ideas and logical structures at the expense of parsimony

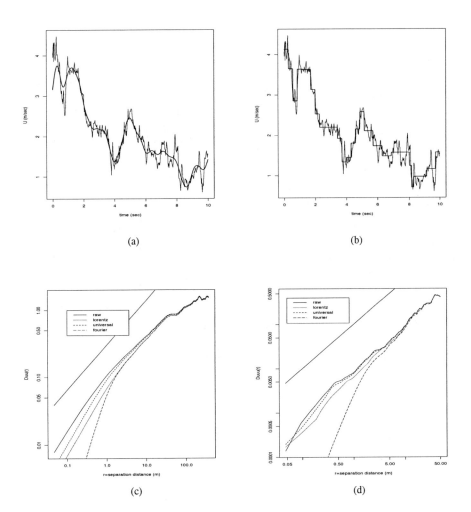

Fig. 11.5 Comparison between the original and the thresholded longitudinal velocity time series. Panel (a) is for the Fourier ranking method (thresholding of associated discrete Fourier transformation). Panel (b) is for the Lorentz thresholding method using the Haar basis (see Section 6.6.6). Panel (c) is for the second-order structure function $D_2(r)$. Panel (d) is for the third order-structure function $D_3(r)$. The solid lines are from the original time series. The power laws are from **K41**.

340 *MISCELLANEOUS STATISTICAL APPLICATIONS*

of programming code. The manual supplied with the program is excellent; not only the use of commands is described, but basic theoretical properties of wavelets are covered as well.

- WAVBOX [http://www.wavbox.com] is a Matlab toolbox for wavelet transforms and adaptive wavelet packet decompositions with the FirWav filter library. Matlab is a technical computing environment available from The MathWorks Inc. WavBox software provides both a function library and a computing environment for wavelet transforms and adaptive wavelet packet decompositions. The software package contains more than 280 Matlab m-files ($\sim$ 1.2 MB uncompressed) implementing an extensive collection of wavelet transforms, expansions, decompositions, wavelet filters, and related functions. This function library performs multiresolution analyses of one-dimensional multichannel signals and images for arbitrary sizes of data. Wavbox was developed by C. Taswell (taswell@wavbox.com). A description of WavBox can be found in [398]. Also, consult the related ToolSmith web page [http://www.toolsmiths.com/].

- WAVELET TOOLBOX [http://www.mathworks.com/products/wavelet] provides a comprehensive collection of Matlab routines for examining local, multiscale, and nonstationary phenomena. It features (i) Complete GUI and command line functionality for analysis, synthesis, de-noising, and compression of signals and images; (ii) continuous wavelet transform; (iii) discrete wavelet transform (DWT); (iv) multiresolution decompositions and analysis of signals and images; (v) wide selection of wavelet basis functions, including several boundary correction methods; and (vi) wavelet packet transform and entropy-based wavelet packet tree pruning for "best-tree" and "best-level" analysis.

11.6.2 Free Wavelet Software

- WAVELAB is a comprehensive library of Matlab routines for wavelet and general time/frequency analysis. It is maintained by the Department of Statistics at Stanford University, and can be downloaded from
   http://www-stat.stanford.edu/~wavelab/ .
The latest version 0.800, compatible with Matlab 5, is developed by The latest version 0.800, compatible with Matlab 5, is developed by Donoho, Huo and Yu. Some contributors to the earlier versions of the software include Scargle, Johnstone, Chen, Buckheit, and Kolaczyk. In addition to routines implementing basic wavelet transforms for finite data sets (both periodic transforms and boundary-corrected transforms), wavelet-packet analysis, cosine-packet analysis and matching pursuit, Wavelab contains scripts that will assist a student in learning the practical aspects of wavelet analysis. Educational add-ons contain: (i) scripts that reproduce the figures

in the authors' published articles, including the de-noising articles of Donoho and Johnstone; and (ii) workouts that give a quick guide to wavelet analysis, wavelet synthesis, wavelet and cosine packets, including the Coifman-Wickerhauser best-basis methodology, and matching pursuit. The software also supports applications such as data expansion, progressive data transmission, image compression, speech segmentation, de-noising, fast matrix multiplication in wavelet bases, etc. See Buckheit et al. [47] for additional information.

- WAVETHRESH is an add-on module for the statistical package S-Plus. The author of the software is Nason from the University of Bristol. WaveThresh contains S-Plus functions for performing one- and two-dimensional wavelet transforms and their inverses for basic wavelet families. It also contains the rudiments of a thresholding scheme for doing wavelet based curve-estimation. See Nason and Silverman [310] for more information. The current version of WaveThresh available on the web is 2.2. A new, more comprehensive version 3.0 exists, but it is not available for the public yet.

- SWAVE by Carmona, Hwang, and Torrésani is an S-Plus add-on module for a general time/frequency analysis. In addition to standard wavelet transformations, the module contains functions and utilities for Gabor-type analyses and for reconstructions of signals from ridges and extrema of moduli of wavelet transformations. The module can be downloaded from:
  http://www.sad.princeton.edu/~rcarmona/TFbook/TFbook.html .
  The description of the package and worked examples can be found in Carmona, Hwang, and Torrésani [64].

- RICE WAVELET TOOLS is a collection of Matlab m-files and mex-files implementing wavelet and filter bank design and analysis. In addition to the design tools, the toolbox provides code for wavelet applications for both uni- and bi-variate denoising, and code for processing of SAR images. The software can be downloaded from [http://www-dsp.rice.edu/software/RWT/].

- MEGAWAVE1 (previously known as "MegaWave") [http://www.ceremade.dauphine.FR/~mw/] is the first environment written and used by the CEREMADE during the years 1988-1992. It is a set of C programs with standardized procedures. It contains algorithms on the wavelet transform, anisotropic diffusion and segmentation. Comprehensive documentation is provided (a volume of 170 pages).

11.6.3 Some WWW Resources

- The Wavelet Digest
  [http://www.wavelet.org/wavelet/index.html] is a free monthly

newsletter edited by Wim Sweldens which contains all kinds of information concerning wavelets: announcements of conferences, preprints, software, questions and answers, etc.

- Mathsoft Wavelet resources

[http://www.mathsoft.com/wavelets.html] is a web page for "one-stop shopping" for wavelet manuscripts. This page contains a list of many preprints on the subject of wavelets and their various applications in the worlds of mathematics, statistics, engineering and physics.

A few other resourceful web pages are:

(i) Wavelet Net Care

[http://www.math.wustl.edu/wavelet/] at Washington University.

(ii) Donoho's web page at Stanford University

[http://www-stat.stanford.edu/reports/donoho/].

and (iii) Blue Devil's Wavelets

[http://www.isds.duke.edu/~brani/wavelet.html] at Duke University; see also the links therein.

An excellent searchable wavelet bibliography can be found at [http://liinwww.ira.uka.de/bibliography/Theory/Wavelets/].

11.7 EXERCISES

11.1. Prove that $\mathfrak{U}_m(x) = \phi(x) - 2^{2m}\phi''(x)$ if the model f is double exponential. [HINT: $\hat{f}(\omega) = 1/(\omega^2 + 1)$]

11.2. Find the vaguellete $u(x)$ and quasi-singular values κ_j for the convolution operator $(Kf)(x) = \int k(x-t)f(t)\,dt$, where $k(x) = \frac{1}{2}e^{-|x|}$. Compare to Exercise 11.1.

11.3. Verify that $[\xi_{jk}, \gamma_{j'k'}] = \delta_{j-j'}\delta_{k-k'}$ for ξ_{jk} and $\gamma_{j'k'}$ from Example 11.2.1.

11.4. Consider Fig. 11.3(b) and the associated discussion on page 326. It is argued that the greedy pursuit unwisely chooses the best function. Even though the two, separately suboptimal, functions exactly describe the signal, greedy pursuit chooses and then compensated the error. The signal is $N(-2, 1) + N(2, 1)$. Illustrate "greediness" of the matching pursuit procedure by finding a single function $c_1 N(0, \sigma^2)$ which better approximates the signal than $c_2 N(-2, 1)$, for any c_2.

11.5. Form a counterpart of Table 11.13 for the SYMM4 wavelet decomposition. Compare the columns "Energy explained" and "Error" and discuss your findings.

11.6. By mimicking Siguira's method propose a wavelet based procedure that calculates moments of order statistics for the standard Cauchy distribution. Note that for the Cauchy distribution, $EX_{r:n}^s$ is finite for all $s < r < n - r$.

References

1. F. Abramovich and Y. Benjamini. Thresholding of wavelet coefficients as a multiple hypotheses testing procedure. In A. Antoniadis and G. Oppenheim, editors, *Wavelets and Statistics*, volume 103 of *Lecture Notes in Statistics*, pages 5–14. Springer-Verlag, New York, 1995.
2. F. Abramovich and Y. Benjamini. Adaptive thresholding of wavelet coefficients. *Comput. Statist. Data Anal.*, 22:351–361, 1996.
3. F. Abramovich, T. Sapatinas, and B. W. Silverman. Wavelet thresholding via a Bayesian approach. *J. Roy. Statist. Soc. Ser. B*, 60(3):725–749, 1998.
4. F. Abramovich and B. W. Silverman. Wavelet decomposition approaches to statistical inverse problems. *Biometrika*, 85:115–129, 1998.
5. P. Abry, P. Gonçlavès, and P. Flandrin. Wavelet-based spectral analysis of 1/f processes. In *Proceedings of the IEEE International Conference on Acoustics, Speech, and Signal Processing*, volume 3, pages 237–240, 1993.
6. P. Abry and D. Veitch. Wavelet analysis of long-range-dependent traffic. *IEEE Trans. Inform. Theory*, 44(1):2–15, 1998.
7. P. Abry, D. Veitch, and P. Flandrin. Long range dependence: Revisiting aggregation with wavelets. *J. Time Ser. Anal.*, 1998.
8. H. Akaike. A new look at statistical model identification. *IEEE Trans. Automat. Control*, 1(9):716–727, 1974.
9. U. Amato and D. T. Vuza. Wavelet regularization for smoothing data. Technical report, Instituto per Applicazioni della Matematica, Napoli, 1994.
10. F. J. Anscombe. The transformation of Poisson, binomial and negative binomial data. *Biometrika*, 35:246–254, 1948.

11. A. Antoniadis. Smoothing noisy data with tapered coiflet series. *Scand. J. Statist.*, 23:313–330, 1996.

12. A. Antoniadis. Wavelets in statistics: A review. To appear in *J. Italian Statist. Soc.*, 1998.

13. A. Antoniadis and R. Carmona. Multiresolution analyses and wavelets for density estimation. Technical report, University of California, Irvine, 1991.

14. A. Antoniadis, I. Gijbels, and G. Grégoire. Model selection using wavelet decomposition and applications. *Biometrika*, 84(4):751–763, 1997.

15. A. Antoniadis, G. Grégoire, and I. McKeague. Wavelet methods for curve estimation. *J. Amer. Statist. Assoc.*, 89:1340–1353, 1994.

16. A. Antoniadis, G. Grégoire, and G. P. Nason. Density and hazard rate estimation for right censored data using wavelet methods. *J. Roy. Statist. Soc. Ser. B*, 61, 1999.

17. A. Antoniadis, G. Grégoire, and P. Vial. Random design wavelet curve smoothing. *Statist. Probab. Lett.*, 35:225–232, 1997.

18. R. Averkamp and C. Houdré. A note on the discrete wavelet transform of second-order processes. Preprint at School of Mathematics, Georgia Institute of Technology, Atlanta. To appear in *IEEE Trans. Inform. Theory*, 1998.

19. R. Averkamp and C. Houdré. Some distributional properties of the continuous wavelet transform of random processes. *IEEE Trans. Inform. Theory*, 44:1111–1124, 1998.

20. R. Averkamp and C. Houdré. Wavelet thresholding for non (necessarily) Gaussian noise: A preliminary report. In *CRM Proceedings and Lecture Notes, vol 18*, pages 347–354. 1999.

21. A. Azzalini. A class of distributions which includes the normal ones. *Scan. J. Statist.*, 12:171–178, 1993.

22. M. Basseville, A. Benveniste, K. C. Chou, S. A. Golden, R. Nikoukhah, and A. S. Willsky. Modeling and estimation of multiresolution stochastic processes. *IEEE Trans. Inform. Theory*, 38(2):766–784, 1992.

23. G. Battle and P. Federbush. Ondelettes and phase cell cluster expansions, a vindication. *Comm. Math. Phys.*, 109:417–419, 1987.

24. A. Benassi. Locally self similar Gaussian processes. In A. Antoniadis and G. Oppenheim, editors, *Wavelets and Statistics*, Lecture Notes in Statistics, pages 43–54. Springer-Verlag, New York, 1995.

25. Y. Benjamini and Y. Hochberg. Controlling the false discovery rate: A practical and powerful approach to multiple testing. *J. Roy. Statist. Soc. Ser. B*, 57(1):289–300, 1995.

26. J. Beran. Reply to comments on *Statistical methods for data with long-range dependence*. *Statist. Sci.*, 7:425–427, 1992.

27. J. Beran. Fitting long-memory models by generalized linear regression. *Biometrika*, 80:817–822, 1993.

28. J. Beran. *Statistical Methods for Long Memory Processes*. Chapman & Hall, London, 1994.

29. J. O. Berger. *Statistical Decision Theory and Bayesian Analysis*. Springer-Verlag, New York, 1985.

30. M. Berliner, C. Wikle, and R. Milliff. Multiresolution wavelet analyses in hierarchical Bayesian turbulence models. In P. Müller and B. Vidakovic, editors, *Bayesian Inference In Wavelet Based Models*, Lecture Notes in Statistics. Springer Verlag, New York, 1999.

31. O. V. Besov. On a family of function spaces: Embeding and extension theorems. *Doklady AN SSSR (In Russian)*, 126:1163–1165, 1959.

32. G. Beylkin. On the representation of the operators in bases of compactly supported wavelets. *SIAM J. Numer. Anal.*, 29:1716–1740, 1992.

33. G. Beylkin and N. Saito. Wavelets, their autocorrelation functions, and multiresolution representation of signals. In *Intelligent Robots and Computer Vision XI: Biological, Neural Net and 3-D Methods*, volume 1826 of *Proceedings of the SPIE*, pages 39–50, 1992.

34. P. Bickel. Minimax estimation of the mean of a normal distribution subject to doing well at the point. In M. H. Rizvi, J. S. Rustagi, and D. Siegmund, editors, *Recent Advances in Statistics*, pages 511–538. Academic Press, New York, 1983.

35. M. E. Bock and G. J. Pliego. Estimating functions with wavelets part II: Using a Daubechies wavelet in nonparametric regression. *Statistical Computing and Statistical Graphics Newsletter*, 3(2):27–34, November 1992.

36. M. E. Bock and G. J. Pliego. Wavelet approach to density estimation through a robust thresholding test. Technical Report 98-10, Department of Statistics, Purdue University, 1998.

37. L. Breiman. Heuristics of instability and stabilization in model selection. *Ann. Statist.*, 24:2350–2383, 1996.

38. D. R. Brillinger. Uses of cumulants in wavelet analysis. Technical report, Department of Statistics, University of California at Berkeley, 1995.

39. P. J. Brockwell and R. A. Davis. *Time Series. Theory and Methods (Second Edition)*. Springer-Verlag, New York, 1991.

40. P. J. Brown, T. Fearn, and M. Vannucci. Bayesian wavelet regression on curves with application to a spectroscopic calibration problem. Technical Report UKC/IMS 98-41, Institute of Maths and Stats, University of Kent at Canterbury, 1998.

41. A. G. Bruce, D. L. Donoho, H. Y. Gao, and R. D. Martin. Smoothing and robust wavelet analysis. In *COMPSTAT. Proceedings in Computational Statistics, 11th Symposium*, pages 531–547, 1994.

42. A. G. Bruce and H-Y. Gao. *Applied Wavelet Analysis with S-PLUS*. Springer-Verlag, New York, 1996.

43. A. G. Bruce and H-Y. Gao. Understanding WaveShrink: Variance and bias estimation. *Biometrika*, 83(4):727–745, 1996.

44. A. G. Bruce, H-Y. Gao, and W. Stuetzle. Subset-selection and ensemble methods for wavelet denoising. *Statistica Sinica*, 9(1):167–182, 1999.

45. H. Brunk. Univariate density estimation by orthogonal series. *Biometrika*, 65:521–528, 1978.

46. R. W. Buccigrossi and E. P. Simoncelli. Embedded wavelet image compression based on a joint probability model. In *Proceedings of 4th IEEE International Conference on Image Processing*, Santa Barbara, CA, October 1997.

348 REFERENCES

47. J. Buckheit, S. S. Chen, D. L. Donoho, I. M. Johnstone, and J. Scargle. *About WaveLab*, Technical report, Department of Statistics, Stanford University, 1995.
48. P. Burman. A comparative study of ordinary cross-validation, ν-fold cross-validation and the repeated learning-testing methods. *Biometrika*, 76:503–514, 1989.
49. C. S. Burrus, R. A. Gopinath, and H. Guo. *Introduction to Wavelets and Wavelet Transforms : A Primer*. Prentice Hall, Englewood Cliffs, NJ, 1998.
50. P. J. Burt and E. H. Adelson. The Laplacian pyramid as a compact image code. *IEEE Trans. Comm.*, COM-31:532–540, 1983.
51. P. J. Burt and E. H. Adelson. A multiresolution spline with applications to image mosaic. *ACM Trans. Graphics*, 2:217–236, 1983.
52. T. Cai. Wavelet regression using block thresholding. Manuscript, Purdue University, 1996.
53. T. Cai. Adaptive wavelet estimation: A block thresholding and oracle inequality approach. Technical Report 98-07, Department of Statistics, Purdue University, 1998.
54. T. Cai and L. Brown. Wavelet shrinkage for nonequispaced samples. Technical Report 97-06, Department of Statistics, Purdue University, 1997.
55. T. Cai and B. W. Silverman. Incorporating information on neighboring coefficients into wavelet estimation. Technical report, Department of Statistics, Purdue University, 1998.
56. S. Cambanis and C. Houdré. On the continuous wavelet transform of second-order random processes. *IEEE Trans. Inform. Theory*, 41(3):628–642, 1995.
57. S. Cambanis and E. Masry. Wavelet approximation of deterministic and random signals: Convergence properties and rates. *IEEE Trans. Inform. Theory*, 40(4):1013–1029, 1994.
58. B. Carlin and S. Chib. Bayesian model choice via Markov chain Monte Carlo. *J. Roy. Statist. Soc. Ser. B*, 57:473–484, 1995.
59. R. Carmona. Wavelet identification of transients in noisy signals. In M. Unser and A. F. Laine, editors, *Mathematical Imaging. Wavelet Applications in Signal and Image Processing, vol 2034*, pages 392–400. SPIE, 1993.
60. R. Carmona. Spline smoothing and extrema representation: Variations on a reconstruction algorithm of Mallat and Zhong. In A. Antoniadis and G. Oppenheim, editors, *Wavelets and Statistics*, Lecture Notes in Statistics, pages 83–94. Springer-Verlag, New York, 1995.
61. R. Carmona and L. Hudgins. Wavelet denoising of EEG signals and identification of evoked response potentials. In M. Unser and A. F. Laine, editors, *Mathematical Imaging. Wavelet Applications in Signal and Image Processing, vol 2304*, pages 91–104. SPIE, 1994.
62. R. Carmona, W-L. Hwang, and B. Torrésani. Identification of chirps with continuous wavelet transformation. In A. Antoniadis and G. Oppenheim, editors, *Wavelets and Statistics*, Lecture Notes in Statistics, pages 95–108. Springer-Verlag, New York, 1995.
63. R. Carmona, W-L. Hwang, and B. Torrésani. Characterization of signals by the ridges of their wavelet transform. *IEEE Trans. Signal Process.*, 45(12):2568–2589, 1997.
64. R. Carmona, W-L. Hwang, and B. Torrésani. *Practical Time–Frequency Analysis*, volume 9 of *Wavelet Analysis and its Applications*. Academic Press, San Diego, 1998.
65. R. Carmona and A. Wang. Comparison tests for the spectra of dependent multivariate time series. In R. J. Adler, P. Müller, and B. Rozovskii, editors, *Stochastic Modeling in*

Physical Oceanography, volume 39 of *Progress in Probability*, pages 69–88. Birkhäuser, Boston, 1996.

66. N. N. Čencov. Evaluation of an unknown distribution density from observations. *Doklady*, (3):1559–1562, 1962.

67. A. Chambolle, R. A. DeVore, N-Y. Lee, and B. J. Lucier. Nonlinear wavelet image processing: Variational problems, compression, and noise removal through wavelet shrinkage. *IEEE Trans. Image Process.*, (7):319–355, 1998.

68. J. Chen and H. Rubin. Drawing a random sample at random. *Comput. Statist. Data Anal.*, 4:219–227, 1986.

69. S. S. Chen. *Basis Pursuit*. Ph.D. thesis, Department of Statistics, Stanford University, 1995.

70. S. S. Chen and D. L. Donoho. On basis pursuit. Technical report, Department of Statistics, Stanford University, 1994.

71. S. S. Chen and D. L. Donoho. Atomic decomposition by basis pursuit. Technical Report 479, Department of Statistics, Stanford University, 1995.

72. M-Y. Cheng. A bandwidth selector for local linear density estimators. *Ann. Statist.*, 25:1001–1013, 1997.

73. C. Chiann and P. A. Morettin. A wavelet analysis for time series. *Journal of Nonparametric Statistics*, 10:1–46, 1998.

74. H. A. Chipman, E. D. Kolaczyk, and R. E. McCulloch. Adaptive Bayesian wavelet shrinkage. *J. Amer. Statist. Assoc.*, 92(440):1413–1421, 1997.

75. C. K. Chui. *An Introduction to Wavelets*, volume 1 of *Wavelet Analysis and its Applications*. Academic Press, San Diego, 1992.

76. C. K. Chui and J. Z. Wang. A cardinal spline approach to wavelets. *Proc. Amer. Math. Soc.*, 113:785–793, 1991.

77. C. K. Chui and J. Z. Wang. A general framework of compactly supported splines and wavelets. *J. Approx. Theory*, 71(3):263–304, 1992.

78. M. Clutton-Brock. Density estimation using exponentials of orthogonal series. *J. Amer. Statist. Assoc.*, 85:760–764, 1990.

79. M. A. Clyde, H. DeSimone, and G. Parmigiani. Prediction via orthogonalized model mixing. Technical Report 94-32, ISDS, Duke University, 1994.

80. M. A. Clyde and E. I. George. Robust empirical Bayes estimation in wavelets. Technical Report 98-21, ISDS, Duke University, 1998.

81. M. A. Clyde and E. I. George. Empirical Bayes for wavelet coefficients. In P. Müller and B. Vidakovic, editors, *Bayesian Inference In Wavelet Based Models*, Lecture Notes in Statistics. Springer Verlag, New York, 1999.

82. M. A. Clyde, G. Parmigiani, and B. Vidakovic. Multiple shrinkage and subset selection in wavelets. *Biometrika*, 85:391–402, 1998.

83. A. Cohen. Biorthogonal wavelets. In C. K. Chui, editor, *Wavelets: A Tutorial in Theory and Applications*, pages 123–152, New York, 1992. Academic Press.

84. A. Cohen, I. Daubechies, and J. Feauveau. Bi-orthogonal bases of compactly supported wavelets. *Comm. Pure Appl. Math.*, 45:485–560, 1992.

85. A. Cohen, I. Daubechies, B. Jawerth, and P. Vial. Multiresolution analysis, wavelets, and fast algorithms on an interval. *C. R. Acad. Sci. Paris Sér. A*, 316:417–421, 1993.

86. A. Cohen, I. Daubechies, and P. Vial. Wavelets on the interval and fast wavelet transforms. *Appl. Comput. Harmon. Anal.*, 1(1):54–81, 1993.

87. A. Cohen and R. D. Ryan. *Wavelets and Multiscale Signal Processing*. Chapman & Hall, London, 1995.

88. R. R. Coifman and D. L. Donoho. Translation-invariant wavelet denoising. In A. Antoniadis and G. Oppenheim, editors, *Wavelets and Statistics*, volume 103 of *Lecture Notes in Statistics*, pages 125–150, New York, 1995. Springer-Verlag.

89. R. R. Coifman and Y. Meyer. Orthonormal wave packet bases. Preprint, 1990.

90. R. R. Coifman, Y. Meyer, S. Quake, and M. V. Wickerhauser. Signal processing and compression with wave packets. In Y. Meyer, editor, *Proceedings of the International Conference on Wavelets, Marseilles*. Masson, Paris, 1989.

91. R. R. Coifman, Y. Meyer, and M. V. Wickerhauser. Wavelet analysis and signal processing. In M. B. Ruskai et al., editor, *Wavelets and Their Applications*, pages 153–178. Jones and Bartlett Publishers, Sudbury, MA, 1992.

92. R. R. Coifman and N. Saito. Constructions of local orthonormal bases for classification and regression. *C. R. Acad. Sci. Paris, Sér. I*, 319(2):191–196, 1994.

93. R. R. Coifman and N. Saito. The local Karhunen-Loève bases. In *Proc. IEEE-SP Intern. Symp. Time-Frequency and Time-Scale Analysis, Jun. 18–21, Paris, France*, pages 129–132. IEEE, 1996.

94. R. R. Coifman and M. V. Wickerhauser. Entropy-based algorithms for best basis selection. *IEEE Trans. Inform. Theory*, 38(2):713–718, 1992.

95. S. Collineau and Y. Brunet. Detection of turbulent coherent motion in a forest canopy: wavelet analysis. *Bound. Layer Meteorol.*, 65:357–379, 1993.

96. P. L. Conti and G. J. Lasinio. Bayesian and linear Bayesian inference on regression functions using orthonormal expansions. Technical report, University of Rome - La Sapienza, 1998. Contact: jona@pow2.sta.uniroma1.it.

97. G. Cristobal, M. Chagoyen, B. Escalante-Ramirez, and J. Lopez. Wavelet based denoising methods: A comparative study with applications in microscopy. In *SPIE conference on wavelet applications, Denver, Colorado*, volume 2825, pages 660–671, 1996.

98. M. Crouse, R. D. Nowak, and R. Baraniuk. Statistical signal processing using wavelet-domain hidden Markov models. In *Wavelet Applications in Signal and Image Processing V*, Proceedings of SPIE, vol. 3169, pages 248–259. SPIE, 1997.

99. R. Dahlhaus. Fitting time series to nonstationary processes. *Ann. Statist.*, 25:1–38, 1997.

100. W. Dahmen and C. A. Micchelli. Using the refinement equation for evaluating integrals of wavelets. *SIAM J. Num. Anal.*, (30):507–537, 1993.

101. A. DasGupta and H. Rubin. Bayes estimators as expanders in one and two dimensions. Technical Report 93-38, Department of Statistics, Purdue University, 1993.

102. I. Daubechies. Orthonormal bases of compactly supported wavelets. *Comm. Pure Appl. Math.*, 41:909–996, 1988.

103. I. Daubechies. Time-frequency localization operators: A geometric phase space approach. *IEEE Trans. Inform. Theory*, 34(4):606–612, July 1988.

104. I. Daubechies. *Ten Lectures on Wavelets*. Number 61 in CBMS-NSF Series in Applied Mathematics. SIAM, Philadelphia, 1992.

105. I. Daubechies. Orthonormal bases of compactly supported wavelets II. Variations on a theme. *SIAM J. Math. Anal.*, 24(2):499–519, 1993.

106. I. Daubechies and J. Lagarias. Two-scale difference equations I. Existence and global regularity of solutions. *SIAM J. Math. Anal.*, 22(5):1388–1410, 1991.

107. I. Daubechies and J. Lagarias. Two-scale difference equations II. Local regularity, infinite products of matrices and fractals. *SIAM J. Math. Anal.*, 23(4):1031–1079, 1992.

108. H. A. David. *Order Statistics (Second Edition)*. Wiley, New York, 1981.

109. G. Davis. *Adaptive Nonlinear Approximations*. Ph.D. thesis, Department of Mathematics - Courant Institute of Mathematical Sciences, 1994.

110. L. Dechevsky and S. Penev. On shape-preserving probabilistic wavelet approximators. *Stochastic Anal. Appl.*, 15:187–215, 1997.

111. L. Dechevsky and S. Penev. On shape-preserving wavelet estimators of cumulative distribution functions and densities. *Stochastic Anal. Appl.*, 16:428–469, 1998.

112. B. Delyon and A. Juditsky. Wavelet estimators, global error measures: Revisited. Technical Report 782, IRISA-INRIA, France, 1993.

113. G. Deslauriers and S. Dubuc. Symmetric iterative interpolation processes. *Constr. Approx.*, 5(1):49–68, 1989.

114. R. A. DeVore and B. J. Lucier. Fast wavelet techniques for near-optimal image processing. In *IEEE Military Communications Conference Record*, pages 1129–1135. Piscataway, NJ, 1992.

115. R. A. DeVore and V. Popov. Interpolation of Besov spaces. *Trans. Amer. Math. Soc.*, 305:397–414, 1988.

116. S. Dharmadhikari and K. Joag-Dev. *Unimodality, Convexity, and Application*. Academic Press, San Diego, 1988.

117. P. J. Diggle and P. Hall. A Fourier approach to nonparametric deconvolution of a density estimate. *J. Roy. Statist. Soc. Ser. B*, 55:523–531, 1993.

118. R. W. Dijkerman and R. R. Mazumdar. Wavelet representations of stochastic processes and multiresolution stochastic models. *IEEE Trans. Signal Process.*, 42(7):1640–1652, 1994.

119. D. L. Donoho. Nonlinear solution of linear inverse problems by wavelet-vaguelette decomposition. Technical Report 403, Stanford University, Department of Statistics, January 1992.

120. D. L. Donoho. Wavelet shrinkage and W.V.D.: A 10-minute tour. In Y. Meyer and S. Roques, editors, *Progress in Wavelet Analysis and Applications*, pages 109–128. Editions Frontieres, 1992. Toulouse, France - June 1992.

121. D. L. Donoho. Nonlinear wavelet methods for recovery of signals, densities, and spectra from indirect and noisy data. In *Proceedings of Symposia in Applied Mathematics*, volume 47, pages 173–205. American Mathematical Society, 1993.

122. D. L. Donoho. De-noising by soft-thresholding. *IEEE Trans. Inform. Theory*, 41(3):613–627, 1995.

123. D. L. Donoho. Cart and best-ortho-basis: A connection. *Ann. Statist.*, 25(5):1870–1911, 1997.

124. D. L. Donoho and I. M. Johnstone. Ideal spatial adaptation via wavelet shrinkage. Technical Report 400, Preprint Department of Statistics, Stanford University, 1992.

125. D. L. Donoho and I. M. Johnstone. Ideal denoising in an orthonormal basis chosen from a library of bases. *C. R. Acad. Sci. Paris Sér. A*, 319:1317–1322, 1994.

126. D. L. Donoho and I. M. Johnstone. Ideal spatial adaptation by wavelet shrinkage. *Biometrika*, 81(3):425–455, 1994.

127. D. L. Donoho and I. M. Johnstone. Minimax risk over l_p-balls for l_q-error. *Probability Theory and Related Fields*, 99:277–304, 1994.

128. D. L. Donoho and I. M. Johnstone. Adapting to unknown smoothness via wavelet shrinkage. *J. Amer. Statist. Assoc.*, 90:1200–1224, 1995.

129. D. L. Donoho and I. M. Johnstone. Neo-classical minimax problems, thresholding and adaptive function estimation. *Bernoulli*, 2(1):39–62, 1996.

130. D. L. Donoho and I. M. Johnstone. Minimax estimation via wavelet shrinkage. *Ann. Statist.*, 26(3):879–921, 1998.

131. D. L. Donoho and I. M. Johnstone. Asymptotic minimaxity of wavelet estimators with sampled data. *Statistica Sinica*, 9(1):1–32, 1999.

132. D. L. Donoho, I. M. Johnstone, G. Kerkyacharian, and D. Picard. Wavelet shrinkage: Asymptopia? (with discussion). *J. Roy. Statist. Soc. Ser. B*, 57(2):301–369, 1995.

133. D. L. Donoho, I. M. Johnstone, G. Kerkyacharian, and D. Picard. Density estimation by wavelet thresholding. *Ann. Statist.*, 24(2):508–539, 1996.

134. D. L. Donoho, S. G. Mallat, and R. von Sachs. Estimating covariances of locally stationary processes: Rates of convergence of best basis methods. 1997.

135. D. L. Donoho and T. P-Y. Yu. Nonlinear "wavelet transforms" based on median thresholding. Technical report, Stanford University, 1998.

136. P. Doukhan. Formes de Töeplitz associées à une analyse multiéchelle. *C. R. Acad. Sci. Paris Sér. A*, 306:663–666, 1988.

137. P. Doukhan and J. R. Léon. Déviation quadratique d'estimateurs de densité par projections orthogonales. *C. R. Acad. Sci. Paris Sér. I Math.*, 310:425–430, 1990.

138. P. Dutilleux. An implementation of the "algorithme à trous" to compute the wavelet transform. In J.-M. Combes, A. Grossmann, and P. Tchamitchian, editors, *Wavelets: Time-Frequency Methods and Phase Space*, Inverse Problems and Theoretical Imaging, pages 298–304, Berlin, 1989. Springer-Verlag.

139. S. Efromovich. Quazi-linear wavelet estimation. Submitted to *J. Amer. Statist. Assoc.*, 1998.

140. B. Efron and C. Morris. Limiting the risk of Bayes and empirical Bayes estimators – part I: The Bayes case. *J. Amer. Statist. Assoc.*, 66:807–815, 1971.

141. J. Q. Fan. On the optimal rates of convergence for nonparametric deconvolution problems. *Ann. Statist.*, 19:1257–1272, 1991.

142. J. Q. Fan. Deconvolution with supersmooth distributions. *The Canadian Journal of Statistics*, 20:155–169, 1992.

143. J. Q. Fan. Adaptively local one-dimensional subproblems with application to a deconvolution problem. *Ann. Statist.*, 21:600–610, 1993.

144. J. Q. Fan. Test of significance based on wavelet thresholding and Neyman's truncation. *J. Amer. Statist. Assoc.*, 91(434):674–688, 1996.

145. J. Q. Fan. Comments on wavelets in statistics: A review by A. Antoniadis. To appear in *J. Italian Statist. Soc.*, 1998.

146. J. Q. Fan and I. Gijbels. *Local Polynomial Modelling and Its Applications*. Chapman & Hall, London, 1996.

147. J. Q. Fan, P. Hall, M. Martin, and P. Patil. Adaptation to high spatial inhomogeneity using wavelet methods. *Statistica Sinica*, 9(1):85–102, 1999.

148. J. Q. Fan, P. Hall, M. A. Martin, and P. Patil. On local smoothing of nonparametric curve estimators. *J. Amer. Statist. Assoc.*, 91(433):258–266, 1996.

149. M. Farge, E. Goirand, Y. Meyer, F. Pascal, and M.V. Wickerhauser. Improved predictability of two-dimensional turbulent flows using wavelet packet compression. *Fluid Dynam. Res.*, 10:229–250, 1992.

150. G. L. Fix and G. Strang. Fourier analysis of the finite element method in Ritz-Galerkin theory. *Stud. Appl. Math*, 48:265–273, 1969.

151. P. Flandrin. On the spectrum of fractional Brownian motions. *IEEE Trans. Inform. Theory*, 35(1):197–199, 1989.

152. P. Flandrin. Some aspects of non-stationary signal processing with emphasis on time-frequency and time-scale methods. In J. M. Combes, A. Grossmann, and Ph. Tchamitchian, editors, *Wavelets Time-Frequency Methods and Phase Space*, pages 68–98, New York, 1989. Springer-Verlag.

153. P. Flandrin. Time-scale analyses and self-similar stochastic processes. In Byrnes et al., editor, *Wavelets and Their Applications*, pages 121 – 142. NATO ASI Series vol. 442, 1992.

154. P. Flandrin. Wavelet analysis and synthesis of fractional Brownian motion. *IEEE Trans. Inform. Theory*, 38(2):910–917, 1992.

155. D. Foster and E. I. George. The risk inflation criterion for multiple regression. *Ann. Statist.*, 22:1947–1975, 1994.

156. G. Foster. Wavelets for period analysis of unequally sampled time series. *Astronomical Journal*, 112:1709–1729, 1996.

157. P. Franklin. A set of continuous orthogonal functions. *Math. Ann.*, 100:522–529, 1928.

158. J. Friedman and W. Stuetzle. Projection pursuit regression. *J. Amer. Statist. Assoc.*, 76:817–823, 1981.

159. J. Friedman, W. Stuetzle, and A. Schroeder. Projection pursuit density estimation. *J. Amer. Statist. Assoc.*, 79:599–608, 1984.

160. U. Frisch. *Turbulence*. Cambridge University Press, New York, 1996.

161. D. Gabor. Theory of communication. *J. Inst. Electr. Engrg.*, 93(3):429–457, 1946.

162. M. Gail and J. Gastwirth. A scale-free goodness of feet test for the exponential distribution based on the Lorentz curve. *J. Amer. Statist. Assoc.*, 73:787–793, 1978.

163. H-Y. Gao. *Estimation of Spectral Densities in Time Series Analysis*. Ph.D. thesis, University of California, Berkeley, 1993.

164. H-Y. Gao. Spectral density estimation via wavelet shrinkage. StatSci Division of MathSoft, Inc., 1996.

165. H-Y. Gao. Choice of thresholds for wavelet shrinkage estimate of the spectrum. *J. Time Ser. Anal.*, 18(3), 1997.

166. H-Y. Gao. Wavelet shrinkage denoising using the non-negative garrote. *J. Comput. Graph. Statist.*, 7(4):469–488, 1998.

167. H-Y. Gao and A. G. Bruce. WaveShrink with firm shrinkage. Technical Report 39, StatSci Division of MathSoft, Inc., 1996.

168. T. Gasser and H-G. Müller. Kernel estimation of wavelet functions. In *Smoothing Techniques for Curve Estimation*, Lecture Notes in Mathematics 757, pages 23–68, New York, 1979. Springer-Verlag.

169. T. Gasser and H-G. Müller. Estimating regression functions and their derivatives by the kernel method. *Scand. J. Statist.*, 11:171–185, 1984.

170. V. Genon-Catalot, C. Laredo, and D. Picard. Estimation non paramétrique de la variance d'une diffusion par méthodes d'ondelettes. *C. R. Acad. Sci. Paris Sér. I Math.*, 311:379–382, 1990.

171. V. Genon-Catalot, C. Laredo, and D. Picard. Nonparametric estimation of variance of a diffusion by wavelet methods. *Scand. J. Statist.*, 19:319–335, 1992.

172. E. I. George and R. E. McCulloch. Approaches to Bayesian variable selection. Technical report, Graduate School of Business, University of Chicago, 1994.

173. J. S. Geronimo, D. P. Hardin, and P. R. Massopust. Fractal functions and wavelet expansions based on several scaling functions. *J. Approx. Theory*, 78(3):373–401, 1994.

174. J. Geweke and S. Porter-Hudak. The estimation and application of long memory time series models. *J. Time Ser. Anal.*, 4:221–238, 1983.

175. S. G. Ghang and M. Vetterli. Spatial adaptive wavelet thresholding for image denoising. In *Proc. IEEE Int. Conf. Image Processing*, volume 2, pages 374–377, 1997.

176. S. G. Ghang, B. Yu, and M. Vetterli. Image denoising via lossy compression and wavelet thresholding. In *Proc. IEEE Int. Conf. Image Processing*, volume 1, pages 604–607, 1997.

177. I. J. Good and R. A. Gaskins. Nonparametric roughness penalties for probability densities. *Biometrika*, 58:255–277, 1971.

178. R. A. Gopinath and C. S. Burrus. Wavelets and filter banks. In C. K. Chui, editor, *Wavelets: A Tutorial in Theory and Applications*, pages 603–654. Academic Press, San Diego, 1992.

179. A. Grossmann and J. Morlet. Decomposition of Hardy functions into square integrable wavelets of constant shape. *SIAM J. Math.*, 15:723–736, 1984.

180. A. Grossmann and J. Morlet. Decomposition of functions into wavelets of constant shape and related transforms. In L. Streit, editor, *Mathematics and physics, lectures on recent results*. World Scientific, River Edge, NJ, 1985.

181. A. Haar. Zur Theorie der orthogonalen Funktionen-Systeme. *Mathematische Annalen*, 69:331–371, 1910. In German.

182. P. Hall and I. Koch. On continuous image models and image analysis in the presence of correlated noise. *Adv. Appl. Probab.*, 22:332–349, 1990.

183. P. Hall and P. Patil. Formulae for mean integrated squared error of nonlinear wavelet-based density estimators. *Ann. Statist.*, 23:905–928, 1995.

184. P. Hall and P. Patil. Effect of thresholding rules on performance of wavelet-based curve estimators. *Statistica Sinica*, 6:331–345, 1996.

185. P. Hall and P. Patil. On the choice of smoothing parameter, threshold and truncation in nonparametric regression by non-linear wavelet methods. *J. Roy. Statist. Soc. Ser. B*, 58(2):361–377, 1996.

186. P. Hall, S. Penev, G. Kerkyacharian, and D. Picard. Numerical performance of block thresholded wavelet estimators. *Statist. Comput.*, 7(2):115–124, 1997.

187. P. Hall and B. A. Turlach. Interpolation methods for nonlinear wavelet regression with irregularly spaced design. *Ann. Statist.*, 25(5):1912–1925, 1997.

188. W. Härdle, G. Kerkyacharian, D. Pickard, and A. Tsybakov. *Wavelets, Approximation, and Statistical Applications*. Lecture Notes in Statistics 129. Springer-Verlag, New York, 1998.

189. T. J. Hastie and R. J. Tibshirani. *Generalized Additive Models*. Chapman & Hall, London, 1990.

190. W. He and M-J. Lai. Examples of bivariate nonseparable compactly supported orthonormal continuous wavelets. Preprint at Department of Mathematics, University of Georgia, Athens, 1998.

191. C. E. Heil and D. F. Walnut. Continuous and discrete wavelet transforms. *SIAM Review*, 31:628–666, 1989.

192. H. Helson. *Harmonic Analysis*. Addison-Wesley, Reading, MA, 1983.

193. E. Hernández and G. Weiss. *A First Course on Wavelets*. CRC Press Inc., Boca Raton, FL, 1996.

194. N. L. Hjort and M. C. Jones. Locally parametric nonparametric density estimation. *Ann. Statist.*, 24:1619–1647, 1996.

195. C. C. Holmes and D. G. T. Denison. Bayesian wavelet analysis with a model complexity prior. Preprint at Imperial College, London, 1998.

196. C. C. Holmes and B. K. Mallick. Bayesian wavelet network for nonparametric regression. Preprint at Imperial College, London, 1998.

197. C. C. Holmes and B. K. Mallick. Perfect simulation for orthogonal mixing. Preprint at Imperial College, London, 1998.

198. M. Holschneider. *Wavelets: An Analysis Tool*. Oxford Science Publications, Oxford, 1995.

199. M. Holschneider, R. Kronland-Martinet, J. Morlet, and P. Tchamitchian. A real-time algorithm for signal analysis with the help of the wavelet transform. In *Wavelets, Time-Frequency Methods and Phase Space*, pages 289–297. Springer-Verlag, New York, 1989.

200. M. Holschneider and P. Tchamitchian. Régularité locale de la fonction 'non-différentiable' de riemann. In P. G. Lemarié, editor, *Les Ondelettes en 1989*, volume 1438 of *Lecture Notes in Math*. Springer-Verlag, New York, 1990. Exposé no. 8.

201. C. Houdré. Wavelets, probability and statistics: Some bridges. In J. J. Benedetto and M. W. Frazier, editors, *Wavelets: Mathematics and Applications*, pages 361–394. CRC Press Inc., Boca Raton, FL, 1993.

202. H-C. Huang and N. Cressie. Deterministic/stochastic wavelet decomposition for recovery of signal from noisy data. Technical Report 97-23, Department of Statistics, Iowa State University, 1997.

203. H-C. Huang and N. Cressie. Empirical Bayesian spatial prediction using wavelets. In P. Müller and B. Vidakovic, editors, *Bayesian Inference In Wavelet Based Models*, Lecture Notes in Statistics. Springer-Verlag, New York, 1999.

204. S-Y. Huang. Wavelet based empirical Bayes estimation for the uniform distribution. *Statist. Probab. Lett.*, 32:141 – 146, 1997.

205. S-Y. Huang. Density estimation by wavelet-based reproducing kernels. *Statistica Sinica*, 9(1):137–151, 1999.

206. L. Hudgins, C. A. Friehe, and M. E. Mayer. Wavelet transforms and atmospheric turbulence. *Physical Review Letters*, 71(20):3279–3282, 1993.

207. G. Huerta. Bayes wavelet shrinkage and applications to data denoising. In electronic proceedings of International Workshop on Wavelets in Statistics, Duke University, 12-13 October, 1997.

208. A. Hyvärinen. Sparse code shrinkage: Denoising of nongaussian data by maximum likelihood estimation. Technical Report A51, Helsinki University of Technology, 1998.

209. D. Rios Insua and B. Vidakovic. Wavelet-based random densities. Technical Report 97-05, ISDS, Duke University, 1997.

210. A. J. Izenman. Recent developments in nonparametric density estimation. *J. Amer. Statist. Assoc.*, 86:205–224, 1991.

211. S. L. Jaffard. Exposants de Hölder en des points donnés et coefficients d'ondelettes. *C. R. Acad. Sci. Paris Sér. I Math.*, 308:79–81, 1989.

212. S. L. Jaffard. Pointwise smoothness, two-microlocalization and wavelet coefficients. *Publ. Mat.*, 35:155–168, 1991.

213. S. L. Jaffard and P. Laurencot. Orthonormal wavelets, analysis of operators, and applications to numerical analysis. In C. K. Chui, editor, *Wavelets: A Tutorial in Theory and Applications*, pages 543–601, San Diego, 1992. Academic Press.

214. S. L. Jaffard and Y. Meyer. Bases d'ondelettes dans des ouverts de $\mathbb{R}^n$. *J. Math. Pures Appl. (9)*, 66:95–108, 1989.

215. A. K. Jain. *Fundamentals of Digital Image Processing*. Prentice Hall, Englewood Cliffs, NJ, 1989.

216. A.J.E.M. Jannsen. The Smith-Barnwell condition and non-negative scaling functions. *IEEE Trans. Inform. Theory*, 38(2):884–885, 1993.

217. M. Jansen and A. Bultheel. Smoothing non-equidistantly sampled data using wavelets and cross validation. In *Proceedings of the IEEE Benelux Signal Processing Symposium*, pages 111–114. Leuven, Belgium, 1998. March 1998.

218. M. Jansen and A. Bultheel. Multiscale hidden Markov models for Bayesian image analysis. In P. Müller and B. Vidakovic, editors, *Bayesian Inference In Wavelet Based Models*, Lecture Notes in Statistics. Springer-Verlag, New York, 1999.

219. M. Jansen, M. Malfait, and A. Bultheel. Generalized cross-validation for wavelet thresholding. *Signal Processing*, 56(1):33–44, January 1997.

220. A. J. E. M. Janssen. The Zak transform: A signal transform for sampled time-continuous signals. *Philips J. Res.*, 43:23–69, 1988.

221. I. M. Johnstone. Wavelet shrinkage for correlated data and inverse problems: Adaptivity results. *Statistica Sinica*, 9(1):51–83, 1999.

222. I. M. Johnstone and B. W. Silverman. Wavelet threshold estimators for data with correlated noise. *J. Roy. Statist. Soc. Ser. B*, 59(2):319–351, 1997.

223. I. M. Johnstone and B. W. Silverman. Empirical Bayes approaches to mixture problems and wavelet regression. Technical report, Department of Statistics, Stanford University, 1998.

224. L. K. Jones. On a conjecture of Huber concerning the convergence of projection pursuit regression. *Ann. Statist.*, 15:880–882, 1987.

225. P. C. Joshi. Bounds and approximations for the moments of order statistics. *J. Amer. Statist. Assoc.*, 64:1617–1624, 1969.

226. L. M. Kaplan and C.-C. J. Kuo. Fractal estimation from noisy data via discrete fractional Gaussian noise (DFGN) and the Haar basis. *IEEE Trans. Signal Process.*, 41(12):3554–3562, 1993.

227. R. E. Kass and A. E. Raftery. Bayes factors. *J. Amer. Statist. Assoc.*, 90:773–795, 1995.

228. G. G. Katul and J. Albertson. Low dimensional turbulent transport mechanics near the forest—atmosphere interface. In P. Müller and B. Vidakovic, editors, *Bayesian Inference In Wavelet Based Models*, Lecture Notes in Statistics. Springer Verlag, New York, 1999.

229. G. G. Katul and M. B. Paralange. Analysis of land-surface heat fluxes using the orthonormal wavelet approach. *Water Resourc. Res.*, 31:2743–2749, 1995.

230. G. G. Katul and B. Vidakovic. The partitioning of attached and detached eddy motion in the atmospheric surface layer using Lorentz wavelet filtering. *Bound. Layer Meteorol.*, 77:153–172, 1996.

231. Y. Katznelson. *An Intoduction to Harmonic Analysis*. Dover Publications Inc., New York, 1976.

232. S. Kelly, M. Kon, and L. Raphael. Local convergence for wavelet expansions. *J. Funct. Anal.*, 126:102–138, 1994.

233. G. Kerkyacharian and D. Picard. Density estimation in Besov spaces. *Statist. Probab. Lett.*, 13:15–24, 1992.

234. G. Kerkyacharian and D. Picard. Density estimation by kernel and wavelets method: Optimality of Besov spaces. *Statist. Probab. Lett.*, 18:327–336, 1993.

235. G. Kerkyacharian, D. Picard, and K. Tribouley. $\mathbb{L}_p$ adaptive density estimation. *Bernoulli*, 2(3):229–247, 1996.

236. V. K. Klonias. Consistency of two nonparametric maximum penalized likelihood estimators of the probability density function. *Ann. Statist.*, 10:811–824, 1982.

237. R. Kohn and S. J. Marron. Bayesian wavelet shrinkage. In electronic proceedings of International Workshop on Wavelets in Statistics, Duke University, 12-13 October, 1997.

238. E. D. Kolaczyk. *Wavelet Methods for the Inversion of Certain Homogeneous Linear Operators in the Presence of Noisy Data*. Ph.D. thesis, Stanford University, 1994.

239. E. D. Kolaczyk. A wavelet shrinkage approach to tomographic image reconstruction. *J. Amer. Statist. Assoc.*, 91(435):1079–1090, 1996.

240. E. D. Kolaczyk. Bayesian multiscale models for Poisson processes. Submitted to the *J. Amer. Statist. Assoc.*, 1998.

241. E. D. Kolaczyk. Poisson data, recursive dyadic partitions, and Bayes. In P. Müller and B. Vidakovic, editors, *Bayesian Inference In Wavelet Based Models*, Lecture Notes in Statistics. Springer-Verlag, New York, 1999.

242. A. N. Kolmogorov. The local structure of turbulence in incompressible viscous fluid for very large Reynolds numbers. *Dokl. Akad. Nauk. SSSR*, 30:301–305, 1941.

243. T. H. Körner. Divergence of decreasing rearranged fourier series. *Ann. of Math.*, 144(1):167–180, 1996.

244. A. Kovac and B. W. Silverman. Extending the scope of wavelet thresholding methods by coefficient-dependent thresholding. Technical report, Dept. of Math., Univ. of Bristol, University Walk, Bristol, BS8 1TW, U.K., 1998.

245. J. Kovačević and M. Vetterli. Nonseparable multidimensional perfect reconstruction filter banks and wavelet bases for $\mathbb{R}^n$. *IEEE Trans. Inform. Theory*, 38(2):533–555, March 1991.

246. J. Kovačević and M. Vetterli. Perfect reconstruction filter banks with rational sampling rate changes. In *IEEE Proc. Int. Conf. Acoust., Speech, Signal Processing*, volume III, pages 1785–1788, Toronto, Canada, May 1991.

247. H. Krim and J.-C. Pesquet. Multiresolution analysis of a class of nonstationary processes. *IEEE Trans. Inform. Theory*, 41(4):1010–1020, 1995.

248. H. Krim and I. C. Schick. Minimax description length for signal denoising and optimized representation. Internal report (LIDS-MIT), to appear in *IEEE Trans. Inform. Theory*, 1998.

249. R. Kronmal and M. Tarter. The estimation of probability densities and cumulatives by Fourier series methods. *J. Amer. Statist. Assoc.*, 63:925–952, 1968.

250. P. Kumar and E. Foufoula-Georgiou. Wavelet analysis for geophysical applications. *Review of Geophysics*, 35(4):385–412, 1997.

251. B. La Borde. Generic explicit wavelet tap derivation. In *SPIE conference on wavelet applications*, volume 2762, pages 94–104, 1996.

252. M. Lang, H. Guo, J. E. Odegard, C. S. Burrus, and R. O. Wells. Noise reduction using an undecimated discrete wavelet transform. *IEEE Signal Process. Lett.*, 3(1):10–12, 1996.

253. W. Lawton. Necessary and sufficient conditions for constructing orthonormal wavelet bases. *J. Math. Phys.*, 32(1):57–61, 1991.

254. N. Lee. *Wavelet-Vaguelette Decompositions and Homogeneous Equations*. Ph.D. thesis, Department of Mathematics, Purdue University, 1997.

255. P. G. Lemarié and Y. Meyer. Ondelettes et bases hilbertiennes. *Rev. Mat. Iberoamericana*, 2:1–18, 1986.

256. L. Lenarduzzi. Denoising non equispaced data with wavelets. Technical Report 97.1, IAMI - CNR Milano, Italy, 1998.

257. T. Leonard. A Bayesian method for histograms. *Biometrika*, 60:297–308, 1973.

258. D. Leporini. Optimized representations of signals using Bayesian decomposition trees. Preprint at Laboratoire des Signaux et Systèmes, CNRS Université Paris-Sud, 1998.

259. D. Leporini and J.-C. Pesquet. Wavelet thresholding for some classes of non-Gaussian noise. Preprint at Laboratoire des Signaux et Systèmes, CNRS Université Paris-Sud, 1998.

260. D. Leporini, J.-C. Pesquet, and H. Krim. Best basis representation s with prior statistical models. In P. Müller and B. Vidakovic, editors, *Bayesian Inference In Wavelet Based Models*, Lecture Notes in Statistics. Springer-Verlag, New York, 1999.

261. Y. Li and Z. Xie. The wavelet detection of hidden periodicities in time series. *Statist. Probab. Lett.*, 35:9–23, 1997.

262. Y. Li and Z. Xie. The wavelet identification of thresholds and time delay of threshold autoregressive models. *Statistica Sinica*, 9(1):153–166, 1999.

263. J. M. Lina and B. MacGibbon. Non-linear shrinkage estimation with complex Daubechies' wavelets. In *Wavelet Applications in Signal and Image Processing V*, Proceedings of SPIE, vol. 3169, pages 67–79. SPIE, 1997.

264. J. E. Littlewood and R. Paley. Theorems on Fourier series and power series. *J. London Math. Soc.*, 42:52–89, 1937.

265. C. R. Loader. Local likelihood density estimation. *Ann. Statist.*, 24:1602–1618, 1996.

266. M. O. Lorentz. Methods of measuring concentration of wealth. *J. Amer. Statist. Assoc.*, 9:209–219, 1905.

267. A. K. Louis, P. Maaß, and A. Reider. *Wavelets: Theory and Applications*. Wiley, Chichester, 1997.

268. H-S. Lu, S-Y. Huang, and Y-C. Tung. Wavelet shrinkage for nonparametric mixed-effects models. Technical report, Institute of Statistics, National Chiao Tung University, 1997. Contact hslu@stat.nctu.edu.tw.

269. B. Lumeau, J.-C. Pesquet, J. F. Bercher, and L. Louveau. Optimization of bias-variance trade-off in nonparametric spectral analysis by decomposition into wavelet packets. In Y. Meyer and S. Roques, editors, *Progress in Wavelet Analysis and Applications*, pages 285–290. Editions Frontieres, 1992. Toulouse, France - June 1992.

270. Y. Ma, G. Strang, and B. Vidakovic. The first moment of wavelet random variables. Technical Report 97-10, ISDS, Duke University, 1997.

271. J. M. Maatta and G. Casella. Developments in decision-theoretic variances estimation. *Statist. Sci.*, 5:90–120, 1990.

272. M. Malfait, M. Jansen, and D. Roose. Bayesian approach to wavelet-based image processing. Presented at the Joint Statistical Meetings, Chicago, August, 1996.

273. M. Malfait and D. Roose. Wavelets and Markov random fields in a Bayesian framework. In A. Antoniadis and G. Oppenheim, editors, *Wavelets and Statistics*, Lecture Notes in Statistics, pages 225–238. Springer-Verlag, New York, 1995.

274. S. G. Mallat. Multiresolution approximations and wavelets. Technical report, GRASP Lab, Dept. of Computer and Information Science, University of Pennsylvania, 1987.

275. S. G. Mallat. Multiresolution approximations and wavelet orthonormal bases of $\mathbb{L}^2(\mathbb{R})$. *Trans. Amer. Math. Soc.*, 315:69–87, 1989.

276. S. G. Mallat. A theory for multiresolution signal decomposition: The wavelet representation. *IEEE Trans. on Patt. Anal. Mach. Intell.*, 11(7):674–693, 1989.

277. S. G. Mallat. *A Wavelet Tour of Signal Processing*. Academic Press, San Diego, 1998.

278. S. G. Mallat and W-L. Hwang. Singularity detection and processing with wavelets. *IEEE Trans. Inform. Theory*, 38(2):617–643, 1992.

279. S. G. Mallat and Z. Zhang. Matching pursuits with time-frequency dictionaries. *IEEE Trans. Signal Process.*, 41(12):3397–3415, 1993.

280. S. G. Mallat and S. Zhong. Characterization of signals from multiscale edges. *IEEE Trans. Patt. Anal. Mach. Intell.*, 14(7):710–732, 1992.

281. S. G. Mallat and S. Zhong. Wavelet transform maxima and multiscale edges. In M. B. Ruskai et. al., editor, *Wavelets and Their Applications*, pages 67–104. Jones and Bartlett Publishers, 1992.

282. D. Marr. *Vision: A Computational Investigation into the Human Representation and Processing of Visual Information*. W.H. Freeman and Company, New York, 1992.

283. J. S. Marron, S. Adak, I. M. Johnstone, M. H. Neumann, and P. Patil. Exact risk analysis of wavelet regression. *J. Comput. Graph. Statist.*, 7(3):278–309, September 1998.

284. S. J. Marron. A comparison of cross-validation techniques in density estimation. *Ann. Statist.*, 15:152–162, 1987.

285. S. J. Marron and M. P. Wand. Exact mean integrated squared error. *Ann. Statist.*, 20:712–736, 1992.

286. A. Marshall and I. Olkin. *Inequalities: Theory of Majorization and Its Applications*. Academic Press, New York, 1979.

287. E. Masry. Asymptotic normality for deconvolution estimators of multivariate densities of stationary processes. *Journal of Multivariate Analysis*, 44:47–68, 1993.

288. E. Masry. Probability density estimation from dependent observations using wavelets orthonormal bases. *Statist. Probab. Lett.*, 21:181–194, 1994.

289. E. J. McCoy. *Some New Statistical Approaches to the Analysis of Long Memory Processes*. Ph.D. thesis, Imperial College, UK, Department of Mathematics, 1994.

290. E. J. McCoy and A. T. Walden. Wavelet analysis and synthesis of stationary long-memory processes. *J. Comput. Graph. Statist.*, 5(1):26–56, 1996.

291. Y. Meyer. Principe d'incertitude, bases Hilbertiennes et algèbres d'opérateurs. *Séminaire Bourbaki*, 662, 1985–86.

292. Y. Meyer. *Ondelettes et Opérateurs II. Opérateurs de Caldéron-Zygmund*. Hermann, 1990.

293. Y. Meyer, editor. *Wavelets and Applications*. Number 20 in Research notes in Applied Mathematics. Springer Verlag, New York, 1991.

294. Y. Meyer. *Wavelets and Operators*. Cambridge Studies in Advanced Mathematics 37. Cambridge University Press, New York, 1992.

295. Y. Meyer. *Wavelets: Algorithms and Applications*. SIAM, Philadelphia, 1993.

296. A. S. Monin and A. M. Yaglom. *Statistical Fluid Mechanics, Volume 1*. MIT Press, Boston, MA, 1971.

297. A. S. Monin and A. M. Yaglom. *Statistical Fluid Mechanics, Volume 2*. MIT Press, Boston, MA, 1975.

298. P. A. Morettin. From Fourier to wavelet analysis of time series. In A. Pratt, editor, *Proceedings in Computational Statistics*, Physica-Verlag, pages 111–122, New York, 1996.

299. P. A. Morettin. Wavelets in statistics. *Resenhas*, 3(2):211–272, 1997.

300. J. Morlet, G. Arens, E. Fourgeau, and D. Giard. Wave propagation and sampling theory. *Geophys.*, 47:203–236, 1982.

301. J. Morlet, G. Arens, I. Fourgeau, and D. Giard. Wave propagation and sampling theory. *Geophysics*, 47:203–236, 1982.

302. P. Moulin. Wavelet thresholding techniques for power spectrum estimation. *IEEE Trans. Signal Process.*, 42(11):3126–3136, 1994.

303. P. Moulin, J. A. O'Sullivan, and D. L. Snyder. A method of sieves for multiresolution spectrum estimation and radar imaging. *IEEE Trans. Inform. Theory*, 38(2):801–813, March 1992.

304. P. Müller and B. Vidakovic. Bayesian inference with wavelets: Density estimation. *J. Comput. Graph. Statist.*, 7(4):456–468, 1998.

305. P. Müller and B. Vidakovic. MCMC methods in wavelet shrinkage: Non-equally spaced regression, density and spectral density estimation. In P. Müller and B. Vidakovic, editors, *Bayesian Inference In Wavelet Based Models*, Lecture Notes in Statistics. Springer Verlag, New York, 1999.

306. E. A. Nadaraya. On estimating regression. *Theory Probab. Appl.*, 9:141–142, 1964.

307. E. A. Nadaraya. *Nonparametric Estimation of Probability Densities and Regression Curves*. Kluwer Academic Publishers Group, Norwell, MA, 1989.

308. G. P. Nason. Choice of the threshold parameter in wavelet function estimation. In A. Antoniadis and G. Oppenheim, editors, *Wavelets and Statistics*, volume 103 of *Lecture Notes in Statistics*, pages 261–280, New York, 1995. Springer-Verlag.

309. G. P. Nason. Wavelet shrinkage by cross-validation. *J. Roy. Statist. Soc. Ser. B*, 58:463–479, 1996.

310. G. P. Nason and B. W. Silverman. The discrete wavelet transform in S. *J. Comput. Graph. Statist.*, 3:163–191, 1994.

311. G. P. Nason and B. W. Silverman. The stationary wavelet transform and some statistical applications. In A. Antoniadis and G. Oppenheim, editors, *Wavelets and Statistics*, volume 103 of *Lecture Notes in Statistics*, pages 281–300, New York, 1995. Springer-Verlag.

312. G. P. Nason, R. von Sachs, and G. Kroisandt. Wavelet processes and adaptive estimation of the evolutionary wavelet spectrum. Technical Report 516, Department of Statistics, Stanford University, 1997.

313. M. H. Neumann. Spectral density estimation via nonlinear wavelet methods for stationary non-gaussian time series. *J. Time Ser. Anal.*, 17(6):601–633, 1996.

314. M. H. Neumann and V. Spokoiny. On the efficiency of wavelet estimators under arbitrary error distributions. *Math. Methods Statist.*, 1(4):137–166, 1995.

315. M. H. Neumann and R. von Sachs. Wavelet thresholding: Beyond the Gaussian i.i.d situation. In A. Antoniadis and G. Oppenheim, editors, *Wavelets and Statistics*, volume 103 of *Lecture Notes in Statistics*, pages 301–329. Springer-Verlag, New York, 1995.

316. M. H. Neumann and R. von Sachs. Wavelet thresholding in anisotropic function classes and application to adaptive estimation of evolutionary spectra. *Ann. Statist.*, 25(1):38–77, 1997.

317. R. D. Nowak. Shift-invariant wavelet-based statistical models and 1/f processes. In *Proceedings of the IEEE DSP Workshop, Bryce Canyon, UT*, 1998. submitted.

362 REFERENCES

318. R. D. Nowak. Multiscale hidden markov models for Bayesian image analysis. In P. Müller and B. Vidakovic, editors, *Bayesian Inference In Wavelet Based Models*, Lecture Notes in Statistics. Springer Verlag, New York, 1999.

319. R. D. Nowak and R. Baraniuk. Wavelet-domain filtering for photon imaging systems. In *Wavelet Applications in Signal and Image Processing V*, Proceedings of SPIE, vol. 3169, pages 55–66. SPIE, 1997. To appear in *IEEE Trans. Image Process.*

320. R. T. Ogden. Wavelets in Bayesian change-point analysis. Technical report, Department of Statistics, University of South Carolina, 1996.

321. R. T. Ogden. *Essential Wavelets for Statistical Applications and Data Analysis*. Birkhäuser, Boston, 1997.

322. R. T. Ogden and M. Hilton. Data analytic wavelet threshold selection in 2-D signal denoising. *IEEE Trans. Signal Process.*, 45(2):496–500, 1997.

323. R. T. Ogden and J. Lynch. Bayesian analysis of change-point models. In P. Müller and B. Vidakovic, editors, *Bayesian Inference In Wavelet Based Models*, Lecture Notes in Statistics. Springer-Verlag, New York, 1999.

324. R. T. Ogden and E. Parzen. Data dependent wavelet thresholding in nonparametric regression with change-point applications. *Comput. Statist. Data Anal.*, 22:53–70, 1996.

325. R. T. Ogden and Emanuel Parzen. Change-point approach to data analytic wavelet thresholding. *Statist. Comput.*, 6(2):93–99, 1996.

326. A. O'Hagan. *Kendall's Advanced Theory of Statistics: Bayesian Inference*. Wiley, New York, 1994.

327. A. O'Hagan and T. Leonard. Bayes estimation subject to uncertainty about parameter constraints. *Biometrika*, 63:201–202, 1976.

328. Y. C. Pati, R. Rezaifar, and P. S. Krishnaprasad. Orthogonal matching pursuit: Recursive function approximation with application to wavelet decomposition. In *Proc. 27th Asilomar Conf. Signals, Systems, and Comp.*, Pacific Grove, CA, November 1993.

329. P. Patil. Nonparametric hazard rate estimation. Technical Report SRR 018-94, Centre for Mathematics and its Applications, Australian National University, 1994.

330. S. Penev and L. Dechevsky. On non-negative wavelet-based density estimators. *Journal of Nonparametric Statistics*, 7:365–394, 1997.

331. M. Pensky. Estimation of a smooth density function using Meyer type wavelets. Submitted to *Statist. Probab. Lett.*, 1998.

332. M. Pensky. Nonparametric empirical Bayes via wavelets. In P. Müller and B. Vidakovic, editors, *Bayesian Inference In Wavelet Based Models*, Lecture Notes in Statistics. Springer-Verlag, New York, 1999.

333. M. Pensky and B. Vidakovic. Adaptive wavelet estimator for nonparametric density deconvolution. Technical Report 98-11, ISDS, Duke University, 1998.

334. M. Pensky and B. Vidakovic. On non-equally spaced wavelet regression. Technical Report 98-06, ISDS, Duke University, 1998.

335. D. B. Percival. On estimation of the wavelet variance. *Biometrika*, 82(3):619–631, 1995.

336. D. B. Percival and A. G. Bruce. Estimation of long memory processes with missing data. Technical Report 64, StatSci Division of MathSoft, Inc., 1997.

337. D. B. Percival and H. O. Mofjeld. Analysis of subtidal coastal sea level fluctuations using wavelets. *J. Amer. Statist. Assoc.*, 92(439):868–880, 1997.

338. D. B. Percival and A. T. Walden. *Wavelet Methods for Time Series Analysis*. Cambridge University Press, London, 1999.

339. J.-C. Pesquet, H. Krim, and H. Carfantan. Time invariant orthonormal wavelet representations. *IEEE Trans. Signal Process.*, Aug. 1996.

340. J.-C. Pesquet, H. Krim, D. Leporini, and E. Hamman. Bayesian approach to best basis selection. In *IEEE International Conference on Acoustics, Speech, and Signal Processing, 7-10 May, Atlanta, GA*, volume 5, pages 2634–2637, 1996.

341. J. Pickands. Maxima of stationary Gaussian process. *Z. Wahrscheinlichkeitstheorie verw. Geb.*, 7:190–223, 1967.

342. A. Pinheiro. *Orthonormal Bases and Statistical Applications; Multiresolution Analysis Function Estimation and Representation of Gaussian Random Fields with Discontinuities*. Ph.D. thesis, University of North Carolina, Chapel Hill, 1997.

343. A. Pinheiro and B. Vidakovic. Estimating the square root of a density via compactly supported wavelets. *Comput. Statist. Data Anal.*, 25:399–415, 1997.

344. D. Pollen. $SU_I(2, F[z, 1/z])$ for F a subfield of $\mathbb{C}$. *J. Amer. Math. Soc.*, 3:611–624, 1990.

345. D. Pollen. Daubechies' scaling function on $[0, 3]$. In C. K. Chui, editor, *Wavelets: A Tutorial in Theory and Applications*, pages 3–15. Academic Press, 1992.

346. M. B. Priestley. *Non-linear and Non-stationary Time Series Analysis*. Academic Press, New York, 1988.

347. M. B. Priestley. Wavelets and time-dependent spectral analysis. *J. Time Ser. Anal.*, 17(1):85–104, 1996.

348. D. L. Ragozin, A. G. Bruce, and H-Y. Gao. Non-smooth wavelets: Graphing functions unbounded on every interval. Preprint at StatSci from the SBIR II Project, 1995.

349. J. Ramanathan and O. Zeitouni. On the wavelet transform of fractional brownian motion. *IEEE Trans. Inform. Theory*, 37:1156–1158, 1991.

350. B. L. S. Prakasa Rao. Nonparametric estimation of the derivatives of a density by the method of wavelets. *Bull. Inform. Cybernet.*, 28:91–100, 1996.

351. B. L. S. Prakasa Rao. Estimation of integral of square of density by wavelets. *Pub. Inst. Stat. Paris*, XXXXI:29–47, 1997.

352. J. M. Restrepo, G. K. Leaf, and G. Schlossnagle. Periodized Daubechies wavelets. Technical Report P423, Math. and Comp. Sci. Div., Argonne National Lab., 1994.

353. C. Robert. *The Bayesian Choice*. Springer-Verlag, New York, 1994.

354. K. Roeder. Density estimation with confidence sets exemplified by superclusters and voids in the galaxies. *J. Amer. Statist. Assoc.*, 85:617–624, 1990.

355. H. Rubin and J. Chen. Some stochastic processes related to random density function. *J. Theoret. Probab.*, 2:227–237, 1988.

356. W. Rudemo. Empirical choice of histograms and kernel density estimators. *Scand. J. Statist.*, 9:65–78, 1982.

357. F. Ruggeri. Robust Bayesian and Bayesian decision theoretic wavelet shrinkage. In P. Müller and B. Vidakovic, editors, *Bayesian Inference In Wavelet Based Models*, Lecture Notes in Statistics. Springer Verlag, New York, 1999.

358. F. Ruggeri and B. Vidakovic. A Bayesian decision theoretic approach to wavelet thresholding. *Statistica Sinica*, 9(1):183–197, 1999.

359. N. Saito. *Local Feature Extraction and Its Applications Using a Library of Bases*. Ph.D. thesis, Department of Mathematics, Yale University, New Haven, CT, 1994.

360. N. Saito. Simultaneous noise suppression and signal compression using a library of orthonormal bases and the minimum description length criterion. In E. Foufoula-Georgiou and P. Kumar, editors, *Wavelets in Geophysics*, pages 299–324. Academic Press, San Diego, 1994.

361. N. Saito. Classification of geophysical acoustic waveforms using time-frequency atoms. In *Amer. Statist. Assoc. 1996 Proceedings of the Statistical Computing Section*, pages 322–327, 1996.

362. N. Saito. Least statistically-dependent basis and its application to image modeling. In A. F. Laine, M. A. Unser, and A. Aldroubi, editors, *Wavelet Applications in Signal and Image Processing VI*, volume 3458 of *Proceedings of the SPIE*, pages 24–37, 1998.

363. N. Saito and G. Beylkin. Multiresolution representations using the autocorrelation functions of compactly supported wavelets. *IEEE Trans. Signal Process.*, 41(12):3584–3590, 1993.

364. N. Saito and R. R. Coifman. Local discriminant bases and their applications. *J. Math. Imaging Vision*, 5(4):337–358, 1995.

365. N. Saito and R. R. Coifman. Improved local discriminant bases using empirical probability density estimation. In *Amer. Statist. Assoc. 1996 Proceedings of the Statistical Computing Section*, pages 312–321, 1996.

366. N. Saito and R. R. Coifman. Extraction of geological information from acoustic welllogging waveforms using time-frequency wavelets. *Geophysics*, 62(6):1921–1930, 1997.

367. S. Sardy, D. B. Percival, A. G. Bruce, H-Y. Gao, and W. Stuetzle. Wavelet denoising for unequally spaced data. Technical report, StatSci Division of MathSoft, Inc., 1997.

368. J. D. Scargle. Wavelet methods in astronomical time series analysis. In T. Subba Rao, M. B. Priestly, and O. Lessi, editors, *Applications of Time Series Analysis in Astronomy and Meteorology*, pages 226–248. Chapman & Hall, London, 1997.

369. M. J. Schauder. Einige Eihenschaft der Haarschen Orthogonalsystems. *Math. Zeit.*, 28:317–320, 1928.

370. I. Schick and H. Krim. Robust wavelet thresholding for noise suppression. In *IEEE International Conference on Acoustics, Speech and Signal Processing*, volume V, Munich, Germany, 1997. IEEE.

371. G. Schwartz. Estimating the dimension of a model. *Ann. Statist.*, 6:461–464, 1978.

372. D. Scott. *Multivariate Density Estimation. Theory, Practice, and Visualization*. Wiley, New York, 1992.

373. W-C. Shan and J-C. Yan. Quadratures involving polynomials and Daubechies' wavelets. Technical Report 9301, Department of Mathematics National Central University, Taiwan, 1994.

374. J. Shen and G. Strang. Asymptotic analysis of Daubechies polynomials. *Proc. Amer. Math. Soc.*, 124:3819–3833, 1996.

375. M. J. Shensa. Wedding the à trous and Mallat algorithms. *IEEE Trans. Signal Process.*, 40(10):2464–2482, 1992.

376. N. Siguira. On the orthogonal inverse expansion with an application to the moments of order statistics. *Osaka Math. J*, 14:253–263, 1962.

377. N. Siguira. The bivariate orthogonal inverse expansion and the moments of order statistics. *Osaka J. Math.*, 1:45–59, 1964.

378. B. W. Silverman. *Density Estimation for Statistics and Data Analysis*. Chapman & Hall, London, 1986.

379. E. P. Simoncelli. Bayesian denoising of visual images in the wavelet domain. In P. Müller and B. Vidakovic, editors, *Bayesian Inference In Wavelet Based Models*, Lecture Notes in Statistics. Springer-Verlag, New York, 1999.

380. E. P. Simoncelli and E. H. Adelson. Noise removal via Bayesian wavelet coring. In *Third International Conference on Image Processing*, Lausanne, Switzerland, September 1996.

381. R. L. Smith. Comment on *chaos, fractals and statistics*. *Statist. Sci.*, 7:109–113, 1992.

382. R. L. Smith. Optimal estimation of fractal dimension. In *Nonlinear Modeling and Forecasting*, pages 115–135, 1992.

383. V. G. Spokoiny. Adaptive hypothesis testing using wavelets. *Ann. Statist.*, 24(6):2477–2498, 1996.

384. L. A. Stefanski and R. J. Carroll. Deconvolution-based score tests in measurement error models. *Ann. Statist.*, 19:249–259, 1991.

385. C. M. Stein. Inadmissibility of the usual estimator for the mean of a multivariate normal distribution. In *Proc. Third Berkeley Symp. Math. Statist. Probab.*, number 1, pages 197–206. University of California Press, Berkeley, 1955.

386. C. M. Stein. Estimation of the mean of a multivariate normal distribution. *Ann. Statist.*, 9:1135–1151, 1981.

387. G. Strang and T. Nguyen. *Wavelets and Filter Banks*. Wellesley-Cambridge Press, Wellesley, MA, 1996.

388. G. Strang and V. Strela. Orthogonal multiwavelets with vanishing moments. *Optical Engineering*, 33(7):2104–2107, 1994.

389. W. E. Strawderman. The James-Stein estimator as an empirical Bayes estimator for an arbitrary location family. In *Bayesian Statistics 4. Proceedings of the Fourth Valencia International Meeting*, pages 821–824, 1992.

390. J. O. Strömberg. A modified Franklin system and higher order spline systems on $\mathbb{R}^n$ as unconditional bases for Hardy spaces. In Beckner et al., editor, *Conference on Harmonic Analysis in Honor of Antony Zygmund*, volume II, pages 475–494. University of Chicago Press, 1981.

391. W. Sweldens. *The Construction and Application of Wavelets in Numerical Analysis*. Ph.D. thesis, University of Leuven, Belgium, 1993.

392. W. Sweldens. The lifting scheme: A custom-design construction of biorthogonal wavelets. *Appl. Comput. Harmon. Anal.*, 3(2):186–200, 1996.

393. W. Sweldens and R. Piessens. Quadrature formulae and asymptotic error expansions for wavelet approximations of smooth functions. *SIAM J. Numer. Anal.*, 31:1240–1264, 1994.

394. M. Talagrand. Sharper bounds for empirical processes. *Ann. Probab.*, 22(1):28–76, 1994.

395. T. Tao. On the almost everywhere convergence of wavelet summation methods. *Appl. Comput. Harmon. Anal.*, 3:384–387, 1996.

396. T. Tao and B. Vidakovic. Almost everywhere convergence of general wavelet shrinkage estimators. Technical Report 98-31, ISDS, Duke University, 1998.

397. R. A. Tapia and J. R. Thompson. *Nonparametric Probability Density Estimation*. The Johns Hopkins University Press, Baltimore, 1978.

398. C. Taswell. WavBox 4: A software toolbox for wavelet transforms and adaptive wavelet packet decompositions. In A. Antoniadis and G. Oppenheim, editors, *Wavelets and Statistics*, Lecture Notes in Statistics, pages 361–376. Springer-Verlag, New York, 1995. Proceedings of the Villard de Lans Conference November 1994.

399. C. Taswell and K. C. McGill. Wavelet transform algorithms for finite-duration discrete-time signals. *ACM Transactions on Mathematical Software*, 20(3):398–412, September 1994.

400. A. H. Tewfik and M. Kim. Correlation structure of the discrete wavelet coefficients of fractional Brownian motion. *IEEE Trans. Inform. Theory*, 38(2):904–909, 1992.

401. A. H. Tewfik, D. Sinha, and P. Jorgensen. On the optimal choice of a wavelet for signal representation. *IEEE Trans. Inform. Theory*, 38(2):747–765, March 1992.

402. J. Tian. *The mathematical theory and applications of biorthogonal Coifman wavelet systems*. Ph.D. thesis, Rice University, Houston, 1996.

403. J. Tian and R. O. Wells, Jr. Vanishing moments and wavelet approximation. Technical Report CML TR95-01, Computational Mathematics Laboratory, Rice University, 1995.

404. R. J. Tibshirani. Regression shrinkage and selection via lasso. *J. Roy. Statist. Soc. Ser. B*, 58:267–288, 1996.

405. L. Tierney. Markov chains for exploring posterior distributions. *Ann. Statist.*, 22:1701–1728, 1994.

406. K. E. Timmermann and R. D. Nowak. Multiscale Bayesian estimation of Poisson intensities. In *Proc. Asilomar Conf. Signals, Systems, and Comp.*, Pacific Grove, CA, pages 85–90. IEEE Computer Society Press, 1997.

407. A. Townsend. *The Structure of Turbulent Shear Flow*. Cambridge University Press, Cambridge, 1976.

408. K. Tribouley. Adaptive density estimation. In A. Antoniadis and G. Oppenheim, editors, *Wavelets and Statistics*, volume 103 of *Lecture Notes in Statistics*, pages 385–395. Springer-Verlag, New York, 1995.

409. K. Tribouley. Practical estimation of multivariate densities using wavelet methods. *Statistica Neerlandica*, 49(1):41–62, 1995.

410. H. Triebel. *Theory of Function Spaces*. Birkhäuser-Verlag, Basel, 1983.

411. H. Triebel. *Theory of Function Spaces, II*. Birkhäuser-Verlag, Basel, 1992.

412. M. Unser, A. Aldroubi, and M. Eden. Polynomial spline approximations: Filter design and asymptotic equivalence with Shannon's sampling theorem. *IEEE Trans. Inform. Theory*, 38:95–103, 1991.

413. M. Unser, A. Aldroubi, and M. Eden. A family of polynomial spline wavelet transforms. *Signal Processing*, 30:141–162, 1993.

414. M. Vannucci. Nonparametric density estimation using wavelets. Technical Report 95-26, ISDS, Duke University, 1995.

415. M. Vannucci and F. Corradi. Some findings on the covariance structure of wavelet coefficients: Theory and models in a Bayesian perspective. Technical Report UKC/IMS 95-05, Institute of Maths and Stats, University of Kent at Canterbury, 1997.

416. M. Vannucci and B. Vidakovic. Preventing the Dirac disaster: Wavelet based density estimation. Technical Report 95-27, ISDS, Duke University, 1995.

417. M. Vetterli and C. Herley. Wavelets and filter banks: Relationships and new results. In *IEEE Proc. Int. Conf. Acoust., Speech, Signal Processing*, Albuquerque, NM, April 1990.

418. M. Vetterli and C. Herley. Wavelets and filter banks: Theory and design. *IEEE Trans. Sig. Proc.*, 40:2207–2232, 1992.

419. M. Vetterli and J. Kovačević. *Wavelets and Subband Coding*. Signal Processing Series. Prentice Hall, Englewood Cliffs, NJ, 1995.

420. B. Vidakovic. Nonlinear wavelet shrinkage with Bayes rules and Bayes factors. Technical Report 94-24, ISDS, Duke University, 1994.

421. B. Vidakovic. A note on random densities via wavelets. *Statist. Probab. Lett.*, 26:315–321, 1996.

422. B. Vidakovic. Nonlinear wavelet shrinkage with Bayes rules and Bayes factors. *J. Amer. Statist. Assoc.*, 93(441):173–179, 1998.

423. B. Vidakovic and C. Bielza. On time-dependent wavelet denoising. *IEEE Trans. Signal Process.*, 46(9):2549–2555, 1998.

424. B. Vidakovic and A. DasGupta. Efficiency of linear rules for estimating a bounded normal mean. *Sankhyā A*, 58:81–100, 1996.

425. B. Vidakovic and G. G. Katul. The filtering of ozone concentration measurements in a turbulent air stream using Bayesian models in the wavelet domain. Technical Report 98–30, ISDS, Duke University, 1998.

426. B. Vidakovic and F. Ruggeri. Expansion estimation by Bayes rules. Technical Report 97-04, ISDS, Duke University, 1997.

427. L. F. Villemoes. Energy moments in time and frequency for two-scale difference equation solutions and wavelets. *SIAM J. Math. Anal.*, 23(6):1119–1543, 1992.

428. H. Volkmer. On the regularity of wavelets. *IEEE Trans. Inform. Theory*, 38:872–876, 1992.

429. R. von Sachs. Modelling and estimation of the time-varying structure of nonstationary time series. Technical report, Department of Statistics, Stanford University, 1996.

430. R. von Sachs, G. P. Nason, and G. Kroisandt. Spectral representation and estimation for locally stationary wavelet processes. FB Mathematik, Universität Kaiserslautern, D-67653 Kaiserslautern, Germany, 1996.

431. R. von Sachs, G. P. Nason, and G. Kroisandt. Adaptive estimation of the evolutionary wavelet spectrum. Technical Report 516, Department of Statistics, Stanford University, 1997.

432. G. Wahba. Automatic smoothing of the log periodogram. *J. Amer. Statist. Assoc.*, 75:122–132, 1980.

433. G. Wahba. Data based optimal smoothing or orthogonal series density estimates. *Ann. Statist.*, 9:146–156, 1981.

434. G. Wahba. Cross-validated spline methods for estimation of multivariate functions from data on functionals. In H. A. David and H. T. David, editors, *Statistics: An Appraisal*, pages 205–235. The Iowa State University Press, 1984.

435. G. Wahba. *Spline Models for Observational Data*. SIAM, Philadelphia, 1990.

436. A. T. Walden and A. C. Cristan. Matching pursuit by undecimated discrete wavelet transform for arbitrary-length time series. Technical Report TR-96-02, Imperial College of Science, Technology and Medicine, Statistics Section, 1996.

437. G. G. Walter. Approximation of the delta function by wavelets. Preprint at Department of Mathematical Sciences, University of Wisconsin-Milwaukee, 1990.

438. G. G. Walter. Approximating of the delta function by wavelets. *J. Approx. Theory*, 71:329–343, 1992.

439. G. G. Walter. *Wavelets and Other Orthogonal Systems with Applications*. CRC Press Inc., Boca Raton, FL, 1994.

440. G. G. Walter. Density estimation in the presence of noise. Preprint at Department of Mathematical Science, University of Wisconsin-Milwaukee, 1998.

441. G. G. Walter and J. Blum. Probability density estimation using delta sequences. *Ann. Statist.*, 7:328–340, 1979.

442. G. G. Walter and L. Cai. Periodic wavelets from scratch. Preprint at Department of Mathematical Science, University of Wisconsin-Milwaukee, 1998.

443. G. G. Walter and J. Ghorai. Advantages and disadvantages of density estimation with wavelets. In *Computing Science and Statistics. Proceedings of the 24rd Symposium on the Interface*, pages 234–243, 1992.

444. G. G. Walter and X. Shen. Continuous non-negative wavelets and their use in density estimation. International Workshop on Wavelets and Statistics, Duke University, 12-13 October, 1997.

445. G. G. Walter and X. Shen. Deconvolution using meyer wavelets. Preprint at Department of Mathematical Science, University of Wisconsin-Milwaukee, 1997.

446. G. G. Walter and A. I. Zayed. Characterization of analytic functions in terms of their wavelet coefficients. *Complex Variables*, 29:265–276, 1996.

447. Y. Wang. Jump and sharp cusp detection by wavelets. *Biometrika*, 82(2):385–397, 1995.

448. Y. Wang. Function estimation via wavelet shrinkage for long-memory data. *Ann. Statist.*, 24(2):466–484, 1996.

449. Y. Wang. Fractal function estimation via wavelet shrinkage. *J. Roy. Statist. Soc. Ser. B*, 59:603–613, 1997.

450. Y. Wang. Minimax estimation via wavelets for indirect long-memory data. *J. Statist. Plann. Inference*, 64:45–55, 1997.

451. Y. Wang. Change curve estimation via wavelets. *J. Amer. Statist. Assoc.*, 93:163–172, 1998.

452. Y. Wang. Change-points via wavelets for indirect data. *Statistica Sinica*, 9(1):103–117, 1999.

453. G. S. Watson. Smooth regression analysis. *Sankhyā, Ser A.*, 26:359–372, 1964.

454. P. D. Welch. The use of fast Fourier transform for the estimation of power spectra: a method based on time averaging over short modified periodograms. *IEEE Trans. on Audio*, AU-15:70–73, 1967.

455. M. West and J. Harrison. *Bayesian Forecasting and Dynamic Models*. Springer-Verlag, New York, 1989.

456. N. Weyrich and G. T. Warhola. De-noising using wavelets and cross validation. In S. P. Singh, editor, *Approximation Theory, Wavelets and Applications*, volume 454 of *NATO ASI Series C*, pages 523–532, 1995.

457. M. V. Wickerhauser. *Adapted Wavelet Analysis from Theory to Software*. A K Peters, Ltd., Wellesley, MA, 1994.

458. M.V. Wickerhauser, M. Farge, E. Goirand, E. Wesfreid, and E. Cubillo. Efficiency comparison of wavelet packet and adapted local cosine bases for compression of a two-dimensional turbulent flow. In C. K. Chui, L. Montefusco, and L. Puccio, editors, *Wavelets: Theory, Algorithms, and Applications*, pages 509–531, San Diego, 1994. Academic Press.

459. P. Wojtaszczyk. *A Mathematical Introduction to Wavelets*. Mathematical Society Student Texts 37. Cambridge University Press, London, 1997.

460. G. W. Wornell. A Karhunen-Loéve-like expansion for 1/f processes via wavelets. *IEEE Trans. Inform. Theory*, 36(4):859–861, 1990.

461. G. W. Wornell. *Signal Processing with Fractals: A Wavelet Based Approach*. Prentice Hall, Englewood Cliffs, NJ, 1996.

462. G. W. Wornell and A. V. Oppenheim. Estimation of fractal signals from noisy measurements using wavelets. *IEEE Trans. Acoust. Signal Speech Process.*, 40(3):611–623, March 1992.

463. D. Wu. *Probability Density Estimation with Wavelets*. Ph.D. thesis, University of Wisconsin-Milwaukee, 1994.

464. D. Wu. Asymptotic normality of the multiscale wavelet density estimator. *Comm. Statist. Theory Methods*, 25(9), 1996.

465. Y. Wu. Wavelet estimation for nonparametric regression: Beyond Gaussian noise, I. Technical report, Department of Statistics, Purdue University, 1998.

466. Y. Wu. Wavelet estimation for nonparametric regression: Beyond Gaussian noise, II. Technical report, Department of Statistics, Purdue University, 1998.

467. X-G. Xia, J. S. Geronimo, D. P. Hardin, and B. W. Suter. Design of prefilters for discrete multiwavelet transforms. *IEEE Trans. Signal Process.*, 44(1):25–35, 1996.

468. X-G. Xia and Z. Zhang. On sampling theorem, wavelets and wavelet transforms. *IEEE Trans. Sig. Proc.*, 41:3524–3535, 1993.

469. M. Yamada and K. Ohkitani. An identification of energy cascade in turbulence by orthonormal wavelet analysis. *Prog. Theor. Phys.*, 86:799–815, 1991.

470. M. Yamada and K. Ohkitani. Orthonormal wavelet analysis of turbulence. *Fluid Dynam. Res.*, 8:101–115, 1991.
471. J. Zak. Finite translations in solid state physics. *Phys. Rev. Lett.*, 19:1385–1397, 1967.
472. F. Zeppenfeldt, J. Börger, and A. Koppes. Optimal thresholding in wavelet image compression. In M. Unser and A. F. Laine, editors, *Mathematical Imaging. Wavelet Applications in Signal and Image Processing, vol 2034*, pages 230–241. SPIE, 1993.
473. J. Zhang and G. G. Walter. A wavelet-based KL-like expansion for wide-sense stationary random processes. *IEEE Trans. Sig. Proc.*, 42:1737–1745, 1994.

Notation Index

$\mathbf{1}(\cdot)$, 3, 23
$\vee$, 23
$\wedge$, 23
$\langle \cdot, \cdot \rangle$, 24
$\perp$, 26
$[\downarrow 2]$, 109
$[\uparrow 2]$, 109
$\prec$, 182
1/f, 290

$\alpha_{j_0,k}$, 225

$\mathbb{B}_{pq}^{\sigma}(\cdot)$, 37, 89

CV, 173
C_ψ, 44
$\mathbb{C}$, 23
$\mathbb{C}^n(\cdot)$, 37
$\mathbb{C}^s(\cdot)$, 37, 88
$\mathcal{C}(\cdot)$, 140
$\mathcal{CWT}_f(a,b)$, 44, 45
c_{jk}, 87

DP, 178
$\mathcal{D}_0$, 145
$\mathcal{D}_1$, 145
$\mathcal{DE}$, 248

d_{jk}, 87
$\Delta_f, \Delta_{\hat{f}}$, 36
$\delta_{u,v}, \delta_u$, 23
$\delta(\cdot)$, 24
$\delta^g(\cdot, \lambda)$, 193
$\delta^h(\cdot, \lambda)$, 176
$\delta^{nng}(\cdot, \lambda)$, 193
$\delta^s(\cdot, \lambda)$, 176
$\delta^{ss}(\cdot, \lambda_1, \lambda_2)$, 191

$\mathcal{E}$, 248
$\mathcal{EPD}$, 194
ess sup, 25

FIR, 39
$\mathcal{F}(\cdot), \hat{f}$, 29
$\mathcal{F}^{-1}(\cdot)$, 29
f_+, 23
f_-, 23
$f \star g$, 31
$\widehat{f}_{n,m}(\cdot)$, 220

GCV, 201
$\mathbf{G}$, 109
$\mathbf{G}^{[r]}$, 147
$\mathcal{G} \equiv [\downarrow 2]\,\mathbf{G}$, 110
$\mathcal{G}^\star$, 111
g_k, g_n, 58

371

$\tilde{g}_k$, 128
$g^{[r]}$, 147

$\mathbf{H}^{[r]}$, 147
$\mathbf{H}$, 109
$\mathcal{H} \equiv [\downarrow 2]\mathbf{H}$, 110
$\mathcal{H}$, 25
$\mathcal{H}^\star$, 111
h_k, h_n, 52
$\tilde{h}_k$, 128
$h^{[r]}$, 147

IIR, 39
iff, 25

$\mathbb{K}(\cdot,\cdot)$, 29, 171

$\mathbb{L}_2(\cdot)$, 25
$\mathbb{L}_p(\cdot)$, 25
$\mathbb{L}_\infty(\cdot)$, 30
$\mathcal{L}(\cdot)$, 73
$\ell_2(\cdot)$, 25
$\ell(2\mathbb{Z})$, 109

MAD, 197
MRA, 51
$\mathcal{M}_k$, 80
$m(\cdot)$, 172
$\tilde{m}(\cdot)$, 173
$m_0(\cdot)$, 52
$\tilde{m}_0(\cdot)$, 127
$m_1(\cdot)$, 58
$\tilde{m}_1(\cdot)$, 128
$\hat{m}(\cdot)$, 172

NES, 209
$\mathbb{N}$, 23
$\mathcal{NIG}$, 260
$\mathcal{N}_k$, 80

$O(\cdot)$, 23
$o(\cdot)$, 23

$\text{Proj}_\mathcal{M}\cdot$, 27
$\mathbf{P}_j f$, 87
$\Phi(\cdot)$, 53
$\tilde{\Phi}(\cdot)$, 127
$\phi(\cdot)$, 52
$\phi_{jk}(\cdot)$, 52
$\phi_{jk}^{per}(\cdot)$, 151

$\tilde{\phi}$, 127
$\Psi(\cdot)$, 58
$\tilde{\Psi}(\cdot)$, 128
$\psi(\cdot)$, 57
$\tilde{\psi}$, 128
$\psi_{a,b}(\cdot)$, 44
$\psi_{jk}(\cdot)$, 50, 57
$\psi_{jk}^{per}(\cdot)$, 151

$\mathbb{Q}(\cdot,\cdot)$, 171

$R_n(\cdot,\cdot)$, 185
$\mathbb{R}$, 23
$\mathbb{R}^n$, 24
$\mathbb{R}^+$, 23
$\mathcal{R}^\star$, 111

SNR, 196
$\mathcal{S}^k$, 145
sinc, 32, 64
$\overline{span}\{\cdot\}$, 26
$\text{supp}(f)$, 23

θ_{jk}, 170, 225
$\Theta_{pq}^\sigma(\cdot)$, 186

$\mathfrak{U}_m$, 321, 342

V_0, 51
V_j, 51
$\tilde{V}_j$, 125
V_j^{per}, 151
$\mathbf{V}_j$, 154
$\mathfrak{V}_m$, 321

W_j, 57
$\tilde{W}_j$, 128
W_j^{per}, 151
$\mathbf{W}_j^{(h)}, \mathbf{W}_j^{(v)}, \mathbf{W}_j^{(d)}$, 155
$\mathbb{W}_2^s(\cdot)$, 37, 88
$\mathbb{W}_p^s(\cdot)$, 37
$\mathcal{W}_{j,n,k}(\cdot)$, 135
$\mathcal{W}_n(\cdot)$, 133
$\widehat{\mathcal{W}}_n(\cdot)$, 138
$w_{j,n,k}$, 138
$w_{r,p}(f;\cdot)$, 38
$\mathbf{w}_{j,n}$, 138

$\mathbb{Z}$, 23

Author Index

Abramovich, F., 202, 256, 257, 324
Abry, P., 290, 293, 295, 296
Adelson, E. H., 104, 194, 265
Akaike, H., 255
Albertson, J., 338
Aldroubi, A., 130
Amato, U., 175
Anscombe, F. J., 239
Antoniadis, A., 120, 170, 172, 173, 175, 207, 209, 217, 224, 225, 239
Averkamp, R., 179, 273
Azzalini, A., 314

Baraniuk, R., 266, 267
Basseville, M., 296, 308
Battle, G., 71
Benassi, A., 292
Benjamini, Y., 202
Beran, J., 292
Berger, J. O., 168
Berliner, M., 338
Besov, O. V., 39
Beylkin, G., 93, 162, 231
Bickel, P., 176
Bielza, C., 266
Bock, M. E., 94, 239
Breiman, L., 193

Brillinger, D. R., 189, 288
Brockwell, P. J., 273, 280
Brown, L., 209
Bruce, A. G., 115, 130, 188, 189, 191, 208, 338
Brunet, Y., 338
Brunk, H., 259
Buccigrossi, R. W., 194
Bultheel, A., 202, 209, 266
Burman, P., 200
Burrus, C. S., 86, 163
Burt, P. J., 104
Börger, J., 194

Cai, L., 153
Cai, T., 206, 207, 209
Cambanis, S., 44, 273
Carfantan, H., 145
Carlin, B., 263
Carmona, R., 211, 212, 217, 225, 290, 296, 341
Carrol, R. J., 319
Casella, G., 168
Chambolle, A., 175, 259
Chen, J., 229, 308
Chen, S. S., 325
Cheng, M-Y., 239

373

Chiann, C., 286, 287
Chib, S., 263
Chipman, H. A., 250, 267
Chui, C. K., 129, 130
Clutton-Brock, M., 229
Clyde, M. A., 251, 253
Cohen, A., 112, 122, 128, 150, 290
Coifman, R. R., 17, 133, 138, 145, 207, 208, 325
Collineau, S., 338
Corradi, F., 262
Cressie, N., 267
Cristan, C. A., 329
Cristobal, G., 212
Crouse, M., 266
Čencov, N. N., 17, 217, 218, 308

Dahlhaus, R., 286
Dahmen, W., 300
DasGupta, A., 248, 314
Daubechies, I., 6, 50, 73, 76, 80, 83, 85, 89, 112, 120–122, 128, 129, 150, 152, 163, 301, 325
David, H. A., 331
Davis, G., 329
Davis, R. A., 273, 280
Dechevsky, L., 225, 229, 232
Delyon, B., 209, 217, 219, 226, 234
Denison, D. G. T., 253
DeSimone, H., 253
Deslauriers, G., 209
DeVore, R. A., 39, 175
Diggle, P. J., 319
Dijkerman, R. W., 296
Donoho, D. L., 17, 142, 145, 150, 168, 170, 175, 178, 185–188, 195, 197, 199–201, 207, 208, 217, 219, 224, 226, 234, 239, 255, 280, 284, 286, 324, 325
Doukhan, P., 217
Dubuc, S., 209
Dutilleux, P., 145

Eden, M., 130
Efroimovich, S., 207
Efron, B., 168, 176

Fan, J. Q., 174, 175, 191, 197, 213, 230, 239, 319, 320
Farge, M., 338

Feauveau, J., 122, 128
Fix, G., 83
Flandrin, P., 286, 290, 292, 293, 295, 296
Foster, D., 209, 255
Foufoula-Georgiou, E., 14
Fourier, J., 1
Franklin, P., 71
Friedman, J., 325
Friehe, C. A., 336
Frisch, U., 14

Gabor, D., 9
Gail, M., 204
Gao, H-Y., 115, 130, 179, 188, 189, 191, 193, 208, 209, 280, 284, 338
Gaskins, R. A., 229
Gasser, T., 172, 173, 242
Gastwirth, J., 204
Genon-Catalot, V., 296
George, E. I., 250, 251, 253, 255
Geronimo, J. S., 131
Geweke, J., 292, 295
Ghang, S. G., 213
Ghorai, J., 217
Gijbels, I., 174, 207, 230
Gonçalvès, P., 290, 293, 295
Good, I. J., 229
Gopinath, R. A., 86, 163
Grossmann, A., 6, 43
Grégoire, G., 170, 172, 173, 207, 209, 224, 239
Guo, H., 86, 163

Haar, A., 2
Hall, P., 175, 200, 207, 209, 227, 319
Hardin, D. P., 131
Harrison, J., 273
Hastie, T. J., 254
He, W, 153
Heil, C. E., 50
Helson, H., 30
Herley, C., 122
Hernández, E., 39, 58
Hilton, M., 203
Hjort, N. L., 239
Hochberg, Y., 202
Holmes, C. C., 253, 255
Holschneider, M., 28, 47, 50
Houdré, C., 44, 179, 273

Huang, H-C., 267
Huang, S-Y., 170, 223, 224, 267
Hudgins, L., 212, 336
Huerta, G., 262
Hwang, W-L., 211, 212, 296, 341
Hyvärinen, A., 258
Härdle, W., 173, 219, 227

Izenman, A. J., 217

Jaffard, S. L., 47, 89, 150, 153
Jain, A. K., 177, 277
Jansen, M., 202, 209, 266
Janssen, A. J. E. M., 299
Johnstone, I. M., 14, 142, 150, 168, 170, 175, 178, 179, 186–188, 195, 197, 199, 200, 207, 208, 253, 255, 284, 324
Jones, L. K., 327
Jones, M. C., 239
Jorgensen, P., 96
Joshi, P. C., 331, 332
Juditsky, A., 209, 217, 219, 226, 234

Kaplan, L. M., 292
Kass, R. E., 255
Katul, G. G., 336, 338
Katznelson, Y., 30
Kelly, S., 171
Kerkyacharian, G., 150, 217, 219, 224, 228
Kim, M., 296
Klonias, V. K., 229
Koch, I., 200
Kohn, R., 267
Kolaczyk, E. D., 201, 250, 267, 324
Kolmogorov, A. N., 14, 335
Kon, M., 171
Koppes, A., 194
Kovac, A., 179
Kovačević, J., 39, 50, 153, 163, 277
Krim, H., 145, 207, 208, 259, 266, 286
Krishnaprasad, P. S., 329
Kroisandt, G., 286
Kronmal, R., 219
Kumar, P., 14
Kuo, C.-C. J., 292
Körner, T. H., 184

La Borde, B., 99

Lagarias, J., 83, 89
Lai, M-J., 153
Lang, M., 208
Laredo, C., 296
Laurencot, P., 89, 150
Lawton, W., 301, 304
Leaf, G. K., 150
Lee, N., 324
Lemarié, P. G., 71
Lenarduzzi, L., 209
Leonard, T., 229, 314
Leporini, D., 259, 266
Li, Y., 296
Lina, J. M., 265
Littlewood, J. E., 5
Loader, C. R., 239
Lorentz, M. O., 180
Louis, A. K., 50
Lu, H-S., 267
Lucier, B. J., 175
Lumeau, B., 280
Lynch, J., 267
Léon, J. R., 217

Ma, Y., 300, 305
Maatta, J. M., 168
Maaß, P., 50
MacGibbon, B., 265
Malfait, M., 202, 266
Mallat, S. G., 6, 14, 17, 51, 104, 145, 153, 194, 211, 286, 290, 325, 329
Mallick, B., 255
Marr, D., 211
Marron, S. J., 188, 189, 231, 240, 267
Marshall, A., 182
Masry, E., 44, 225, 273, 275, 319
Massopust, P. R., 131
Mayer, M. E., 336
Mazumdar, R. R., 296
McCoy, E. J., 290, 295
McCulloch, R. E., 250, 251, 267
McGill, K. C., 115
McKeague, I., 170, 172, 173, 209, 224
Meyer, Y., 6, 38, 39, 50, 65, 88, 133, 138, 153, 187
Micchelli, C. A., 300
Milliff, R., 338
Mofjeld, H. O., 143
Monin, A. S., 333, 338

Morettin, P., 286, 287, 290
Morlet, J., 6, 43
Morris, C., 168, 176
Moulin, P., 280
Müller, H-G., 172, 173, 242
Müller, P., 262, 264

Nadaraya, E. A., 173
Nason, G. P., 145, 194, 197, 200, 201, 239, 286, 341
Neumann, M. H., 208, 284, 286, 288
Nguyen, T., 39, 78, 109, 133, 304, 305
Nowak, R. D., 266, 267, 296

O'Hagan, A., 260, 314
O'Sullivan, J. A., 280
Ogden, R. T., 200, 203, 267
Ohkitani, K., 336, 338
Olkin, I., 182
Oppenheim, A. V., 292

Paley, R., 5
Paralange, M. B., 336
Parmigiani, G., 251, 253
Parzen, E., 203
Pati, Y. C., 329
Patil, P., 175, 209, 227
Penev, S., 225, 229, 232
Pensky, M., 207, 209, 227, 267, 320, 323
Percival, D. B., 143, 290, 296
Pesquet, J.-C., 145, 207, 259, 266, 286
Picard, D., 150, 217, 219, 224, 228, 296
Pickands, J., 196
Piessens, R., 94
Pinheiro, A., 229–231
Pliego, G., 94, 239
Pollen, D., 94
Popov, V., 39
Porter-Hudak, S., 292, 295
Prakasa Rao, B. L. S., 241, 242
Priestley, M. B., 273, 290

Raftery, A. E., 255
Ragozin, D. L., 130
Raphael, L., 171
Reider, A., 50
Restrepo, J. M., 150
Rezaifar, R., 329
Rios, D., 308
Robert, C., 168, 314

Roeder, K., 261, 264
Roose, D., 266
Rubin, H., 229, 248, 308
Rudemo, W., 240
Ruggeri, F., 267
Ryan, R. D., 290

Saito, N., 93, 162, 207
Sapatinas, T., 256, 257
Sardy, S., 209
Scargle, J. D., 289
Schauder, M. J., 3
Schick, I. C., 208, 259
Schlossnagle, G., 150
Schroeder, A., 325
Schwartz, G., 255
Scott, D., 217
Shann, W-C., 300
Shen, J., 78
Shen, X., 222, 229, 235, 236, 300, 323
Shensa, M. J., 145
Siguira, N., 329, 332
Silverman, B. W., 14, 145, 179, 207, 208, 217, 253, 256, 257, 324, 341
Simoncelli, E. P., 194, 259, 265
Sinha, D., 96
Smith, R., 292, 295
Snyder, D. L., 280
Spokoiny, V., 208, 213
Stefansky, L. A., 319
Stein, C., 167, 199
Strang, G., 39, 78, 83, 109, 133, 304, 305
Strawderman, W. E., 168
Strela, V., 133
Strömberg, J. O., 5
Stuetzle, W., 208, 325
Suter, B. W., 133
Sweldens, W., 94, 130, 160

Talagrand, M., 228
Tao, T., 184
Tapia, R. A., 229
Tarter, M., 219
Taswell, C., 115, 340
Tchamitchian, P., 47
Tewfik, A. H., 96, 296
Thompson, J. R., 229
Tian, J., 130
Tibshirani, R. J., 177, 254

Tierney, L., 264
Timmermann, K. E., 267
Torrésani, B., 211, 212, 296, 341
Townsend, A., 336, 337
Tribouley, K., 228, 237, 240
Triebel, H., 39, 187
Tung, Y-C., 267
Turlach, B. A., 209

Unser, M., 130

Vannucci, M., 231, 238, 262
Veitch, D., 296
Vetterli, M., 39, 50, 122, 153, 163, 213, 277
Vial, P., 112, 150, 209
Vidakovic, B., 184, 209, 229–231, 251, 262, 264, 266, 267, 305, 308, 314, 320, 323, 336, 338
Villemoes, L., 41, 300, 305
Volkmer, H., 83
von Sachs, R., 208, 286, 288
Vuza, D. T., 175

Wahba, G., 175, 200, 201, 219, 260, 281
Walden, A. T., 290, 295, 296, 329
Walnut, D. F., 50
Walter, G. G., 14, 18, 65, 69, 70, 94, 125, 153, 170, 217, 219, 221, 222, 229, 235, 236, 275, 277, 279, 280, 300, 323
Wand, M. P., 231

Wang, A., 290
Wang, J. Z., 130
Wang, Y., 201, 212, 324
Warhola, G. T., 201
Watson, G. S., 173
Weiss, G., 39, 58
Welch, P. D., 293
Wells, Jr., R. O., 130
West, M., 273
Weyrich, N., 201
Wickerhauser, M. V., 17, 110, 115, 133, 138, 325, 336
Wikle, C., 338
Wojtaszczyk, P., 39
Wornell, G. W., 14, 277, 290, 292
Wu, D., 240, 241
Wu, Y., 209

Xia, X-G., 93, 133
Xie, Z., 296

Yaglom, A. M., 333, 338
Yamada, M., 336, 338
Yan, J-C., 300
Yu, B., 213
Yu, T. P-Y., 208

Zayed, A. I., 70
Zeppenfeldt, F., 194
Zhang, J., 277, 279
Zhang, Z., 93, 325
Zhong, S., 211

Subject Index

adaptive estimator, 322
admissibility condition, 44
aliasing error, 33, 93
autocorrelation shell, 93, 162
autocovariance function, 272

bandlimited function, 32
basis
    best, 140
    biorthogonal, 124
    Daubechies, 6
    Gabor, 9
    Haar, 60, 133
    Hermite, 9
    Legendre, 9
    orthonormal, 27
    Riesz, 28, 52, 278
    Schauder, 3
    Shannon, 63
    unconditional, 28
    Walsh, 133
basis pursuit, 325
Bayesian paradigm, 247
Besov body, 186
Besov seminorm, 186
Bessel inequality, 27
best orthogonal basis, 325
Bezout's lemma, 76
block thresholding, 322
BlockJS estimator, 207
Bonferroni procedure, 202

California earthquakes, 19
cascade algorithm, 104, 152
Cauchy sequence, 25
change point, 212
change point for indirect data, 324
closed span, 26
"comb"-function, 225
convolution, 31
cost
    ℓ_2, 142
    entropy, 142
    Sure, 142
    threshold, 142
cost measures, 141
critical sampling, 50
cummulants, 288
cycle-spin method, 208
cyclostationary process, 297

data-set
    `armadillo`, 135
    `Duke Forest turbulence`, 333

380 SUBJECT INDEX

earthquake, 19
galaxy, 231, 264
motorcycle accident, 174
Old Faithful geyser, 238
sea-level, 143
Wolf's sunspot numbers, 282
Daubechies-Lagarias algorithm, 89
decimation, 109
decimation operator, 146
decomposition algorithm, 115
deconvolution, 319
deconwavelet, 321
δ-sequence, 219
diagonal projection estimator, 178
dictionaries of functions, 325
dilation, 109
Dirac function, 24
Dirichlet kernel, 220
downsampling, *see* dcimation, 109

eddy motion, 336
energy, 10
energy conservation, 46
energy spectrum, 31
exact risk analysis, 188
exponential power distribution, 194, 259
extremal phase, 86

filter, 39
    à trous, 55, 93
    band, 39
    casual, 39
    coiflet (COIF), 121
    Daubechies (DAUB), 80
    finite impulse response (FIR), 39
    halfband, 305
    high-pass, 40, 59
    infinite impulse response (IIR), 39, 65
    low-pass, 40, 52
    quadrature mirror, 59
    symmlet (SYMM), 86
    taps, 39
fixed design, 172
flow
    laminar, 333
    turbulent, 333
Fourier series, 33
frame, 28, 41, 50
    tight, 28, 41
Franklin system, 5

Γ-minimax, 314
Gasser-Müller kernel estimator, 172
generalized cross-validation function, 202
generalized moments, 300
Gibbs effect, 34, 41
global block thresholding, 227

Hausdorff dimension, 291
Heaviside function, 24, 218
Heisenberg's uncertainty principle, 7, 35
Hurst exponent, 291, 296
hyperbole rule, 193

incomplete Beta function, 331
indicator, 3

jig-saw effect, 97

K41 theory, 335
Karhunen-Loève expansions, 276
klipping, 177
Kronecker symbol, 23

lasso method, 177
Lawton's matrix, 301
Lebesgue point, 24, 184
level-wise energies, 10
lifting scheme, 160
Littlewood-Paley technique, 5
local polynomial models, 174
localization, 47
log-periodogram, 281
Lorentz curve, 10, 180

magnitude spectrum, 30
Mallat's model, 194
MAP principle, 177, 257
matching pursuit, 325, 329
maximum modulus reconstruction, 211
median absolute deviation (MAD), 197
method of frames, 325
minimax risk, 187
minimum description length, 207
"$-\frac{5}{3}$ law", 336
moment conditions, 91
moments of order statistics, 329
mother wavelet, 44, 57

MRA, 51
multiresolution analysis, 51
  d-dimensional, 153
multiscale wavelet density estimator, 241

Nadaraya-Watson estimator, 173
Navier-Stokes equations, 335
nn-garrote rule, 191
non-linear density estimators, 225
non-linear estimator, 322
norm, 24
normalization property, 53, 92
NP-hard problem, 325
Nyquist rate, 33

O-notation, 23
Ockham's razor principle, 17
operator notation of DWT, 109
oracular risk, 177
orthogonal projection, 26
orthogonality property, 54, 92
orthonormal set, 27
oscillation index, 135

Parseval's identity, 27, 125, 229, 308
percentile thresholding, 194
periodogram, 280
phase spectrum, 30
Plancherel's identity, 31
Poisson summation formula, 32
Pollen-type parameterization, 94
power low, 292
pretest estimator, 176
process
  $1/f$, 290
  autoregressive, 272
  Brownian bridge, 277
  fractional Brownian motion (fBm), 290
  fractional Gaussian noise, 292
  long-memory, 290
  moving average, 272
  non Gaussian stationary, 284
  stationary, 272, 274
  white noise, 272
  Wiener, 277
projection estimator, 170, 218
pursuit methods, 325

Rademacher system, 164

random density, 308
random design, 172
reproducing kernel, 29, 47, 171
resolution of identity, 44
Reynolds number, 333, 335
Riesz lemma, 77

scaling equation, 52
scaling function, 51
scalogram, 232, 243, 289
Schur number, 182
Schur order, 181
self-similarity, 14, 291
semisoft shrinkage, 191
Shannon's entropy, 142
shift operator, 145
shrinkage estimation, 167
signal-to-noise ratio (SNR), 196, 329
Siguira's method for order statistics, 329
singular value decomposition, 323
smoother-cleaner, 208
SNR, 197
Sobolev norm, 221
software
  free, 340
  Megawave1, 341
  Rice Wavelet Tools, 341
  S+Wavelets, 104, 338
  S-Plus, 338
  Swave, 341
  WavBox, 340
  Wavelab, 340
  Wavelet Explorer, 338
  Wavethresh, 341
space
  $\mathbb{C}^n(\mathbb{R})$, 37
  $\mathbb{L}_2(\mathbb{R})$, 25
  ℓ_2, 25
  Besov, 37, 88, 169, 185
  Hardy, 5
  Hilbert, 24
    separable, 28
  Hölder, 88
  inner product, 24
  Lebesgue, 25
  reproducing kernel, 29
  Sobolev, 37, 88, 322
spectral density, 279
spectral exponent, 292

spectral factorization, 76
steerable pyramid, 266
Strang-Fix condition, 83
structure functions, 335
supersmooth density, 320
support, 23
SureShrink, 200

test function
    `blocks`, 149, 150
    `doppler`, 7, 197
    `heavisine`, 210
theorem
    Balian-Low, 9
    Poisson, 32
    projection, 26
    sampling, 32
theory
    Calderón-Zygmund, 5
threshold
    block, 206
    cross-validation, 200
    false discovery rate, 202
    Lorentz, 205
    optimal minimax, 197
    sigmoidal, 213
    SURE, 200
    "top n", 206
    universal, 195
Townsend's decomposition, 336
transformation
    à trous, 145
    continuous wavelet, 43
    discrete Fourier, 34
    discrete wavelet, 101
    ϵ-decimated, 145
    fast Fourier, 101, 104
    Fourier, 29
    Karhunen-Loève, 10, 125, 276
    non-decimated, *see* stationary
    stationary, 50, 145, 147
    Zak, 221
translation invariant shrinkage, 208
turbulence, 14, 333
    attached, 336
    detached, 336
    locally isotropic, 335
two-scale equation, 52

unitary matrix, 115

vanishing moments, 73, 80, 92
vertical-line rule, 140
viscosity, 335
VisuShrink method, 197
Wahba's model, 282
wavelet
    r-regular, 82, 88
    atoms, 138
    B-spline, 128
    bandlimited, 70
    Battle-Lemarié, 71
    biorthogonal, 124
    cardinal, 93
    coiflets, 120
    cross spectrum, 289
    crystals, 138
    Daubechies', 73
    digest, web resource, 341
    filter, 52
    Franklin, 70, 71
    Haar, 47, 60
    indexing, 52
    interpolating, 93
    least asymmetric (symmlet), 86
    Littlewood-Paley, 63
    Marr's, 49
    Mexican hat, 49
    Meyer, 65
    Meyer-type, 70, 277
    Morlet, 46
    packet table, 135
    packets, 133
    periodized, 150
    Poisson, 49
    regularity, 80
    second generation, 160
    semi-orthogonal, 130
    Shannon, 63
    shrinkage, 167
    spectrum, 14, 286
wavelet autocorrelation function, 287
wavelet-vaguelette decomposition, 323
Welch's estimator, 293
whitening property, 295
Wolf's sunspot numbers, 284

WILEY SERIES IN PROBABILITY AND STATISTICS
ESTABLISHED BY WALTER A. SHEWHART AND SAMUEL S. WILKS

Editors
Vic Barnett, Noel A. C. Cressie, Nicholas I. Fisher,
Iain M. Johnstone, J. B. Kadane, David G. Kendall, David W. Scott,
Bernard W. Silverman, Adrian F. M. Smith, Jozef L. Teugels;
Ralph A. Bradley, Emeritus, J. Stuart Hunter, Emeritus

Probability and Statistics Section

*ANDERSON · The Statistical Analysis of Time Series
ARNOLD, BALAKRISHNAN, and NAGARAJA · A First Course in Order Statistics
ARNOLD, BALAKRISHNAN, and NAGARAJA · Records
BACCELLI, COHEN, OLSDER, and QUADRAT · Synchronization and Linearity: An Algebra for Discrete Event Systems
BASILEVSKY · Statistical Factor Analysis and Related Methods: Theory and Applications
BERNARDO and SMITH · Bayesian Statistical Concepts and Theory
BILLINGSLEY · Convergence of Probability Measures
BOROVKOV · Asymptotic Methods in Queuing Theory
BOROVKOV · Ergodicity and Stability of Stochastic Processes
BRANDT, FRANKEN, and LISEK · Stationary Stochastic Models
CAINES · Linear Stochastic Systems
CAIROLI and DALANG · Sequential Stochastic Optimization
CONSTANTINE · Combinatorial Theory and Statistical Design
COOK · Regression Graphics
COVER and THOMAS · Elements of Information Theory
CSÖRGŐ and HORVÁTH · Weighted Approximations in Probability Statistics
CSÖRGŐ and HORVÁTH · Limit Theorems in Change Point Analysis
DETTE and STUDDEN · The Theory of Canonical Moments with Applications in Statistics, Probability, and Analysis
DEY and MUKERJEE · Fractional Factorial Plans
*DOOB · Stochastic Processes
DRYDEN and MARDIA · Statistical Analysis of Shape
DUPUIS and ELLIS · A Weak Convergence Approach to the Theory of Large Deviations
ETHIER and KURTZ · Markov Processes: Characterization and Convergence
FELLER · An Introduction to Probability Theory and Its Applications, Volume 1, *Third Edition,* Revised; Volume II, *Second Edition*
FULLER · Introduction to Statistical Time Series, *Second Edition*
FULLER · Measurement Error Models
GHOSH, MUKHOPADHYAY, and SEN · Sequential Estimation
GIFI · Nonlinear Multivariate Analysis
GUTTORP · Statistical Inference for Branching Processes
HALL · Introduction to the Theory of Coverage Processes
HAMPEL · Robust Statistics: The Approach Based on Influence Functions
HANNAN and DEISTLER · The Statistical Theory of Linear Systems
HUBER · Robust Statistics
IMAN and CONOVER · A Modern Approach to Statistics
JUREK and MASON · Operator-Limit Distributions in Probability Theory
KASS and VOS · Geometrical Foundations of Asymptotic Inference

*Now available in a lower priced paperback edition in the Wiley Classics Library.

Probability and Statistics (Continued)

    KAUFMAN and ROUSSEEUW · Finding Groups in Data: An Introduction to Cluster Analysis
    KELLY · Probability, Statistics, and Optimization
    LINDVALL · Lectures on the Coupling Method
    McFADDEN · Management of Data in Clinical Trials
    MANTON, WOODBURY, and TOLLEY · Statistical Applications Using Fuzzy Sets
    MORGENTHALER and TUKEY · Configural Polysampling: A Route to Practical Robustness
    MUIRHEAD · Aspects of Multivariate Statistical Theory
    OLIVER and SMITH · Influence Diagrams, Belief Nets and Decision Analysis
    *PARZEN · Modern Probability Theory and Its Applications
    PRESS · Bayesian Statistics: Principles, Models, and Applications
    PUKELSHEIM · Optimal Experimental Design
    RAO · Asymptotic Theory of Statistical Inference
    RAO · Linear Statistical Inference and Its Applications, *Second Edition*
    RAO and SHANBHAG · Choquet-Deny Type Functional Equations with Applications to Stochastic Models
    ROBERTSON, WRIGHT, and DYKSTRA · Order Restricted Statistical Inference
    ROGERS and WILLIAMS · Diffusions, Markov Processes, and Martingales, Volume I: Foundations, *Second Edition;* Volume II: Îto Calculus
    RUBINSTEIN and SHAPIRO · Discrete Event Systems: Sensitivity Analysis and Stochastic Optimization by the Score Function Method
    RUZSA and SZEKELY · Algebraic Probability Theory
    SCHEFFE · The Analysis of Variance
    SEBER · Linear Regression Analysis
    SEBER · Multivariate Observations
    SEBER and WILD · Nonlinear Regression
    SERFLING · Approximation Theorems of Mathematical Statistics
    SHORACK and WELLNER · Empirical Processes with Applications to Statistics
    SMALL and McLEISH · Hilbert Space Methods in Probability and Statistical Inference
    STAPLETON · Linear Statistical Models
    STAUDTE and SHEATHER · Robust Estimation and Testing
    STOYANOV · Counterexamples in Probability
    TANAKA · Time Series Analysis: Nonstationary and Noninvertible Distribution Theory
    THOMPSON and SEBER · Adaptive Sampling
    WELSH · Aspects of Statistical Inference
    WHITTAKER · Graphical Models in Applied Multivariate Statistics
    YANG · The Construction Theory of Denumerable Markov Processes

Applied Probability and Statistics Section

    ABRAHAM and LEDOLTER · Statistical Methods for Forecasting
    AGRESTI · Analysis of Ordinal Categorical Data
    AGRESTI · Categorical Data Analysis
    ANDERSON, AUQUIER, HAUCK, OAKES, VANDAELE, and WEISBERG · Statistical Methods for Comparative Studies
    ARMITAGE and DAVID (editors) · Advances in Biometry
    *ARTHANARI and DODGE · Mathematical Programming in Statistics
    ASMUSSEN · Applied Probability and Queues
    *BAILEY · The Elements of Stochastic Processes with Applications to the Natural Sciences
    BARNETT and LEWIS · Outliers in Statistical Data, *Third Edition*

*Now available in a lower priced paperback edition in the Wiley Classics Library.

Applied Probability and Statistics (Continued)
   BARTHOLOMEW, FORBES, and McLEAN · Statistical Techniques for Manpower Planning, *Second Edition*
   BATES and WATTS · Nonlinear Regression Analysis and Its Applications
   BECHHOFER, SANTNER, and GOLDSMAN · Design and Analysis of Experiments for Statistical Selection, Screening, and Multiple Comparisons
   BELSLEY · Conditioning Diagnostics: Collinearity and Weak Data in Regression
   BELSLEY, KUH, and WELSCH · Regression Diagnostics: Identifying Influential Data and Sources of Collinearity
   BHAT · Elements of Applied Stochastic Processes, *Second Edition*
   BHATTACHARYA and WAYMIRE · Stochastic Processes with Applications
   BIRKES and DODGE · Alternative Methods of Regression
   BLOOMFIELD · Fourier Analysis of Time Series: An Introduction
   BOLLEN · Structural Equations with Latent Variables
   BOULEAU · Numerical Methods for Stochastic Processes
   BOX · Bayesian Inference in Statistical Analysis
   BOX and DRAPER · Empirical Model-Building and Response Surfaces
   BOX and DRAPER · Evolutionary Operation: A Statistical Method for Process Improvement
   BUCKLEW · Large Deviation Techniques in Decision, Simulation, and Estimation
   BUNKE and BUNKE · Nonlinear Regression, Functional Relations and Robust Methods: Statistical Methods of Model Building
   CHATTERJEE and HADI · Sensitivity Analysis in Linear Regression
   CHILÈS and DELFINER · Geostatistics: Modeling Spatial Uncertainty
   CHOW and LIU · Design and Analysis of Clinical Trials: Concepts and Methodologies
   CLARKE and DISNEY · Probability and Random Processes: A First Course with Applications, *Second Edition*
   *COCHRAN and COX · Experimental Designs, *Second Edition*
   CONOVER · Practical Nonparametric Statistics, *Second Edition*
   CORNELL · Experiments with Mixtures, Designs, Models, and the Analysis of Mixture Data, *Second Edition*
   *COX · Planning of Experiments
   CRESSIE · Statistics for Spatial Data, *Revised Edition*
   DANIEL · Applications of Statistics to Industrial Experimentation
   DANIEL · Biostatistics: A Foundation for Analysis in the Health Sciences, *Sixth Edition*
   DAVID · Order Statistics, *Second Edition*
   *DEGROOT, FIENBERG, and KADANE · Statistics and the Law
   DODGE · Alternative Methods of Regression
   DOWDY and WEARDEN · Statistics for Research, *Second Edition*
   DRYDEN and MARDIA · Statistical Shape Analysis
   DUNN and CLARK · Applied Statistics: Analysis of Variance and Regression, *Second Edition*
   ELANDT-JOHNSON and JOHNSON · Survival Models and Data Analysis
   EVANS, PEACOCK, and HASTINGS · Statistical Distributions, *Second Edition*
   FLEISS · The Design and Analysis of Clinical Experiments
   FLEISS · Statistical Methods for Rates and Proportions, *Second Edition*
   FLEMING and HARRINGTON · Counting Processes and Survival Analysis
   GALLANT · Nonlinear Statistical Models
   GLASSERMAN and YAO · Monotone Structure in Discrete-Event Systems
   GNANADESIKAN · Methods for Statistical Data Analysis of Multivariate Observations, *Second Edition*
   GOLDSTEIN and LEWIS · Assessment: Problems, Development, and Statistical Issues
   GREENWOOD and NIKULIN · A Guide to Chi-Squared Testing
   *HAHN · Statistical Models in Engineering

*Now available in a lower priced paperback edition in the Wiley Classics Library.

Applied Probability and Statistics (Continued)

HAHN and MEEKER · Statistical Intervals: A Guide for Practitioners
HAND · Construction and Assessment of Classification Rules
HAND · Discrimination and Classification
HEIBERGER · Computation for the Analysis of Designed Experiments
HINKELMAN and KEMPTHORNE: · Design and Analysis of Experiments, Volume 1: Introduction to Experimental Design
HOAGLIN, MOSTELLER, and TUKEY · Exploratory Approach to Analysis of Variance
HOAGLIN, MOSTELLER, and TUKEY · Exploring Data Tables, Trends and Shapes
HOAGLIN, MOSTELLER, and TUKEY · Understanding Robust and Exploratory Data Analysis
HOCHBERG and TAMHANE · Multiple Comparison Procedures
HOCKING · Methods and Applications of Linear Models: Regression and the Analysis of Variables
HOGG and KLUGMAN · Loss Distributions
HOSMER and LEMESHOW · Applied Logistic Regression
HØYLAND and RAUSAND · System Reliability Theory: Models and Statistical Methods
HUBERTY · Applied Discriminant Analysis
JACKSON · A User's Guide to Principle Components
JOHN · Statistical Methods in Engineering and Quality Assurance
JOHNSON · Multivariate Statistical Simulation
JOHNSON and KOTZ · Distributions in Statistics
    Continuous Multivariate Distributions
JOHNSON, KOTZ, and BALAKRISHNAN · Continuous Univariate Distributions, Volume 1, *Second Edition*
JOHNSON, KOTZ, and BALAKRISHNAN · Continuous Univariate Distributions, Volume 2, *Second Edition*
JOHNSON, KOTZ, and BALAKRISHNAN · Discrete Multivariate Distributions
JOHNSON, KOTZ, and KEMP · Univariate Discrete Distributions, *Second Edition*
JUREČKOVÁ and SEN · Robust Statistical Procedures: Aymptotics and Interrelations
KADANE · Bayesian Methods and Ethics in a Clinical Trial Design
KADANE AND SCHUM · A Probabilistic Analysis of the Sacco and Vanzetti Evidence
KALBFLEISCH and PRENTICE · The Statistical Analysis of Failure Time Data
KELLY · Reversability and Stochastic Networks
KHURI, MATHEW, and SINHA · Statistical Tests for Mixed Linear Models
KLUGMAN, PANJER, and WILLMOT · Loss Models: From Data to Decisions
KLUGMAN, PANJER, and WILLMOT · Solutions Manual to Accompany Loss Models: From Data to Decisions
KOVALENKO, KUZNETZOV, and PEGG · Mathematical Theory of Reliability of Time-Dependent Systems with Practical Applications
LAD · Operational Subjective Statistical Methods: A Mathematical, Philosophical, and Historical Introduction
LANGE, RYAN, BILLARD, BRILLINGER, CONQUEST, and GREENHOUSE · Case Studies in Biometry
LAWLESS · Statistical Models and Methods for Lifetime Data
LEE · Statistical Methods for Survival Data Analysis, *Second Edition*
LePAGE and BILLARD · Exploring the Limits of Bootstrap
LINHART and ZUCCHINI · Model Selection
LITTLE and RUBIN · Statistical Analysis with Missing Data
LLOYD · The Statistical Analysis of Categorical Data
MAGNUS and NEUDECKER · Matrix Differential Calculus with Applications in Statistics and Econometrics
MALLER and ZHOU · Survival Analysis with Long Term Survivors
MANN, SCHAFER, and SINGPURWALLA · Methods for Statistical Analysis of Reliability and Life Data

*Now available in a lower priced paperback edition in the Wiley Classics Library.

Applied Probability and Statistics (Continued)

McLACHLAN and KRISHNAN · The EM Algorithm and Extensions
McLACHLAN · Discriminant Analysis and Statistical Pattern Recognition
McNEIL · Epidemiological Research Methods
MEEKER and ESCOBAR · Statistical Methods for Reliability Data
MILLER · Survival Analysis
MONTGOMERY and PECK · Introduction to Linear Regression Analysis, *Second Edition*
MYERS and MONTGOMERY · Response Surface Methodology: Process and Product in Optimization Using Designed Experiments
NELSON · Accelerated Testing, Statistical Models, Test Plans, and Data Analyses
NELSON · Applied Life Data Analysis
OCHI · Applied Probability and Stochastic Processes in Engineering and Physical Sciences
OKABE, BOOTS, and SUGIHARA · Spatial Tesselations: Concepts and Applications of Voronoi Diagrams
PANKRATZ · Forecasting with Dynamic Regression Models
PANKRATZ · Forecasting with Univariate Box-Jenkins Models: Concepts and Cases
PIANTADOSI · Clinical Trials: A Methodologic Perspective
PORT · Theoretical Probability for Applications
PUTERMAN · Markov Decision Processes: Discrete Stochastic Dynamic Programming
RACHEV · Probability Metrics and the Stability of Stochastic Models
RÉNYI · A Diary on Information Theory
RIPLEY · Spatial Statistics
RIPLEY · Stochastic Simulation
ROUSSEEUW and LEROY · Robust Regression and Outlier Detection
RUBIN · Multiple Imputation for Nonresponse in Surveys
RUBINSTEIN · Simulation and the Monte Carlo Method
RUBINSTEIN and MELAMED · Modern Simulation and Modeling
RYAN · Statistical Methods for Quality Improvement
SCHUSS · Theory and Applications of Stochastic Differential Equations
SCOTT · Multivariate Density Estimation: Theory, Practice, and Visualization
*SEARLE · Linear Models
SEARLE · Linear Models for Unbalanced Data
SEARLE, CASELLA, and McCULLOCH · Variance Components
SENNOTT · Stochastic Dynamic Programming and the Control of Queueing Systems
STOYAN, KENDALL, and MECKE · Stochastic Geometry and Its Applications, *Second Edition*
STOYAN and STOYAN · Fractals, Random Shapes and Point Fields: Methods of Geometrical Statistics
THOMPSON · Empirical Model Building
THOMPSON · Sampling
TIJMS · Stochastic Modeling and Analysis: A Computational Approach
TIJMS · Stochastic Models: An Algorithmic Approach
TITTERINGTON, SMITH, and MAKOV · Statistical Analysis of Finite Mixture Distributions
UPTON and FINGLETON · Spatial Data Analysis by Example, Volume 1: Point Pattern and Quantitative Data
UPTON and FINGLETON · Spatial Data Analysis by Example, Volume II: Categorical and Directional Data
VAN RIJCKEVORSEL and DE LEEUW · Component and Correspondence Analysis
VIDAKOVIC · Statistical Modeling by Wavelets
WEISBERG · Applied Linear Regression, *Second Edition*
WESTFALL and YOUNG · Resampling-Based Multiple Testing: Examples and Methods for *p*-Value Adjustment
WHITTLE · Systems in Stochastic Equilibrium

*Now available in a lower priced paperback edition in the Wiley Classics Library.

Applied Probability and Statistics (Continued)

WOODING · Planning Pharmaceutical Clinical Trials: Basic Statistical Principles
WOOLSON · Statistical Methods for the Analysis of Biomedical Data
*ZELLNER · An Introduction to Bayesian Inference in Econometrics

Texts and References Section

AGRESTI · An Introduction to Categorical Data Analysis
ANDERSON · An Introduction to Multivariate Statistical Analysis, *Second Edition*
ANDERSON and LOYNES · The Teaching of Practical Statistics
ARMITAGE and COLTON · Encyclopedia of Biostatistics: Volumes 1 to 6 with Index
BARTOSZYNSKI and NIEWIADOMSKA-BUGAJ · Probability and Statistical Inference
BERRY, CHALONER, and GEWEKE · Bayesian Analysis in Statistics and Econometrics: Essays in Honor of Arnold Zellner
BHATTACHARYA and JOHNSON · Statistical Concepts and Methods
BILLINGSLEY · Probability and Measure, *Second Edition*
BOX · R. A. Fisher, the Life of a Scientist
BOX, HUNTER, and HUNTER · Statistics for Experimenters: An Introduction to Design, Data Analysis, and Model Building
BOX and LUCEÑO · Statistical Control by Monitoring and Feedback Adjustment
BROWN and HOLLANDER · Statistics: A Biomedical Introduction
CHATTERJEE and PRICE · Regression Analysis by Example, *Second Edition*
COOK and WEISBERG · An Introduction to Regression Graphics
COX · A Handbook of Introductory Statistical Methods
DILLON and GOLDSTEIN · Multivariate Analysis: Methods and Applications
DODGE and ROMIG · Sampling Inspection Tables, *Second Edition*
DRAPER and SMITH · Applied Regression Analysis, *Third Edition*
DUDEWICZ and MISHRA · Modern Mathematical Statistics
DUNN · Basic Statistics: A Primer for the Biomedical Sciences, *Second Edition*
FISHER and VAN BELLE · Biostatistics: A Methodology for the Health Sciences
FREEMAN and SMITH · Aspects of Uncertainty: A Tribute to D. V. Lindley
GROSS and HARRIS · Fundamentals of Queueing Theory, *Third Edition*
HALD · A History of Probability and Statistics and their Applications Before 1750
HALD · A History of Mathematical Statistics from 1750 to 1930
HELLER · MACSYMA for Statisticians
HOEL · Introduction to Mathematical Statistics, *Fifth Edition*
HOLLANDER and WOLFE · Nonparametric Statistical Methods, *Second Edition*
HOSMER and LEMESHOW · Applied Survival Analysis: Regression Modeling of Time to Event Data
JOHNSON and BALAKRISHNAN · Advances in the Theory and Practice of Statistics: A Volume in Honor of Samuel Kotz
JOHNSON and KOTZ (editors) · Leading Personalities in Statistical Sciences: From the Seventeenth Century to the Present
JUDGE, GRIFFITHS, HILL, LÜTKEPOHL, and LEE · The Theory and Practice of Econometrics, *Second Edition*
KHURI · Advanced Calculus with Applications in Statistics
KOTZ and JOHNSON (editors) · Encyclopedia of Statistical Sciences: Volumes 1 to 9 wtih Index
KOTZ and JOHNSON (editors) · Encyclopedia of Statistical Sciences: Supplement Volume
KOTZ, REED, and BANKS (editors) · Encyclopedia of Statistical Sciences: Update Volume 1
KOTZ, REED, and BANKS (editors) · Encyclopedia of Statistical Sciences: Update Volume 2

*Now available in a lower priced paperback edition in the Wiley Classics Library.

Texts and References (Continued)

LAMPERTI · Probability: A Survey of the Mathematical Theory, *Second Edition*
LARSON · Introduction to Probability Theory and Statistical Inference, *Third Edition*
LE · Applied Categorical Data Analysis
LE · Applied Survival Analysis
MALLOWS · Design, Data, and Analysis by Some Friends of Cuthbert Daniel
MARDIA · The Art of Statistical Science: A Tribute to G. S. Watson
MASON, GUNST, and HESS · Statistical Design and Analysis of Experiments with Applications to Engineering and Science
MURRAY · X-STAT 2.0 Statistical Experimentation, Design Data Analysis, and Nonlinear Optimization
PURI, VILAPLANA, and WERTZ · New Perspectives in Theoretical and Applied Statistics
RENCHER · Methods of Multivariate Analysis
RENCHER · Multivariate Statistical Inference with Applications
ROSS · Introduction to Probability and Statistics for Engineers and Scientists
ROHATGI · An Introduction to Probability Theory and Mathematical Statistics
RYAN · Modern Regression Methods
SCHOTT · Matrix Analysis for Statistics
SEARLE · Matrix Algebra Useful for Statistics
STYAN · The Collected Papers of T. W. Anderson: 1943–1985
TIERNEY · LISP-STAT: An Object-Oriented Environment for Statistical Computing and Dynamic Graphics
WONNACOTT and WONNACOTT · Econometrics, *Second Edition*

WILEY SERIES IN PROBABILITY AND STATISTICS
ESTABLISHED BY WALTER A. SHEWHART AND SAMUEL S. WILKS

Editors
Robert M. Groves, Graham Kalton, J. N. K. Rao, Norbert Schwarz, Christopher Skinner

Survey Methodology Section

BIEMER, GROVES, LYBERG, MATHIOWETZ, and SUDMAN · Measurement Errors in Surveys
COCHRAN · Sampling Techniques, *Third Edition*
COUPER, BAKER, BETHLEHEM, CLARK, MARTIN, NICHOLLS, and O'REILLY (editors) · Computer Assisted Survey Information Collection
COX, BINDER, CHINNAPPA, CHRISTIANSON, COLLEDGE, and KOTT (editors) · Business Survey Methods
*DEMING · Sample Design in Business Research
DILLMAN · Mail and Telephone Surveys: The Total Design Method
GROVES and COUPER · Nonresponse in Household Interview Surveys
GROVES · Survey Errors and Survey Costs
GROVES, BIEMER, LYBERG, MASSEY, NICHOLLS, and WAKSBERG · Telephone Survey Methodology
*HANSEN, HURWITZ, and MADOW · Sample Survey Methods and Theory, Volume 1: Methods and Applications

*Now available in a lower priced paperback edition in the Wiley Classics Library.

Survey Methodology (Continued)
 *HANSEN, HURWITZ, and MADOW · Sample Survey Methods and Theory,
  Volume II: Theory
 KISH · Statistical Design for Research
 *KISH · Survey Sampling
 LESSLER and KALSBEEK · Nonsampling Error in Surveys
 LEVY and LEMESHOW · Sampling of Populations: Methods and Applications,
  Third Edition
 LYBERG, BIEMER, COLLINS, de LEEUW, DIPPO, SCHWARZ, TREWIN (editors) ·
  Survey Measurement and Process Quality
 SIRKEN, HERRMANN, SCHECHTER, SCHWARZ, TANUR, and TOURANGEAU
  (editors) · Cognition and Survey Research
 SKINNER, HOLT, and SMITH · Analysis of Complex Surveys

*Now available in a lower priced paperback edition in the Wiley Classics Library.